DEEP IMPACT MISSION: LOOKING BENEATH THE SURFACE OF A COMETARY NUCLEUS

Front cover:

Comet Tempel 1 and the Flyby Spacecraft, artist: Pat Rawlings

Back cover:

The Deep Impact Flyby Spacecraft and released Impact Spacecraft heading toward comet Tempel 1, graphic rendering: Ball Aerospace & Technology Corporation.

DEEP IMPACT MISSION: LOOKING BENEATH THE SURFACE OF A COMETARY NUCLEUS

Edited by

CHRISTOPHER T. RUSSELL
University of California,
California, USA

Reprinted from *Space Science Reviews*, Volume 117, Nos. 1–2, 2005

🐎 Springer

A.C.I.P. Catalogue record for this book is available from the Library of Congress

ISBN: 1-4020-3488-1

Published by Springer
P.O. Box 990, 3300 AZ Dordrecht, The Netherlands

Sold and distributed in North, Central and South America
by Springer,
101 Philip Drive, Norwell, MA 02061, U.S.A.

In all other countries, sold and distributed
by Springer,
P.O. Box 322, 3300 AH Dordrecht, The Netherlands

Printed on acid-free paper

TABLE OF CONTENTS

FOREWORD

Deep Impact is NASA's answer to the media's concern that robotic exploration is becoming boring. Deep Impact does more than just address its scientific objectives. It attacks them. It is an active mission, an energetic mission, a mission on a mission! It is a risky mission whose architecture is complex. The two spacecraft must work together, communicate, co-navigate. The flyby spacecraft must see the impact site. The impactor is aimed at the lit surface for better imagery; perhaps it will hit in a shadowed area. We do not know well the physical properties of the cometary nucleus. Is it dense and rigid or is it friable with many voids like pumice? Will there be a deep crater or a shallow crater? Will the nucleus breakup? Will the impactor go right through the nucleus? Will the comet suddenly become active? There are models, theories, ideas and wild guesses as to what will happen but at this writing they can be all lumped together as scientific speculation, as we have never been this close to a comet nucleus before. It is new territory for space exploration and the planetary scientist.

This volume provides an in depth examination of the Deep Impact mission: the system architecture, the scientific payload, the history and dynamics of the target, 9 P/Tempel 1, and the expected properties of the nucleus and the coma. It also describes the expected results of the mission: remote sensing of the geology of the surface, the size of the crater and the spectroscopy of the materials ejected. The auxiliary observations from 1 AU are described as are the returned data and the ground system that processes it. Lastly there is a discussion of the Education and Public Outreach Program for Deep Impact.

Assembling this volume required the efforts of many individuals who graciously donated their time to this effort. The referees and authors deserve much appreciation as they proceeded with this project in near record time, producing a most readable and complete description of the mission. We wish to thank especially Marjorie Sowmendran who assisted in the editorial office, and kept the process moving smoothly and promptly to completion.

C. T. Russell
University of California, Los Angeles
February, 2005

DEEP IMPACT: A LARGE-SCALE ACTIVE EXPERIMENT ON A COMETARY NUCLEUS

MICHAEL F. A'HEARN[1,*], MICHAEL J. S. BELTON[2], ALAN DELAMERE[3]
and WILLIAM H. BLUME[4]

[1] *Department of Astronomy, University of Maryland, College Park, MD 20742-2421, U.S.A.*
[2] *Belton Space Exploration Initiatives, LLC, Tucson, AZ, U.S.A.*
[3] *Delamere Support Services, Boulder, CO, U.S.A.*
[4] *Jet Propulsion Laboratory, California Institute of Technology, Pasadena, CA, U.S.A.*
(*Author for correspondence; e-mail: ma@astro.umd.edu)

(Received 20 August 2004; Accepted in final form 14 January 2005)

Abstract. The Deep Impact mission will provide the first data on the interior of a cometary nucleus and a comparison of those data with data on the surface. Two spacecraft, an impactor and a flyby spacecraft, will arrive at comet 9P/Tempel 1 on 4 July 2005 to create and observe the formation and final properties of a large crater that is predicted to be approximately 30-m deep with the dimensions of a football stadium. The flyby and impactor instruments will yield images and near infrared spectra (1–5 μm) of the surface at unprecedented spatial resolutions both before and after the impact of a 350-kg spacecraft at 10.2 km/s. These data will provide unique information on the structure of the nucleus near the surface and its chemical composition. They will also used to interpret the evolutionary effects on remote sensing data and will indicate how those data can be used to better constrain conditions in the early solar system.

Keywords: comets, space missions, solar system formation, craters, chemical composition

1. Introduction and History

The Deep Impact mission was conceived as a proposal to NASA's Discovery Program because we know so little about cometary nuclei and because missions that either land or do remote sensing cannot easily make measurements far enough below the surface to have a chance to characterize primitive cometary material. When the comets formed 4.5 Gy ago, they formed at very low temperatures from a mixture of different ices that is expected to be very sensitive to the actual temperatures at which they formed (e.g., Bar-Nun and Laufer, 2003; and references therein). The comets also included a mixture of more refactory materials including a broad range of both organics and silicates (e.g., Langevin *et al.*, 1987), Because the comets are small (<100 km), whether they formed at their present sizes or represent fragments of somewhat larger Trans-Neptunian objects, they have not been subject to much internal heating and they therefore preserve a record of conditions in the outer half of the protoplanetary disk. On the other hand, the surface layers of comets have evolved (see Section 2.3), in some cases due to solar heating at previous perihelion passages and in other cases due to irradiation by galactic cosmic rays. The European Space Agency's Rosetta mission is now on its way to place a lander on the nucleus of comet 67P/Churyumov-Gerasimenko (up-to-date information on the status of this mission can be obtained in the internet at http://www.esa.int/export/esaMI/Rosetta). The Rosetta lander, which will probe into the uppermost meter of the nucleus, was an enormous engineering challenge because of the uncertainty in the cometary properties. The concept of Deep Impact arose while thinking about how to make measurements far enough below the evolved surface materials to have a substantial chance to probe cometary material that is essentially unchanged since the formation of the solar system 4.5 Gy ago. With current technology, the only method for sampling this material is to excavate a crater and this is the experiment of Deep Impact.

The primary goal of Deep Impact is to understand the differences between the material at the surface of a cometary nucleus and the material in the interior in order to understand the evolutionary processes that have taken place in the surface layers. These processes occur typically at previous perihelion passages (the companion paper by Yeomans et al. (2005) provides a detailed discussion of the orbital dynamics of our target comet). The material in the deep interior is expected to retain much of the original molecular abundances from the formation of the comet 4.5 Gy ago but theoreticians (Section 2.3) disagree about the depth to which this evolution has penetrated.

The general concept of Deep Impact is to recreate, using an artificial impactor, a process that occurs regularly throughout the solar system, namely the impact of one body into another. The impactor is a fully functional spacecraft that flies attached to the flyby spacecraft until 1 day prior to impact. When comet D/Shoemaker-Levy 9 (hereafter S-L9) was about to impact Jupiter (Chodas and Yeomans, 1996), astronomers worldwide made a wide range of predictions and carried out extensive observational programs aimed at better understanding both the nature of the comet and the nature of the Jovian atmosphere below the clouds. The biggest limitation in studying the Jovian atmosphere was in fact the huge lack of knowledge about the impacting bodies, not only the chemical composition (e.g., Crovisier, 1996; Lellouch, 1996) but even the size (Sekanina, 1996) of the impactors. Deep Impact is intended to impact a comet such that everything is known about the impacting body and the only unknowns are the properties of the comet itself. The scientific basis for the mission was first laid out by Belton and A'Hearn (1999).

Prior to the Apollo program, the series of Ranger missions impacted the moon in order to provide close-up photographs of the surface prior to manned missions to the moon. Deep Impact's impactor will take an analogous series of images. During the Apollo program, nine large impact experiments were carried out to study the seismic properties of Moon (Nakamura et al., 1982; see also the summary by Cook, 1980). These experiments delivered kinetic energies to the moon that range from 20–200% of the energy that Deep Impact will deliver. A similar mission concept to an asteroid was described by Clarke (1968), although this was not in our thinking at the time

Deep Impact was conceived. The actual heritage of Deep Impact, came in part from an early, unpublished, concept study led by M. Neugebauer (M.J.S. Belton, personal communication) for JPL as part of the work for the Comet Rendezvous Asteroid Flyby (CRAF) mission that was subsequently cancelled. Although, in that study, a hypersonic impact was not envisioned. Prior to the selection of Deep Impact by NASA, other proposals for impact experiments had been rejected on technical feasibility grounds or have failed. Since the selection of Deep Impact by NASA, there have been additional proposals to NASA's Discovery Program for other types of impact experiments on asteroids.

2. What We Don't Know

2.1. MASS AND RELATED PARAMETERS

It is not widely realized outside the community of cometary scientists, that we do not have a single, direct measurement of the mass of a cometary nucleus. That means, of course, that we do not have a single direct measurement of the density. Several investigators, beginning with Rickman et al. (1987), have cleverly used the measured, non-gravitational acceleration of comets together with models for the outgassing to deduce the masses of some nuclei, but the results are still very model dependent (a more detailed discussion of nuclear mass and density is given in a companion paper by Belton et al., 2005). In the most recent case of comet 19P/Borrelly, for example, the location of the strongest active area on the surface and the orientation of the rotational axis are reasonably well known. Nevertheless, two independent determinations of the mass of Borrelly based on the same measured accelerations yield very different numerical results with reasonably large and only partially overlapping error bars (Farnham and Cochran, 2002; Davidsson and Guitérrez, 2004). All of the recent determinations, however, yield densities between 0.1 and 1.0 g cm^{-3}, suggesting that cometary nuclei are porous, unless there is some still unidentified flaw in the approach using non-gravitational accelerations. The degree of porosity, however, depends critically on the ratio of ices to silicates, since these two components have different bulk densities. The ice-to-silicate ratio is also unknown. Estimates of dust-to-gas ratios by mass in cometary comae are currently tending to be of order unity or larger for typical comets, but this is model dependent, and the ratio in the coma is not necessarily representative of the ratio in the nucleus. The gas and dust coma of 9P/Tempel 1 is discussed in detailed in the companion paper by Lisse et al., 2005.

2.2. STRUCTURAL PROPERTIES

The structural strength is also unknown with one prominent exception. The tensile strength, at least on spatial scales of a kilometer or so, must be $< 10^3$ dyn cm^{-2} on the basis of the tidal fragmentation of S-L9 (Sekanina, 1996). The distribution of the fragments of S-L9 has been very well described by models that assume reaccretion of 100-m fragments, suggesting but not proving that the tensile strength is comparably small at 100-m spatial scales. It remains to be seen whether the strength becomes large for scales somewhat below 100 m, or for scales of 1 m or 1 cm or even less. The presence of a large quantity of debris detected during radar observations of near-Earth-approaching comets IRAS-Araki-Alcock (Harmon et al., 1989) and Hyakutake (Harmon et al., 1997) suggests that chunks of order 10 cm must have sufficient strength to be lifted by hydrodynamic or other forces. The spontaneous fragmentation of comets when far from any source of tidal stress (Sekanina, 1997) indicates that they are generally very weak on some scales, but without understanding the mechanism for spontaneous splitting, it is impossible to obtain strengths quantitatively. These numbers might be compared with numbers like 10^6 dyn cm^{-2} for solid ice and 10^8 dyn cm^{-2} for rock measured in the laboratory (both materials are usually much weaker on large geophysical scales). In addition to tensile strength, of course, there are other strengths that matter, including shear strength and strength against compression. Again, these are not known, although one commonly assumes that these strengths are also small in the case of cometary material. Further discussion of this topic can be found in Belton et al. (2005).

Closely related to this question is whether the structure, and thus the strength, has a characteristic size resulting from the original formation process. For example, Weidenschilling (1997) has suggested that cometary nuclei are made up of primordial cometesimals with preferred sizes in the range of 10–100 m. If this is correct, one expects significant changes in strength at this characteristic scale.

2.3. DIFFERENTIATION AND EVOLUTION

Another interesting characteristic of cometary nuclei is that it is widely assumed that they must be differentiated, either primordially from extinct radio-nuclides (not very likely in our view based on the available evidence) or more recently from the effect of insolation at previous perihelion passages (see Belton et al., 2005 for further discussion). Regrettably there are essentially no data to show this differentiation. In fact, the only differentiation that is well documented is in the outermost layer of Oort-cloud comets arriving in the inner solar system for the first time. These dynamically recognizable comets are also photometricly recognizable, brightening as r^{-2} as they approach the sun, a much shallower variation than exhibited by any other comets, including these same dynamically new comets as they recede from the sun (Whipple, 1978). This photometric behavior is generally attributed to the irradiation by galactic cosmic rays of the outermost layer of cometary nuclei beyond the heliopause, resulting in a highly chemically unstable layer that is blown off at some large heliocentric distance on the first approach to the inner solar system.

Evolution and differentiation during previous perihelion passages have been extensively studied with numerical simulations, but there are few data to constrain the models (see also the companion paper by Belton *et al.*, 2005). As a result, the simulations exhibit a wide range of depths to which the evolution proceeds and also a wide range in the variation of chemical and physical properties with depth. The evolution and differentiation are sensitive to the ice-to-dust ratio, the porosity of that mixture, the volatility of the ices, the mean free path of a vaporized ice molecule inside the pores, and so on. There are relatively few constraints on the models although some investigators have, for example, tried to reproduce the variation of brightness with heliocentric distance for Jupiter-family comets (e.g., Benkhoff and Huebner, 1995; Prialnik, 2002). The observations at radio wavelengths of many different species in comet Hale-Bopp show dramatic and systematic variations in relative release/production rates for different species as shown in Figure 1 (Biver

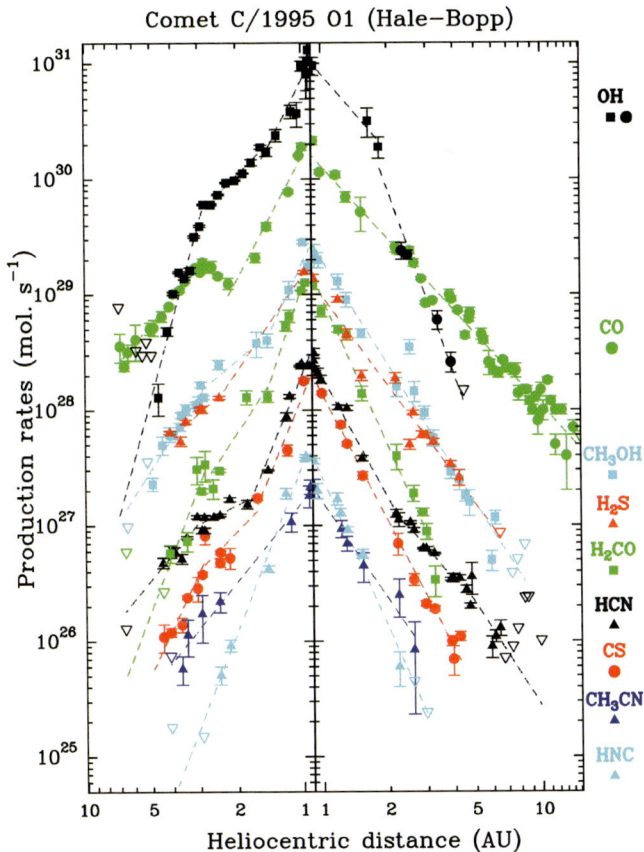

Figure 1. Molecular production rates by comet Hale-Bopp as a function of heliocentric distance both pre- (left half of diagram) and post-perihelion. Relative abundances vary systematically with distance suggesting either differentiation or processes in the coma. Diagram courtesy of Nicolas Biver (Biver *et al.*, 2002).

et al., 2002). This could be due to chemical reactions in the coma (proposed for HNC relative to HCN by Irvine *et al.*, 1997), or due to differentiation of volatiles in the sub-surface at some time prior to the observations, such as at the previous perihelion passage, or due to a differentiating process that was active at the time of the observations. These data might provide useful constraints on the differentiation in the nucleus, but detailed models to fit these data are still lacking.

A key issue in the modeling is the depth to which the evolution has penetrated. The cosmic ray irradiation of dynamically new comets from the Oort cloud is thought to have penetrated tens of meters (Moore *et al.*, 1983). For the comets of interest to Deep Impact (Jupiter-family comets, originally from the Kuiper belt), the depth of evolution is more uncertain. At the shallow extreme is the prediction of Kouchi and Sirono (2001).

2.4. END STATES

The ultimate evolutionary fate of comets after many perihelion passages is not known. Statistical studies of dynamical evolution suggest that some comets, particularly Oort-cloud comets, must disappear due to physical evolution before dynamical processes have time to eject them either out of the solar system or into a planet or the sun, but the nature of that evolutionary end-state is not known. The nuclei could either dissipate like comet LINEAR (C/1999 S4; see, e.g., Weaver *et al.*, 2001) or they could become inert and thus apparently asteroidal, much like the apparently dead cometary nuclei discussed by Fernández *et al.* (2001). If the latter is the dominant mechanism, there is a further question, namely whether the nuclei become inert because they have exhausted their supply of volatile ices or because they have developed a mantle that seals the volatile ices inside.

3. What Other Missions Are Doing

The 1990s led to a wide suite of missions to comets (and asteroids), ranging from the technology demonstration mission Deep Space 1, which flew past 19P/Borrelly, to ESA's mission that is *en route* to a rendezvous with 67P/Churyumov-Gerasimenko in 2014. The Stardust mission flew through the coma of 81P/Wild 2 and will return dust grains from the coma to Earth in January 2006. The ill-fated CONTOUR (the final status of the CONTOUR mission is available on the web at http://discovery.nasa.gov/contour.html) was slated to fly past two (a third was reduced in scope due to cost) comets with very different dynamical histories and possibly fly to a 'new' comet to assess what diversity exists in their physical and chemical properties and to put quantitative constraints on the origins of such diversity. The Rosetta Lander will sample material to depths of about a meter at the landing site and the CONSERT experiment (Barbin *et al.*, 1999). will perform radioabsorption sounding through the nucleus between the orbiter and the lander. The surface experiments are clearly important but they are limited in depth ($\sim$1 m) and/or spatial ($\sim$2 m) resolution and may not fully characterize the evolutionary processes of chemical and physical differentiation that are expected to occur. Most importantly, they will not be carried out for another decade. Deep Impact will be the first mission to sample to substantial depths (10–30 m) into the subsurface of a cometary nucleus. It is striking that a wide variety of missions to comets still allows each mission to address quite different scientific goals, a testimony to the progress that can be made in the field by means of a series of narrow missions, particularly when we know so little about cometary nuclei.

Differences among the topographies of cometary nuclei are obvious from the very limited sample of three comets for which we have *in situ* imaging. Wild 2 is clearly more nearly spherical, although the deduced semi-axes imply that it is really best described as a triaxial body (Duxbury, cited by Brownlee *et al.*, 2004) than either 19P/Borrelly (Oberst *et al.*, 2004; Kirk *et al.*, 2004) or 1P/Halley (Merényi *et al.*, 1990), and it also has much greater vertical relief at scales <1 km than does Borrelly (Brownlee *et al.*, 2004). Whether this is an evolutionary effect related to lowering of the perihelion distance of Wild 2 only a few apparitions earlier is unknown. Borrelly and Halley have both had many perihelion passages well within the orbit of Mars and they show substantial differences, Halley being nearly convex except for a narrowing at what might be termed the waist (Merényi *et al.*, 1990) while Borrelly has large-scale concavity on the side that was imaged. Thus the one technique, optical imaging, that is common to essentially all missions, shows substantial differences among comets that are not readily explained. Variations in gross shape for other comets are also inferred from the wide range of amplitudes observed for lightcurves if they are interpreted as being due to the varying cross-section of the nucleus as it rotates.

Finally, we note that Wild 2, for which we have the most complete set of images, shows many topographic features, some of which are nearly circular and may be impact craters, but which do not look like impact craters elsewhere in the solar system (a discussion of topography including our views of the enigmatic circular features on Wild 2 can be found in the companion paper by Thomas *et al.*, 2005). Deep Impact will provide at least one example of a crater that is known to be an impact crater and thus provide an important point of comparison for the craters on Wild 2.

4. Overview of Deep Impact

Deep Impact is the eighth mission in NASA's Discovery Program. It was proposed and accepted as a partnership between the University of Maryland, which provides the scientific direction and manages the science and the outreach, the Jet Propulsion Laboratory, which manages the project development and carries out the operations, and Ball Aerospace and Technologies Corp., which provides the spacecraft and instruments, other than some components that are provided by JPL.

4.1. SCIENTIFIC OBJECTIVES AND SUCCESS CRITERIA

The scientific objectives of the mission, as described in the original proposal and as quoted here from the relevant portion of the Discovery Program Plan, are

The Deep Impact mission will fly to and impact a short-period comet understood to have a nuclear radius>2 km (large enough so that it will sustain a crater of cometesimal size and ensure reliable targeting). The direct intent of the impact is to excavate a crater of approximately 100 m in diameter and 25 m in depth. The overall scientific objectives are to

1. Dramatically improve the knowledge of key properties of a cometary nucleus and, for the first time, assess directly the interior of a cometary nucleus by means of a massive impactor hitting the surface of the nucleus at high velocity.
2. Determine properties of the surface layers such as density, porosity, strength, and composition from the resultant crater and its formation.
3. Study the relationship between the surface layers of a cometary nucleus and the possibly pristine materials of the interior by comparison of the interior of the crater with the pre-impact surface.
4. Improve our understanding of the evolution of cometary nuclei, particularly their approach to dormancy, from the comparison between interior and surface.

The conversion of these goals into success criteria is more complicated than in many missions because of the very large uncertainty in what it would take to produce a crater of the size mentioned in the objectives, as discussed earlier in Section 2 and in more detail in the companion articles by Richardson *et al.* (2005) and by Schultz and Ernst (2005). For this reason, the success criteria are stated in terms of delivering a minimum mass at a minimum velocity, followed by success criteria based on the scale and sensitivity of images and spectra. The baseline success criteria are taken from the relevant appendix to the Discovery Program Plan (section numbering omitted) and are as follows:

  i. Target a short period comet understood to have a nuclear radius >2 km.
 ii. Deliver an impactor of mass > 350 kg to an impact on the cometary nucleus at a velocity > 10 km/s. The impact event and crater formation shall be visible from the flyby spacecraft and observable from Earth.
iii. Obtain pre-impact visible-wavelength images of the impact site including one with resolution < 3 m and *FOV* > 50 pixels.
iv. Obtain three visible-wavelength images, using at least two different filters, of the entire comet, pre-impact, with resolution < 50 m and average S/N >50 for the illuminated portion of the nucleus.

  v Obtain five visible-wavelength images containing the impact site with resolution < 50 m and showing the crater evolution from within 3 s of time of impact until full crater development (assumed to take less than 660 s).

  vi. Obtain five visible-wavelength images of the ejecta cone, showing the ejecta cone evolution at a resolution <50 m from within 1 s of impact until late in the cone evolution (assumed to take less than 60 s).

  vii. Obtain five near-infrared (1.1 to 4.8 μm), long-slit spectra of the ejecta cone, showing the ejecta cone evolution with spectral resolving power > 200 from within 2 s of time of impact until late in the cone evolution (assumed to take less than 60 s).

  viii. Obtain one image of the final crater with a resolution <7 m.

  ix. Obtain one near-infrared (1.1 to 4.8 μm), long-slit spectrum of the impact region pre-impact and one post impact, both with spectral resolving power >200 and with noise-equivalent-surface-brightness <150 k Rayleigh per spectral resolution element at 3.5 μm.

  x Obtain two near-infrared (2.0 to 4.8 μm), long-slit spectra of the coma, one before impact and one after formation of the crater (assumed to take <660 s), with spectral resolving power >200 and Noise-equivalent surface brightness <500 k Rayleigh per spectral resolution element at 4.7 μm.

  xi. Obtain at least three Earth-orbital or ground-based datasets of two different types of data complementary to the data from the spacecraft.

The original baseline success criteria included a requirement to deliver an impactor of mass 500 kg, coupled with a minimum requirement of 300 kg. The descope to 350 kg was approved by NASA before CDR (Critical Design Review) in order to save considerable funds by using a smaller launch vehicle. If the project's favored scenario for the impact, gravitational control of the cratering, is correct, the difference in the size of the crater due to the reduction in scope will be very small, as one can see from the discussion of scaling laws in the accompanying papers by Richardson *et al.* (2005) and by Schultz and Ernst (2005). The baseline success criteria are otherwise unchanged since selection, despite numerous other reductions in scope taken between PDR (Preliminary Design Review) and CDR. The only other requirement from the original proposal that has been waived is the window for the launch date. Due in largest part to difficulty in developing the spacecraft's computer system, the launch was allowed to slip from the original window (opening on 2 Jan 2004) to the backup window (opening on 31 Dec 2004), which was outside the originally defined window for launch, but this had no effect on the encounter with the comet.

4.2. OVERVIEW OF THE INSTRUMENTS AND THE MISSION

The details of the flight mission are described in the companion paper by Blume (2005), details of the scientific instruments in the companion paper by Hampton *et al.* (2005), and details of the data expected to be returned and its archival disposition in the companion paper by Klaasen *et al.* (2005). This section provides only an introduction to those topics.

Deep Impact, which consists of two spacecraft – an impactor vehicle and a flyby vehicle that are initially mated and launched together, will reproduce the impact of a boulder into a cometary nucleus at a speed characteristic of collisions in the asteroid belt, delivering an impactor of 363 kg (plus whatever remains of the initial 8 kg of hydrazine fuel) onto the nucleus of comet 9P/Tempel 1 at $10.2 \, \text{km s}^{-1}$. This kinetic energy, about 19 GJ, corresponds to the explosive power of 4.5 tons of TNT. We note that the speed is such that the kinetic energy per unit mass substantially exceeds the chemical energy per unit mass of the most efficient chemical explosives. Also, the localized and explosive liberation of this energy, which initially serves to accelerate the subsurface material in the vicinity of the explosion, ultimately causes the excavation of a large volume of material at much larger depths. The material left in the crater, which may have seen the passage of one or more shock waves, is not expected to have its chemical composition changed. The effect should be to produce, in roughly 200 s, a crater about 100 m in diameter and 25 m deep, although there is a large uncertainty in this prediction (see the companion papers by Schultz and Ernst, 2005 and by Richardson *et al.*, 2005). The impactor spacecraft has a camera for scientific imaging and autonavigation (see Mastrodemos *et al.*, 2005), a complete attitude control system using gyros and thrusters and a complete propulsion system using hydrazine. In order to minimize chemical reactions that might lead to species with bright lines in the spectrum, the use of copper, from the noble metals column of the periodic table, was maximized, comprising nearly half the mass of the impactor. The camera on the impactor has no filter wheel, taking only white-light images, which are both used on board for autonavigation and transmitted to the flyby spacecraft for retransmission to Earth (see companion paper by Hampton *et al.*, 2005 for details). As the impactor approaches the comet, we expect that dust in the coma will sandblast the primary mirror of the camera in the last minute before impact, while a single, major dust impact closer to the nucleus could destroy the camera. In either case, this is long after the last navigation maneuver (Figure 3). The last image, presumably only partially transmitted, will have a scale of 20 cm per pixel.

The flyby spacecraft is launched mated to the impactor, Figure 2, in a 30-day launch window beginning 30 December 2004, and remains joined to the impactor until it releases the impactor with a spring mechanism 24 h before impact onto a collision course with the nucleus. The flyby spacecraft diverts by about 2 arcmin in order to miss the nucleus by 500 km, our best estimate of the radius of the Hill sphere for the comet. (The Hill sphere is the volume within which orbits around the nucleus are stable against solar perturbations.) It also decelerates by about $100 \, \text{m s}^{-1}$, passing closest to the nucleus 850 s after impact and providing an 800-s window for making all of our observations of the crater and its formation [our predictions of the formation time of the crater range up to 700 s for the lowest

Figure 2. The flyby spacecraft being lowered onto the impactor spacecraft in the clean room at Ball Aerospace and Technologies Corp. prior to system environmental tests. When the spacecraft are joined, the pentagonal gold-colored panels of the impactor are recessed entirely inside the flyby spacecraft. The white ring at the bottom of the impactor, here bolted to the top of a test stand, is the fixture that will join the impactor to the launch vehicle. Photo courtesy of Ball Aerospace and Technology Corporation.

assumed density of the nucleus]. The mission design, and particularly the short (800 s) window for observation after impact, require very intelligent auto-navigation as described by Mastrodemos *et al.* (2005). Instruments include a high-resolution camera (2 μrad per pixel) with a series of intermediate-band filters, an infrared spectrometer covering the range from 1.05 to 4.8 μm, and a medium-resolution camera (10 μrad per pixel) that is identical to that on the impactor except for the addition of a filter wheel. The medium resolution camera is intended to have a field of view large enough to image the entire nucleus near closest approach and provide geophysical context for the high resolution frames. The flyby instruments are body mounted and co-aligned so that tracking is achieved by rotating the spacecraft.

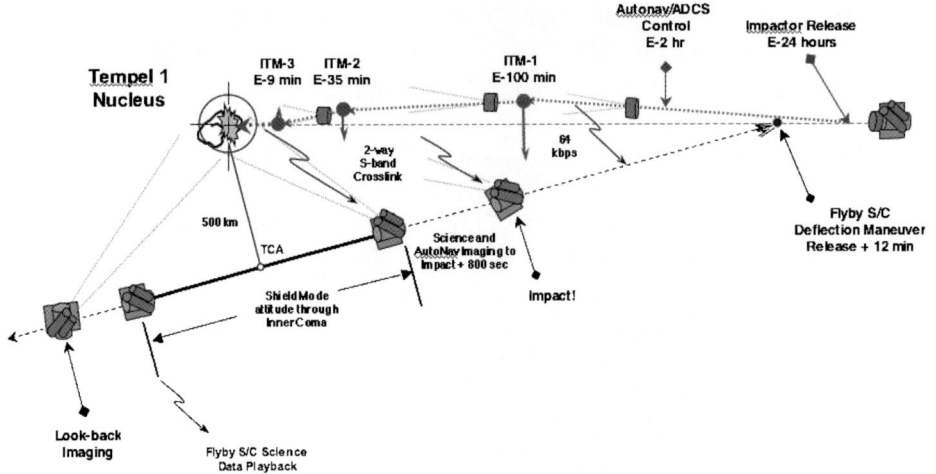

Figure 3. The encounter sequence. In this view, which is in the rest frame of the comet, the sun is behind the page at an angle of 63° to the plane of the paper and generally toward the upper right. In the heliocentric frame, the comet is moving down and to the right at about 30 km/s. Earth is also to the right and behind the page. The encounter sequence is described in detail in the companion paper by Blume (2005).

As the spacecraft approaches the comet, the spacecraft-fixed cameras are pointed by turning the spacecraft. The maximum spacecraft attitude turn rate of 0.6 deg/s occurs at a range of 700 km when the spacecraft has rotated 45° from the approach asymptote. Most of the shielding against dust being placed for this orientation.

The time of impact, on 4 July 2004, is determined to within a small window by the requirement for redundant linking of data from the spacecraft to two stations of the Deep Space Network coupled with the desire to observe the event in darkness from one of the major astronomical sites. The first constraint forces the astronomical site to be Hawaii and the onset of darkness there sets the earliest time for the impact. The window is roughly 05:50 to 06:30 UTC on 4 July, with the choice of time within that window to be made 1–2 months before impact in order to ensure that HST (Hubble Space Telescope) is on the correct side of the Earth to make observations.

The science team has developed a baseline scenario for the impact, and thus for its observation, assuming that the growth of the crater will be controlled by the low gravity of the cometary nucleus rather than, e.g., by the strength of the material. Depending on the size (effective radius probably near 3.3 km; please refer to the companion paper by Belton *et al.* (2005) for our latest estimates of the parameters that describe the nucleus of 9P/Tempel 1) and density (unknown) of the nucleus, the crater might be between 100 and 200 m in diameter (for reasonable assumptions about density) and 25 to 30 m deep. However, some colleagues have argued that the crater will be controlled by the strength of porous ice with the penetration being limited by the development of instabilities (Rayleigh-Taylor and Kelvin-Helmholtz) at the sides of the impactor (O'Keefe *et al.*, 2001), while others might argue that the energy will go primarily into compaction of the porous nucleus against moderate compressive strength as has been argued for Mathilde (Housen *et al.*, 1999). These scenarios lead to very different final craters. Many cometary scientists, on the other hand, have suggested that the impact might either break a piece off the nucleus or even shatter it to many pieces. Finally, although it seems very unlikely, the impactor might just bury itself very deeply if the density of the comet is much lower than currently accepted values, just as Stardust gently captures dust particles in ultra-low-density aerogel (Tsou *et al.*, 2003). The key point in designing the mission was to optimize our measurements for the baseline scenario but to ensure that our data would be robust in providing useful information about the other scenarios. Several of the alternative scenarios would require substantial revisions to prevailing ideas about the structure of comets such that even minimal data for the unlikely scenarios, as long as the data identify the cratering mode, will lead to fundamental conclusions about cometary structure.

An important aspect of the mission is the role of remote sensing from ground-based and space-based telescopes. The impact event is expected to be readily observable from Earth, and it was designed to allow convenient observation with many different techniques. On a spacecraft one is severely limited in the nature of the instrumentation one can carry to the target and, because of the fast flyby, the Deep Impact instruments are limited to a rather short observing window. Thus remote sensing will be important for applying the full range of techniques that astronomers use, from high-speed photometry through high-resolution spectroscopy and at wavelengths from X-ray to radio. The details of the Earth-based program are described by Meech *et al.* (2005) and the role of amateurs in this program is also discussed by McFadden *et al.* (2005).

4.3. MEASUREMENTS TO BE MADE

The instruments take a suite of measurements at different times (see Klaasen *et al.*, 2005) that we summarize here. The relevant scientific phases used here (different from the mission phases described by Blume, 2005) are approach, encounter, and lookback. Approach is from 60 days before impact (E-60d; the point at which we expect to be able to reliably detect the comet) until release of the impactor (E-24h). Encounter is the day from release of the impactor until closest approach of the flyby spacecraft to the nucleus (at E+850s). Lookback is from closest approach until end of observations of Tempel 1, nominally at E+2.5d. We summarize the types of data here as a function of instrument and phase indicating how they address our scientific goals. Many of the imaging data are also used for navigation, as described by Mastrodemos *et al.* (2005) but we discuss here the scientific use only. Details of the spectroscopic plans are given by Sunshine *et al.* (2005).

4.3.1. *Impactor Targeting Sensor (ITS)*

During encounter the ITS takes images at steadily increasing frequency. In order to maintain a minimum sampling interval corresponding to $\sqrt{2}$ changes in distance while not exceeding the telemetry bit rate to the flyby spacecraft, the images are gradually decreased in size from 1024×1024 by taking central subframes down to 64×64. These images provide the highest resolution ever obtained on a cometary nucleus and provide the context images of the surface immediately prior to impact which provide valuable input to simulations of the cratering process.

4.3.2. *IR Spectrometer (HRI-IR)*

During approach the spectrometer is used to study the coma, allowing us to determine the spatial distribution and abundances of numerous molecules as well as characteristics of the dust. Toward the end of the approach phase, the spectrometer is also used to study the rotational variation of the reflectivity characteristics of the spatially unresolved nucleus. During encounter, the spectrometer is used to study in detail the distribution of molecules in the coma and the variations in the dust with location in the coma. It is also used to study the rotational variation of the reflectivity characteristics of the cometary nucleus. As the range shortens prior to impact, the spectrometer maps the reflectivity of the spatially resolved nucleus. During crater formation, the spectrometer monitors the evolution of the spectra of the ejecta as material from successively greater depths is ejected from the crater, thus revealing compositional variations. Shortly before closest approach, the spectrometer studies the coma in detail to determine what the differences are relative to the pre-impact composition. The spectrometer is also used to make a spatial map of the reflectivity of the region including the crater to understand the differences between the ambient surface and the crater floor. During lookback, the spectrometer is used to study any continuing outgassing from the crater as well as to map the coma from a different direction thus allowing resolution of the three-dimensional structure and compositional variation.

4.3.3. *High Resolution Imager (HRI-Vis)*

During approach the HRI-Vis imager is used to photometrically monitor the comet's variations, both with rotation and with orbital position. Regular observations of the structures in the coma will provide the best data on the rotational state prior to impact. Once the nucleus is photometrically resolved in the central pixel, the rotational variation of the nucleus is monitored to study lateral heterogeneity of the surface. The nucleus becomes spatially resolved before the release of the impactor spacecraft. During the encounter phase, imaging is used to obtain colorimetry of the nuclear surface at steadily improving spatial resolution over half a rotation of the nucleus. This provides considerable information on lateral variation of the compostion of the surface. At the time of impact, images are taken as rapidly as possible in white light to study the evolution of the ejecta and determine the morphology of the ejecta cone. As the evolution proceeds, the image rate gradually decreases since things change more slowly at the later times. These images of the ejecta cone will allow determination of the cratering mode as well as providing estimates of the amount of ejecta from optical depth consideration. Clumps of ejecta will be tracked in order to study the distribution of ejecta velocities thus providing a good test of structural properties that are important to the simulations of the cratering process. Very slowly moving clumps may allow an estimate of the local gravity if tracking continues through lookback. During the latest portion of the encounter phase, the HRI-Vis is used to study the properties of the crater and its surroundings, providing information on vertical stratigraphy, on any boulders that may have survived the shock and the excavation, and on colorimetric differences between the floor of the crater and the ambient surface. This will allow determination of differences between the surface and the interior. If natural outgassing persists from the crater, producing a new active area, this will indicate that comets become dormant by sealing ice in the interior rather than by exhausting the supply of ice. The last images will provide by far ($5\times$) the highest resolution images ever of a cometary nucleus other than in those from the impactor. During the lookback phase, the HRI-Vis will take images to determine the three-dimensional shape, and thus the total volume, of the nucleus. The images will also track any continuing outgassing from a possible new active area by monitoring the jets at the limb of the nucleus (the crater itself will be on the far side at this point). Details of the geological approach are given by Thomas *et al.* (2005), while details of the interpretation of the cratering data are given by Richardson *et al.* (2005) and by Schultz and Ernst (2005).

4.3.4. *Medium Resolution Imager (MRI)*

The MRI is used during approach primarily to supplement the observations with HRI-Vis. At the end of approach and during encounter the MRI is used to take deep images of the coma in narrow-bands to isolate the gaseous and dusty structures in the coma, using filters that are not available in HRI-Vis as well as some of the same filters that are available in the HRI-Vis. At the time of impact, MRI provides higher-speed imaging of the impact than does HRI-Vis, although the crater itself is spatially unresolved by MRI at the time of impact. MRI provides a wide field of view for tracking clumps of ejecta that are seen initially in the HRI-Vis. The last images before closest approach image the entire nucleus and thus, combined with earlier images, provide the stereoscopic information to allow reconstruction of the three-dimensional shape of the nucleus. During lookback, the MRI will provide the wide field of view for studying the entire nucleus and much of the coma. As before impact, deep exposures in the narrow-band filters can isolate different gases and dust in the coma for detailed study.

5. Choosing the Target

The choice of a target for the Deep Impact mission was constrained by many factors including

- Launch in the window defined by NASA's Announcement of Opportunity (AO).
- A trajectory with sufficiently small launch energy per unit mass that a large mass could be delivered to the comet.
- Encounter at hypervelocity, i.e., more than a few kilometers per second.
- Encounter at low enough velocity that the flyby spacecraft can realistically decelerate enough to observe the entire process of crater formation.
- Approach from a moderate to small phase angle for approach navigation and crater illumination.
- Impact event readily observable from Earth, i.e., large solar elongation and moderate to small geocentric distance.
- Nuclear size large enough that self-gravitational energy should substantially exceed the delivered kinetic energy and large enough that targeting to hit the nucleus would not present a major challenge.

With these constraints there is a "good" target available for launches every few years and an acceptable target that meets most constraints available every year. Tempel 1 was chosen as the best target available for launches in the launch window allowed by the AO from NASA. To minimize the launch energy one should encounter the target near one of its orbital nodes and the descending node occurs on 7 July 2004, which is fortuitously close to the comet's perihelion on 5 July 2004. Thus impacts between late June and mid-July are energetically best and 4 July was selected. The dynamical history as well as the observational history of the comet are described in a companion paper by Yeomans *et al.* (2005). Details of what is known about the nucleus parameters and of the coma of comet Tempel 1 are given in the companion papers by Belton *et al.* (2005) and by Lisse *et al.* (2005), respectively.

Some of the alternative targets that were rejected for one reason or another included

- 4P/Faye – larger phase angle on approach, no backup launch window, higher launch energy,
- 58P/Jackson-Neujmin – launch too early without backup window, larger phase angle on approach, small nucleus,
- 10P/Tempel 2 – impact not observable from Earth, large phase angle on approach for backup launches,
- 41P/Tuttle-Giacobini-Kresak – erratic outbursts, small nucleus, higher flyby speed,
- 78P/Schwassmann-Wachmann 3 – small nucleus, long flight time,
- 2P/Encke – high launch energy,
- 37P/Forbes – high launch energy, high phase angle on approach.

Of all these targets, Schwassmann-Wachmann 3 was the most promising alternative and this was planned as one of the targets of the CONTOUR mission (primarily because it had recently undergone a major splitting event and a "young" surface might be observed). However, it was dropped from prime consideration for Deep

Impact because of the increased difficulty of hitting a nucleus with radius thought to be less than 1 km. The point of this discussion, of course, is to emphasize the fact that many targets are available for missions such as Deep Impact, and the choice of target is mostly a tradeoff between optimizing the science and minimizing the risk.

The choice of Schwassmann-Wachmann 3 would also have resulted in two missions to the same comet. In our view, the ideal mission, which is clearly not doable under the Discovery Program, would send two separate spacecraft to the same comet. An orbiter would arrive first and it would map the comet in detail, determining the mass and developing complete maps while the impactor is *en route*. The orbiter might then be used to adjust the targeting and the time of impact to hit a certain portion of the surface and the orbiter would also be used to study the impact process.

6. What We Should Learn

This section explains, in order, how we answer each of the questions raised in Section 2. However, we certainly won't answer all the questions raised there. Nevertheless, we organize the answers in the same way as the questions. As emphasized by Harwit (1984), and as exemplified in observations from Halley to Borrelly to Wild 2, the surprising results usually come from measurements in a new regime, in this case first an entirely new regime of experimentation, but also even traditional measurements in an entirely new regime of spatial resolution. We could only speculate on what might be learned from increasing the spatial resolution by an order of magnitude. We provide here only an outline. More details are in some of the companion papers in this volume.

6.1. MASS AND RELATED PARAMETERS

The mass, while of fundamental importance, will probably not be determined by the Deep Impact project, although we will certainly attempt to do so. Tracking of the spacecraft during a cometary flyby, as has been clear in previous flyby missions, is incapable of deducing a mass because of the combination of the small mass of the nucleus with the relatively fast flyby speed. Our hope for determining the mass lies in tracking clumps of ejecta that emerge at relatively low velocity and end up orbiting the nucleus or at least showing significant deceleration. A second possibility, if the geometry turns out to allow us a good view from the side of the ejecta cone, is to carefully determine the shape at the base of the cone when it expands beyond the edge of the crater and the base of the cone is falling back onto the surface Detailed discussion on the processes involved in crater formation and the resultant ejecta will be found in the companion papers by Schultz and Ernst (2005) and Richardson *et al.* (2005). Neither of these approaches can be characterized as having a high probability of success and this is not among our scientific requirements for success.

On the other hand, the morphology of the impact event may make it possible to constrain the porosity of the material being excavated, i.e., the porosity of the outermost 20–30 m. Extrapolating this deeper into the interior, however, would be inadvisable. Our detailed measurements of both the ambient coma and the ejecta, both gaseous and solid, may also allow us to better understand whether the ice-to-silicate ratio of the bulk material is the same as that of the solid nucleus.

6.2. STRUCTURAL PROPERTIES

Perhaps the first step in understanding the results from Deep Impact will come from studying the morphology of the impact phenomena. We need to determine which physics is relevant to the event, i.e., to determine which physical processes control the formation of the crater and the distribution of the ejecta. Fortunately, the behavior of the ejecta cone with time is very different for gravitationally controlled craters (remaining in contact with the surface throughout the event) and strength-controlled craters (lifting entirely off the surface at an early stage) and the ejecta cone will be much weaker for a compression-controlled crater. All other things being equal, the opening angle of the ejecta cone is sensitive to the porosity of the target. Filling in of the conical shell occurs, for example, due to enhancement by buried volatiles. Thus simple morphology will be the key to understanding which physics is relevant to the cratering process and thus to the structure of the nucleus in its outer layers. Simply resolving which scenario of crater formation is qualitatively correct will dramatically reduce the qualitative uncertainty in models of the surface layers. The final crater is also diagnostic not only of the process but also of the values of certain parameters depending on which physical processes dominate. The challenge will be to use different types of measurements to separate the various parameters that we would like to determine. Details of our expectations are given by Schultz and Ernst (2005) and by Richardson *et al.* (2005).

6.3. DIFFERENTIATION AND EVOLUTION

The ejecta from the crater are likely to have a composition that varies with time as material is excavated from deeper and deeper within the crater. This will be diagnostic of the differentiation in the outer layers. As noted earlier, the morphology of the ejecta is sensitive to whether volatiles are buried beneath the surfaces.

Spectroscopy and photogeology of the resultant crater will also be important in allowing us to investigate differences between the ambient surface of the nucleus and the material below the ambient surface – can we see layering in the walls of the crater? Can we see differences in spectral properties? And so on.

Spectroscopy of the coma, comparing the outgassing from the crater with the ambient outgassing, should allow us to determine whether the relative abundances of various volatiles is different in the interior and thus whether or not we are approaching primitive material. While it will be nearly impossible to determine whether we have gotten deep enough to reach primordial material, the differences should enable us to estimate the differences between primordial abundances and the abundances observed in the coma of many comets.

6.4. END STATES

Since we calculate that the nucleus has an active surface fraction of only a few to 10% (cf. the discussion in Belton *et al.*, 2005), the impactor will hit with very high probability in an inactive area. If the process of cometary inactivation is primarily one of sealing the ice inside by developing a thermally insulating and/or impenetrable mantle of refractory material, then one might expect the crater to become a new, active area, producing a jet with rather different ratios of volatiles than are observed in the ambient outgassing. Study of such a jet would be a key project for Earth-based observatories since the narrow window of the flyby observations limits *in situ* observations to the initiation of the jet and such a jet might last anywhere from hours to months.

6.5. OTHER POSSIBLE RESULTS

The puzzling topographic features seen from Stardust at comet Wild 2 lead us to expect great advances in understanding the surface features with our higher spatial resolution. If nothing else, we will observe a feature that is unambiguously an impact crater that can then be compared with features seen at Wild 2 that might be impact craters. Details of these investigations are given by Thomas *et al.* (2005). We will also obtain unprecedented spatial resolution on the coma near the nucleus. This will enable us to better understand the processes by which both gas and dust leave the nucleus and expand into the coma where they can subsequently be observed from Earth. There should be advances, for example, in understanding the acceleration in the inner coma and in understanding the chemical changes that take place in the inner coma.

7. Summary

Deep Impact is unusual among space missions in several ways. It will conduct one of the very few active, *in situ*, experiments ever done and our range of predictions of the outcome of that experiment are so qualitatively different that simple, qualitative results can lead to a major advance in our understanding of cometary nuclei. Furthermore, the rich variety of phenomena that could occur makes Earth-based observations a crucial part of the experiment.

Acknowledgements

This work was made possible by support from NASA for the Deep Impact project under contract NASW00004. We also acknowledge support from the Deep Impact Team.

References

Barbin, Y., Kofman, W., Nielson, E., Hagfors, T., Sui, R., Picardi, G., *et al.*: 1999, *Adv. Space Res.* **24**, 1127.

Bar-Nun, A. and Laufer, D.: 2003, *Icarus* **161**, 157.

Belton, M. J. S. and A'Hearn, M. F.: 1999, *Adv. Space Res.* **24**, 1167.

Belton, M. J. S., Meech, K. J., A'Hearn, M. F., Groussin, O., McFadden, L., Lisse, C., *et al.*: 2005, *Space Sci. Rev.* **117**, xxx.

Benkoff, J. and Huebner, W.:1995, *Icarus* **114**, 348.

Biver, N., Bockelée-Morvan, D., Colom, P., Crovisier, J., Henry, F., Lellouch, E., *et al.*: 2002, *Earth, Moon, Planets* **90**, 5.

Blume, W. H.: 2005, *Space Sci. Rev.* **117**, xxx.

Brownlee, D. E., Horz, F., Newburn, R. L., Zolensky, M., Duxbury, T. C., Sanford, S., *et al.*: 2004, *Science* **304**, 1676.

Chodas, P. W. and Yeomans, D. K.: 1996, in Noll, K. S., Weaver, H. A., and Feldman, P. D. (eds.) *Proc. IAU Coll. 156,* Cambridge University Press, New York.

Clarke, A. C.: 1968, 2001, *A Space Odyssey,* Signet Books, New York, Chap. 18.

Cook, A. H.: 1980, *Interiors of the Planets,* Cambridge University Press, Cambridge, Section 5.6.

Crovisier, J.: 1996, in Noll, K. S., Weaver, H. A., and Feldman, P. D. (eds.) *The Collision of Comet Shoemaker-Levy 9 and Jupiter,* Cambridge University Press, Cambridge, UK, p. 31.

Davidsson, B. J. R. and Gutiérrez, P. J.: 2004, *Icarus* **168**, 392.

Farnham, T. L. and Cochran, A. L.: 2002, *Icarus* **160**, 398.

Fernández, Y. R., Jewitt, D. C., and Sheppard, S. S.: 2001, *Astrophys. J.* **553**, L197.

Hampton, D. L., Baer, J. W., Huisjen, M. A., Varner, C. C., Delamere, A., Wellnitz, D. D., *et al.*: 2005, *Space Sci. Rev.* **117**, xxx.

Harmon, J. K., Campbell, D. B., Hine, A. A., Shapiro, I. I., and Marsden, B. G.: 1989, *Astrophys. J.* **338**, 1071.

Harmon, J. K., Ostro, S. J., Benner, L. A. M., Rosema, K. D., Jurgens, R. F., Winkler, R., *et al.*: 1997, *Science* **278**, 1921.

Harwit, M.: 1984, *Cosmic Discovery: The Search, Scope, and Heritage of Astronomy,* MIT Press, Cambridge, MA.

Housen, K. R., Holsapple, K. A., and Voss, M. E.: 1999, *Nature* **402**, 155.

Irvine, W. M., Dickens, J. E., Lovell, A. J., Schloerb, F. P., Senay, M., Bergin, E. A., *et al.*: 1997, *Earth, Moon, Planets* **78**, 29.

Kirk, R. L., Howington-Kraus, E., Soderblom, L. A., Giese, B., and Oberst, J.: 2004, *Icarus* **167**, 154.

Klaasen, K. P., Carcich, B., Carcich, G., Grayzeck, E. J., and McLaughlin, S.: 2005, *Space Sci. Rev.* **117**, xxx.

Kouchi, A. and Sirono, S.: 2001, *Geophys. Res. Lett.* **28**, 827.

Langevin, Y., Kissel, J., Bertaux, J.-L. and Chassefiere, E.: 1987, *Astron. Astrophys.* **187**, 761.

Lellouch, E.: 1996, in Noll, K. S., Weaver, H. A., and Feldman, P. D. (eds.) *The Collision of Comet Shoemaker-Levy 9 and Jupiter,* Cambridge University Press, Cambridge, UK, p. 31.

Lisse, C. M., A'Hearn, M. F., Farnham, T. L., Groussin, O., Meech, K. J., Fink, U., *et al.*: 2005, *Space Sci. Rev.* **117**, xxx.

Mastrodemos, N., Kubitschek, D. G., and Synnott, S. P.: 2005, *Space Sci. Rev.* **117**, xxx.

McFadden, L., Rountree-Brown, M., Warner, E., McLaughlin, S., Behne, J., Ristvey, J., *et al.*: 2005, *Space Sci. Rev.* **117**, xxx.

Meech, K. J., A'Hearn, M. F., Lisse, C. M., Weaver, H. A., and Biver, N.: 2005, *Space Sci. Rev.* **117**, xxx.

Merényi, E., Földy, L., Szegö, K., Tóth, I., and Kondor, A.: 1990, *Icarus* **86**, 9.

Moore, M. H., Donn, B., Khanna, R., and A'Hearn, M. F.: 1983, *Icarus* **54**, 388.

Nakamura, Y., Latnam, G. V., and Dorman, H. J.: 1982, *J. Geophys. Res.* **87**, A117.

Oberst, J., Giese, B., Howington-Kraus, E., Kirk, R., Soderblom, L., Buratti, B., *et al.*: 2004, *Icarus* **167**, 70.

O'Keefe, J. D., Stewart, S. T., and Ahrens, T. J.: 2001, *32nd Lunar and Plan. Sci. Conf. Abstract* #2002.

Prialnik, D.: 2002, *Earth, Moon, Planets* **89**, 27.

Richardson, J., Melosh, H. J., Artemeiva, N. A., and Pierazzo, E.: 2005, *Space Sci. Rev.* **117**, xxx.

Rickman, H., Kamel, L., Festou, M. C., and Froeschle, C.: 1987, Rolfe, E. J., and Battrick, B. (eds.), *Diversity and Similarity of Comets*, ESA SP278, European Space Agency, Noordwijk.

Schultz, P. H. and Ernst, C.: 2005, *Space Sci. Rev.* **117**, xxx.

Sekanina, Z.: 1996, in Noll, K. S., Weaver, H. A., and Feldman, P. D. (eds.) *The Collision of Comet Shoemaker-Levy 9 and Jupiter,* Cambridge University Press, Cambridge, UK, p. 55.

Sekanina, Z.: 1997, *Astron. Astrophys.* **318**, l5.

Sunshine, J. M., A'Hearn, M. F., Groussin, O., McFadden, L., Klaasen, K. P., Schultz, P. H., *et al.*: 2005, *Space Sci. Rev.* **117**, xxx.

Thomas, P. C., Veverka, J., A'Hearn, M. F., McFadden, L., and Belton, M. J. S.: 2005, *Space Sci. Rev.* **117**, xxx.

Tsou, P., Brownlee, D. E., Sandford, S. A., Hőrz, F., and Zolensky, M. E.: 2003, *J. Geophy. Res.* **108**, 8113.

Weaver, H. A., Sekanina, Z., Toth, I., Delahodde, C. E., Hainaut, O. R., Lamy, P., *et al.*: 2001, *Science* **292**, 1329.

Weidenschilling, S. J.: 1997, *Icarus* **127**, 290.

Whipple, F. L.: 1978, *Moon Planets* **18**, 343.

Yeomans, D. K., Giorgini, J. D. and Chesley, S. R.: 2005, *Space Sci. Rev.* **117**, xxx.

DEEP IMPACT MISSION DESIGN

WILLIAM H. BLUME

*Jet Propulsion Laboratory, California Institute of Technology, 4800 Oak Grove Drive,
Mail Stop 301-140L, Pasadena, CA 91109, U.S.A.*
(e-mail: wblume@jpl.nasa.gov)

(Received 10 September 2004; Accepted in final form 28 December 2004)

Abstract. The Deep Impact mission is designed to provide the first opportunity to probe below the surface of a comet nucleus by a high-speed impact. This requires finding a suitable comet with launch and encounter conditions that allow a meaningful scientific experiment. The overall design requires the consideration of many factors ranging from environmental characteristics of the comet (nucleus size, dust levels, etc.), to launch dates fitting within the NASA Discovery program opportunities, to launch vehicle capability for a large impactor, to the observational conditions for the two approaching spacecraft and for telescopes on Earth.

Keywords: comets, space missions, mission design, 9P/Tempel 1

1. Mission Summary

The Deep Impact mission explores the interior of comet 9P/Tempel 1 by using a 364-kg impactor to excavate a crater in the comet's surface and collecting observations of the ejecta and newly exposed cometary interior with a companion flyby spacecraft. Deep Impact is the eighth mission in NASA's Discovery Program, following NEAR-Shoemaker, Mars Pathfinder, Lunar Prospector, Stardust, Genesis, CONTOUR, and MESSENGER. The project is organized as a team between the principal investigator, Dr. Michael A'Hearn of the University of Maryland; the science team of 11 other prominent experts on comets, remote sensing, and impact physics; the industrial partner, Ball Aerospace and Technologies Corp.; and the Jet Propulsion Laboratory as the NASA lead center.

The Deep Impact cratering experiment targets the nucleus of comet Tempel 1 near the time of perihelion for its 2005 apparition. This is accomplished by launching two joined spacecraft (flyby spacecraft + impactor) in December 2004/January 2005 to approach the comet in early July 2005. Using spacecraft optical observations of the comet and conventional ground-based navigation techniques, the joined spacecraft are maneuvered as close as possible to a collision trajectory with the nucleus of Tempel 1, and the impactor is released 24 h before impact. Figure 1 shows the two spacecraft at impactor release.

The impactor, a battery-powered spacecraft with a dry mass of 364 kg, observes the approaching nucleus with an optical camera and maneuvers itself to a collision course toward the lighted portion of the nucleus. After separation from the impactor,

Space Science Reviews (2005) 117: 23–42
DOI: 10.1007/s11214-005-3386-4

the flyby spacecraft maneuvers to delay and deflect its flight path toward the nucleus so that it can observe the impact, ejecta, crater development, and crater interior during a 500-km flyby of the nucleus that occurs about 14 min after the impact. The flyby spacecraft carries a remote sensing payload of two instruments for imaging and infrared spectroscopy. Close-in observations of the nucleus by the impactor camera are sent to the flyby spacecraft by a radio link in the last minutes before impact. The flyby spacecraft sends the highest priority scientific and engineering data to the ground in realtime during the encounter and also records the primary data sets for later playback. Simultaneous observations of the comet before, during, and after the impact are also conducted from ground and space-based observatories as an essential part of the total experiment. All scientific and supporting engineering data is archived for future use by the scientific community. Figure 2 shows an overview timeline for the mission.

The Deep Impact mission was originally proposed to launch in January 2004, using a 1-year Earth-to-Earth trajectory and an Earth flyby to initiate a direct 6-month transfer to intercept the comet. In March 2003, NASA approved a 1-year launch delay to allow more time for delivery of the spacecraft hardware and system-level testing. The mission now uses essentially the same 6-month direct trajectory to the comet that was the final trajectory segment of the 2004 mission. Although some of the launch conditions now change, the launch energy and launch mass capability are nearly identical for use of the Delta II 7925 launch vehicle. As well, the approach conditions at Tempel 1 are unchanged, so the original designs for the approach and encounter phases of the mission are unaffected.

2. The Target: Comet Tempel 1

Tempel 1, officially designated 9P/Tempel 1, is a periodic comet with a current orbital period of 5.5 years and an inclination of 10.5°. It was first discovered in 1867 by Ernst Tempel in Marseilles, France. Gravitational interactions with Jupiter near aphelion have changed the comet's orbit over time, and it was not observed between its 1879 and 1967 apparitions. Orbital elements for the 2005 apparition are given in Table I (Yeomans *et al.*, 2005). The descending nodal crossing (1.506 AU on 7 July 2005) is quite close to perihelion and allows a fairly low-energy intercept trajectory to depart Earth in late December 2004/January 2005, with favorable geometry for Earth-based observing as the comet approaches perihelion. Tempel 1 is observable from Earth throughout the mission, with the solar elongation never less than 95° (on the first launch date) and reaching opposition on 4 April 2005.

Regular observations of the comet have been conducted by the Science Team and cooperating astronomers since before the 2000 apparition (perihelion 3 Jan 2000) to gather information on characteristics of the comet to support the engineering design of the mission and plans for in-flight observations (Meech *et al.*, 2005). Analysis of these data is homing in on estimates of the nucleus albedo, mean radius and axial

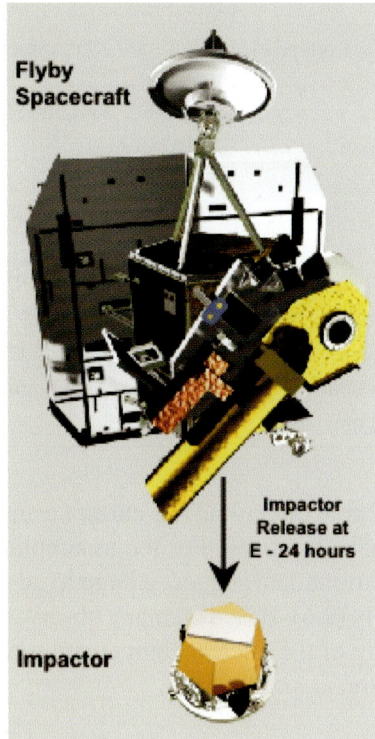

Figure 1. The Deep Impact flyby spacecraft and impactor shown at the time of impactor release, 24 h before the intended impact on comet 9P/Tempel 1.

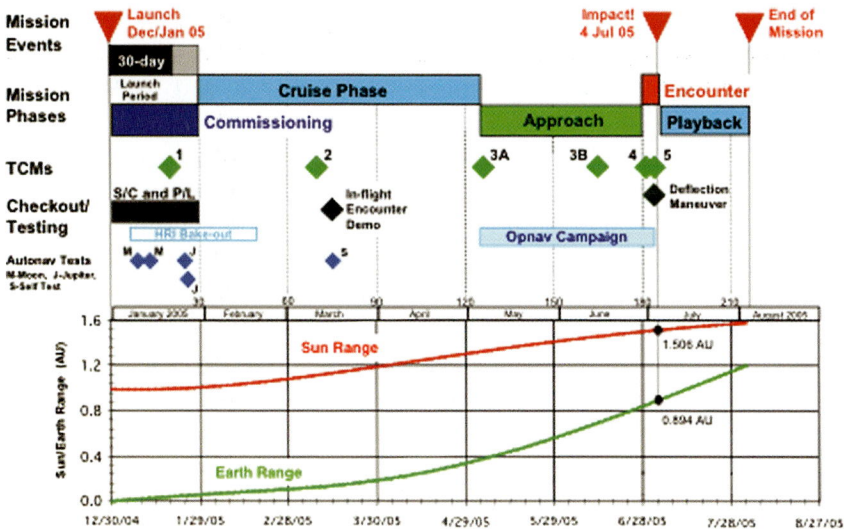

Figure 2. Timeline for the Deep Impact mission, based on the earliest launch date of 30 December 2004.

TABLE I

Tempel 1 orbital elements for the 2005 apparition.

Epoch	9 July 2005, 00:00:00 ET
Semi-major axis	3.1215288 AU
Eccentricity	0.5174906
Inclination	10.53009°
Ascending node	68.93732°
Argument of perihelion	178.83893°
Perihelion distance	1.5061670 AU
Perihelion epoch	5 July 2005, 07:34:02 ET
Coordinate system	Ecliptic and Mean Equinox of J2000
Astronomical unit (IAU)	149597870 km

ratio, rotation period, and pole direction. The current estimate of the nucleus mean radius is 3.25 km for an albedo of 4%. For the assumption that the nucleus has a prolate spheroid shape, current data suggest a large axial ratio of 3.2, as compared to Halley (2.0) and Borrelly (2.6). Based on many observations attempting to define the magnitude and phasing of the nuclear light curve, a rotational period of about 41.85 h is suggested (Belton $et\ al.$, 2005).

3. Alternate Targets

Comet Tempel 1 was selected as the target for the Deep Impact mission after a study of several potential targets that were available for launches from mid-2003 to September 2004, the time period required by the third NASA announcement of opportunity for the Discovery Program. Beginning in early 1998, this study looked for comets that could be reached for low launch energies ($\leq 15\ \mathrm{km}^2/\mathrm{s}^2$, twice the injection energy per unit mass), in order to fly the greatest impactor mass. Favorable approach phase angle (require $< 70°$, desire $< 45°$), approach speed (require 10–15 km/s), Earth viewing conditions (desire distance < 1 AU and elongation $> 90°$) at impact, encounter solar distance (desire 1.0–1.5 AU), and nucleus size (require radius ≥ 2 km) were also important considerations.

Two of the targets studied, asteroids 3200 Phaethon and 4015 Wilson-Harrington, are considered probable extinct or dormant comet nuclei, but these were rejected because of the uncertainty on the result of an impact experiment if there may be no volatile material. Two comet targets, 2P/Encke and 73P/Schwassmann-Wachmann 3, were studied, but these were already targets of the ill-fated CONTOUR mission and were rejected. Comet 9P/Tempel 1 became an early contender in this study because it had favorable launch and encounter conditions and, at the time, it was the target for the US-French comet rendezvous mission study Deep Space 4/Champollion (later called Space Technology 4 or ST4), which was planned to reach the comet after perihelion in December 2005. The possibility that this rendezvous/sample-return mission, under the NASA New Millennium space technology program, could arrive to study the newly formed crater made the synergy between these missions very attractive. Unfortunately, the ST4 mission study was cancelled only weeks before Deep Impact was selected as the eighth Discovery mission.

Tempel 1 is the selected target for the mission because the 2005 apparition allows a spacecraft of over 1000 kg to be launched to meet the comet near its perihelion at a distance of less than one AU from Earth. The Sun–Earth–Comet angle of 104° at the time of impact is favorable for simultaneous viewing from Earth. The encounter conditions give an approach speed of 10.2 km/s to provide high momentum for the impactor and a phase angle of about 63°, which is acceptable for approach imaging to target the nucleus. The comet activity is considered modest and predictable, so that the dust hazard could be well characterized for the design of spacecraft shielding. Some of the alternate targets considered for the Deep Impact mission and the reasons that they were not selected are listed in Table II. Although trajectory

TABLE II

Alternate targets considered for the Deep Impact mission.

Comet	Reasons not selected
(3200) Phaethon (launch Feb 2004, encounter Jan 2005)	Comet heritage isn't definite, very high flyby speed (32 km/s)
(4015) 107P/Wilson-Harrington (launch Nov 2002, encounter May 2005)	Comet heritage isn't definite, launch date too early, long flight time
2P/Encke (launch Feb 2003, encounter Nov 2003)	Launch date too early, high launch energy $(34\,km^2/s^2)$, very high flyby speed (28 km/s), CONTOUR target
73P/Schwassmann-Wachmann 3 (launch Mar 2004, encounter May 2006)	Small nucleus, uncertain dust environment after comet split in 1995, CONTOUR target
4P/Faye (launch March 2004, encounter Oct 2006)	Larger phase angle on approach (74°), higher launch energy, long flight time, high Sun range (1.7 AU)
10P/Tempel 2 (launch Feb–May 2003, encounter Jan–Feb 2005)	Impact near conjunction, not observable from Earth, launch date too early
37P/Forbes (launch Mar 2004 with Earth flyby, encounter Sep 2005)	Higher launch energy $(17\,km^2/s^2)$, quite high phase angle (86°)
41P/Tuttle-Giacobini-Kresak (launch Feb–Mar 2004, encounter Apr 2006)	Small nucleus with erratic outbursts, higher flyby speed (15 km/s), longer flight time
49P/Arend-Rigaux (launch Jul 2004, encounter Apr 2005)	Higher phase angle, poorer Earth viewing (1.4 AU, elongation 70°), very low activity
58P/Jackson-Neujmin (launch Jan 2003, encounter Dec 2003)	Launch date too early, larger phase angle on approach (84°), small nucleus, poorer Earth viewing (1.9 AU)

opportunities to a number of comets existed in the time period originally considered for launch, a number of factors, including high launch energy, poor approach phase angle, poor Earth viewing, small nucleus size, and uncertain dust environment eliminated most of the potential targets.

4. System Overview

4.1. LAUNCH VEHICLE

The launch vehicle for Deep Impact is the Boeing Delta II three-stage rocket. Launch occurs from Cape Canaveral from late December 2004 to January 2005. The Delta vehicle provides a reliable capability and a moderate cost for delivering a sufficient impactor mass (require >350 kg) at the Tempel 1 launch opportunity. For the required launch energy to reach Tempel 1 near perihelion in 2005 (see trajectory and launch phase descriptions below), the Delta II can carry a spacecraft mass of approximately 1020 kg.

The project originally planned to use a "heavy" version of the Delta II with larger strap-on rockets to fly an impactor mass of up to 500 kg, but elected to save about $6 million in launch vehicle cost with a lighter impactor. The heavy version had the greatest payload capability of the vehicles allowed by the Discovery guidelines. Launch vehicles smaller than the Delta were available, but could not carry sufficient payload.

4.2. FLIGHT SYSTEM (FLYBY SPACECRAFT AND IMPACTOR)

The primary element of the flight system (Figure 3) is the flyby spacecraft, which carries and targets the impactor up to 24 h before the impact; provides the stable platform for remote sensing of the comet nucleus before, during, and after the impact event; and provides the communications node for storage and return of the scientific data to Earth.

As seen in Figure 3, the flyby spacecraft is made up of the five-sided bus, which houses most of the subsystems; the instrument platform; and the two panels of the solar array, which fold against the bus in the launch configuration. The impactor is attached partially inside the bottom side of the flyby and serves as the launch vehicle adapter, mounting on the Delta II third stage so that the flyby $+X$ axis is upward at launch. After separation from the third stage, the solar panels are deployed to form a flat array normal to the flyby $+Y$ axis. The instrument platform mounts on the $-Y$ face of the flyby bus, with the instrument boresights aimed at a 45° angle between the $+Z$ and $-X$ axes. When the impactor is released, springs push it away in the $-X$ direction, and most of the shielding to protect against comet dust impacts is located on the flyby bottom face.

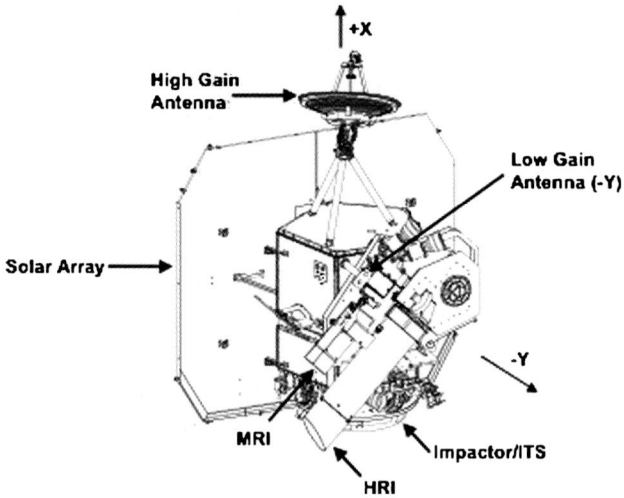

Figure 3. The Deep Impact flyby spacecraft and impactor shown mated in the cruise configuration.

The bus structure is made up of two modules: the propulsion module and the instrument module. The propulsion module is the backbone of the structure and houses all the components required for the propulsion subsystem and the components required on the flyby for the flyby/impactor separation. The instrument module houses all the components and wire harnesses required for other engineering subsystems. Most of the electronic components of the flyby spacecraft engineering subsystems (computers, star trackers, transponders, etc.) are redundant.

The solar array has two deployable wings, each one requiring a 36° deployment to reach the fully deployed position. The instrument platform provides the thermally and structurally stable mounting for all the alignment-critical components, including the flyby instrument complement, the two star trackers, and the inertial reference unit. The mechanisms on the flyby spacecraft are two gimbals for the high gain antenna, deployment and latch mechanisms for the solar arrays, and the flyby/impactor separation system. The mass of the flyby spacecraft at launch is about 601 kg, with a total flight system mass (fully fueled) of 973 kg. The flyby spacecraft carries about 86 kg of hydrazine propellant (including pressurant) for propulsive maneuvers, momentum control, and damping high rotational rates, such as at separation from the launch vehicle. Reaction wheels provide fine pointing control for imaging and in the cruise attitude for transit to Tempel 1. An S-band radio relay system provides communication with the impactor after release.

4.3. THE IMPACTOR

The impactor is a fully functional battery-powered spacecraft that operates on its own for 24 h after release from the flyby spacecraft. Its primary purpose is to open a large crater on the nucleus of Comet Tempel 1. To accomplish this, the impactor uses its autonomous navigation software to guide itself to an impact in a lighted portion of the nucleus using images from the impactor targeting sensor (ITS) to estimate targeting errors. Hydrazine thrusters are used for pointing and to correct the flight path. Because of its short operational life, the impactor has a single-string design, with many of its components being identical to those on the flyby spacecraft for cost efficiency. This includes the flight computer, inertial measurement unit, star tracker, and S-band radio relay. The impactor has a mass at launch of about 372 kg, including the hydrazine propellants (about 8 kg, including pressurant), and it also serves as the launch vehicle adapter, providing the mechanical interface to the Delta third stage.

4.4. THE PAYLOAD

Three remote sensing instruments collect the scientific observations of Tempel 1 and the cratering event. The instruments also serve as navigation tools for setting up the impactor trajectory toward the comet, by ground navigation, and for pointing of the spacecraft toward the comet (Hampton *et al.*, 2005). The high-resolution instrument (HRI) combines a visual imager and infrared spectrometer with a 30-cm mirror in a Cassegrain telescope to provide high-resolution images and spectral maps of the nucleus and coma (2-mrad FOV, 2-μrad IFOV). The HRI is also the primary optical navigation sensor. The medium-resolution instrument (MRI) is a visual imager with a 12-cm mirror in a Cassegrain telescope to provide wider-field images, including full-nucleus context views at the closest distances (10-mrad FOV, 10-μrad IFOV). It also serves as the autonomous navigation sensor to support pointing of the flyby spacecraft on approach to the nucleus. The impactor target sensor (ITS) is identical to the MRI, except without a filter wheel, and serves as the autonomous navigation sensor on the impactor. In the final minutes before impact, it also provides close-in images of the nucleus to capture the topography at the impact site.

4.5. AUTONOMOUS NAVIGATION SOFTWARE

The autonomous navigation software (AutoNav) provides the capability on the impactor and flyby spacecraft to process images of the comet and update the ephemerides of the nucleus and the spacecraft (Mastrodemos *et al.*, 2005). Because the spacecraft ephemeris is well known from ground navigation processes, this effectively provides a target-relative ephemeris. This information is used to control timing and pointing on both spacecraft, and additionally to compute maneuvers to accomplish an intercepting trajectory for the impactor. AutoNav uses information from the attitude control system sensors (star trackers and inertial measurement units) on both the impactor and the flyby spacecraft, the MRI and HRI on the flyby and the ITS on the impactor.

The AutoNav software requires images from the MRI or ITS at 15-s intervals and attitude information associated with each image from the attitude control sensors. In return, AutoNav performs image processing, orbit determination, and maneuver computations to provide target relative position and velocity, ΔV magnitude and inertial direction for impactor targeting maneuvers, and timing adjustments for image sequence optimization on both spacecraft. The AutoNav capability is based on heritage from the Deep Space 1 and Stardust missions, which successfully encountered comets Borrelly (2001) and Wild 2 (2004), respectively (Bhaskaran et al., 2004). On Deep Space 1, autonomous interplanetary navigation was also successfully demonstrated (Bhaskaran et al., 2000).

Because of the large uncertainty in the comet's downtrack position, which cannot be resolved by ground-based optical navigation, autonomous navigation is an essential capability on the flyby spacecraft for pointing the instruments. Because of the cost of a "smart" self-guided impactor, a recurring issue on the project was whether a "dumb" impactor could be substituted for a lower cost. This question was always resolved in favor of the self-guided impactor, because numerous studies showed that late maneuvers by the impactor were essential to having a high confidence of hitting in a lighted area of an irregular nucleus, where the crater could be observed by instruments on the flyby spacecraft (Mastrodemos et al., 2005).

5. Trajectory Description

5.1. INTERPLANETARY TRAJECTORY

A direct trajectory strategy, with about 6 months between launch and encounter, is used to launch Deep Impact in late December 2004 to January 2005. Figure 4 shows the interplanetary trajectory for Deep Impact for the first launch date. The original project baseline was to launch a year earlier in January 2004 and use a 1-year Earth-to-Earth trajectory segment to link up to the 2005 direct trajectory opportunity. The January 2004 launch strategy fit a shorter development schedule and provided more time in space for payload and spacecraft testing, but the project encountered hardware delivery delays that required a slip in the launch schedule by 1 year.

In terms of delivering a substantial science payload, the best opportunity for an intercept mission of Comet Tempel 1 is to arrive near the descending node of the comet's orbit (1.506 AU on 7 July 2005), which is very close to perihelion (5 July 2005). However, 4 July 2005 is a rough optimum for the encounter date when considering all the mission parameters, as it provides nearly the maximum injected mass (to allow a larger impacting mass). Later arrival dates do not provide much additional mass, and the approach phase angle and Earth range are degraded (therefore degrading approach observations, telecommunication rates, and Earth-based

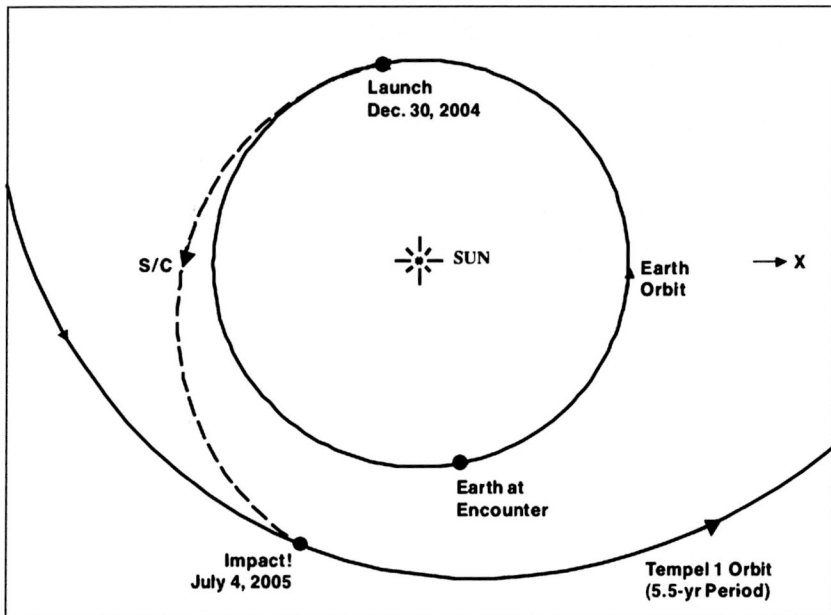

Figure 4. The Deep Impact interplanetary trajectory projected into the Ecliptic. A launch opportunity of 30 days begins on 30 December 2005, with a fixed arrival date at the comet of 4 July 2005.

observing). Earlier arrival dates provide a better approach phase angle and Earth range, but the mass capability falls off significantly and the encounter speed increases (resulting in either a larger deflection maneuver or less flyby observing time between impact and entering a shielded attitude for the coma crossing). Figure 5 shows the primary trade for the arrival date between the launch energy (for the optimal launch date) and the approach phase angle. While a lower phase angle is highly desirable for approach targeting, the autonomous navigation capability allowed selection of a later arrival date to provide for a larger impactor mass, to open a larger crater on the nucleus. For a 4 July encounter date, the optimum launch date is 9 January 2005 (minimum energy), and the best 21-day launch period extends from 30 December 2004 through 19 January 2005. Table III shows characteristics of the trajectory for the beginning, middle, and end of this primary launch period.

5.2. ENCOUNTER TRAJECTORY

Although the Deep Impact spacecraft and comet P/Tempel 1 are both in curved orbits around the Sun, their high-speed intersection results in nearly straight-line relative trajectories, as shown in Figure 6. Two final targeting maneuvers by the flyby spacecraft place the impactor on a trajectory that ideally impacts the comet after

TABLE III

Deep Impact mission parameters 21-day primary launch period – 30 December 2004 to 19 January 2005.

Trajectory parameter	Open	Middle	Close
Launch (L)			
Launch date (2004/2005)	30 Dec	9 Jan	19 Jan
Launch energy (km/s)^2	11.58	10.79	11.75
Launch declination	$-4.37°$	$-3.48°$	$-1.70°$
Launch mass capability[a] (kg)	1018	1035	1015
Encounter (E)			
Encounter date (2005)	4 July	4 July	4 July
Impact speed (km/s)	10.20	10.28	10.36
Approach phase angle[b]	63.4°	62.5°	61.9°
Sun range (AU)	1.506	1.506	1.506
Earth range (AU)	0.894	0.894	0.894
Sun–earth-S/C angle	103.9°	103.9°	103.9°
Sun–S/C-earth angle	40.9°	40.9°	40.9°
Deflection ΔV[b] (m/s)	100.0	100.8	101.6
End of mission (EOM)			
EOM date (2005)[c]	3 August	3 August	3 August
EOM Sun range (AU)	1.570	1.573	1.576
EOM Earth range (AU)	1.203	1.206	1.208

[a]Single fixed flight system mass (976 kg, with ballasting on the third stage as required) will be used over the 21-day launch period (now extended to 30 days through 28 Jan 2005).
[b]At E−24 h. Deflection ΔV for post-impact imaging time of 800 s.
[c]At E+30 days.

release at E−24 h, but an accurate impact on the sunward side of the nucleus cannot be guaranteed without maneuvers closer to the target. After release, the impactor makes up to three small adjustments to this trajectory to correct any errors in the initial targeting and to aim for an impact on the lighted side of the nucleus. Each adjustment (impactor targeting maneuver or ITM) is computed by the AutoNav software. The impactor approaches the comet from a 63° phase angle, so the ITS does not see a fully illuminated nucleus.

After releasing the impactor, the flyby spacecraft deflects and delays its trajectory in order to safely pass by the nucleus and to have time to observe the impact and resulting crater. This deflection maneuver targets for a 500-km miss distance, which is selected to provide a survivable path through the inner coma dust environment while still allowing the HRI to meet the total resolution objective (7-m requirement, 3.4-m goal) for the last crater images. The flyby spacecraft configuration and dust

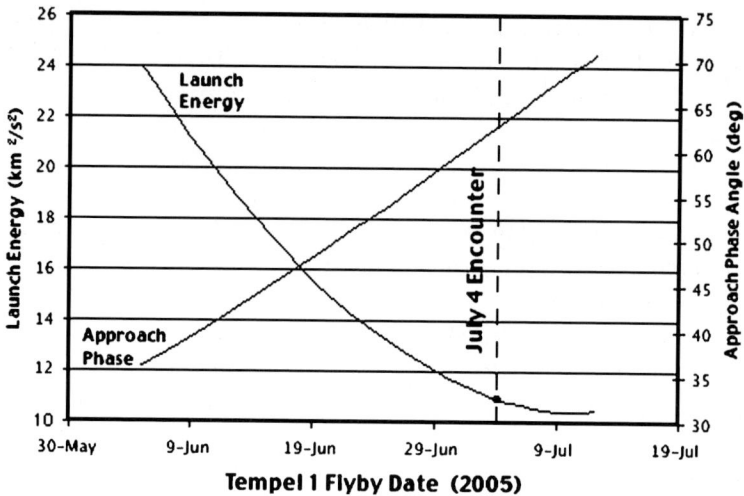

Figure 5. Arrival date trade space between launch energy for the optimal launch date and approach phase angle. The 4 July encounter date was selected to allow a massive impactor (originally 500 kg) with adequate margin for developing the flight hardware within the launch vehicle capability.

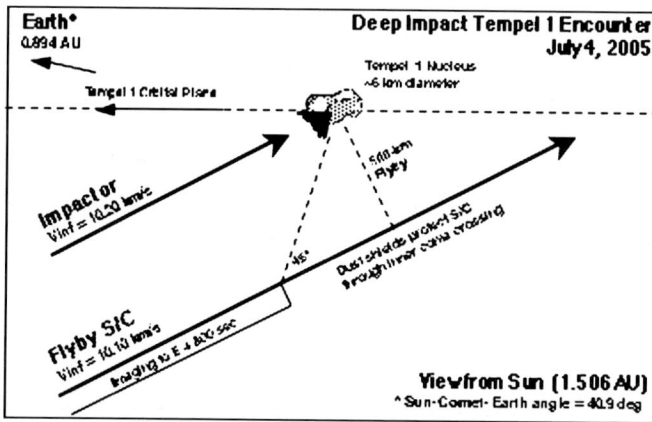

Figure 6. The flight path of the Deep Impact flyby spacecraft and impactor approaching the comet nucleus, as viewed from the Sun. The flyby spacecraft trajectory is targeted "below" the nucleus at a distance of 500 km to cross the comet's orbital plane after the closest approach.

shielding are designed so that MRI and HRI imaging can continue on approach to the nucleus until the spacecraft rotation reaches 45° to the relative velocity vector at about 700 km from the nucleus. At this orientation, the spacecraft control system then holds the spacecraft inertially in this "shield mode" for the passage past the nucleus, with the dust shielding providing protection to the critical subsystem components.

The deflection maneuver also slows down the flyby spacecraft, delaying closest approach until 850 s after impact, to allow time for imaging the impact event. Because imaging continues until the flyby spacecraft reaches a turn angle of 45° on the ideal trajectory, this strategy provides the science requirement of 800 s of observation time from impact until the shield mode attitude is reached. The science expectation is that the crater formation time is about 200 s, but with a high uncertainty, so the 800-s delay time provides a factor of four of conservatism.

The third component of the deflection maneuver is determined by the aiming point for the flyby trajectory, which determines its geometry relative to the Sun and Earth and to the orbit plane. For some small-body missions, a subsolar flyby is selected to provide the widest range of illumination angles, but the Deep Impact aimpoint is selected 90° to the Sun direction on the south ecliptic side of the nucleus. This provides two advantages: First, for a fixed solar array oriented parallel to the relative approach velocity, the direction of rotation to keep the imaging instruments pointed at the nucleus also keeps the Sun angle to the solar array constant at a maximum value during the flyby. Second, because the flyby spacecraft approaches the nucleus from "below," in the sense of its angular momentum vector around the Sun, this aimpoint means that the flyby spacecraft will pass through the comet's orbit plane after closest approach, as seen in Figure 6. This is considered to be an advantage in mitigating the possibility of impacting larger debris from the nucleus that may be co-orbiting around the Sun. The Giotto spacecraft suffered a large impact on approach to comet Halley in 1986, on a trajectory that crossed the orbit plane before closest approach. Although there is no good model for the size, volume density, and extent of such debris, it is thought to be more concentrated near the orbital plane. This approach geometry provides the greatest time for imaging and realtime data return before the crossing, and the crossing is at nearly the greatest distance possible from the nucleus for a 500-km closest approach.

The time of impact on 4 July 2005 is selected to provide highly reliable reception of the critical telemetry immediately before and after impact and excellent observing of the event from Earth-based observatories. These considerations favor use of the observatories at Mauna Kea, Hawaii, which lie between the NASA Deep Space Network (DSN) tracking sites at Goldstone, California and Canberra, Australia. The time of impact is scheduled after darkness at Mauna Kea and in the tracking overlap between the DSN tracking sites in a window between 0540 and 0635 UTC (Earth received time). This window allows for the possibility to further optimize the time of impact after launch to allow simultaneous observations by the Hubble Space Telescope (HST), which is periodically occulted from viewing the comet on each orbit. Because of variable atmospheric drag, the timing of HST orbital events cannot be predicted reliably months in advance, so the 55-min window allows the time of impact to be adjusted by maneuvers in the last 60 days before the encounter. The science team selects an impact time in this window that allows the best combination of observing by Earth-based and space-based observatories.

6. Mission Phases and Key Events

Six mission phases are defined to simplify description of the different periods of activity during the mission. As shown in Figure 2, these are the launch, commissioning, cruise, approach, encounter, and playback phases. Two time epochs are useful for defining activities and the boundaries of some of the mission phases. Launch (L) is the time of liftoff of the Delta II launch vehicle. Encounter (E) is the time of impact with Tempel 1.

6.1. LAUNCH PHASE

The launch phase begins with the start of the launch countdown and ends with stabilization of the flight system in a sun-pointed attitude under three-axis attitude control, which is expected within 27 h after liftoff. This phase includes the important event of initial acquisition of the spacecraft radio signal by the DSN. Launch occurs from the Cape Canaveral Air Force Station during a 21-day primary launch period beginning 30 December 2004, with a 9-day secondary launch period through 28 January 2005 providing additional launch opportunities. The primary launch period is selected to maximize the injected mass by minimizing the launch energy ($11.75 \, \text{km}^2/\text{s}^2$ on 19 Jan 2005). Because the flight system mass came in lighter than expected, nine secondary launch dates with reduced, but acceptable margin were added with a maximum launch energy of $14.24 \, \text{km}^2/\text{s}^2$ on 28 January 2005. Two instantaneous liftoff times, separated by 39–40 min, are provided on each launch date to accommodate short interruptions that may occur late in the countdown.

6.2. COMMISSIONING PHASE

The commissioning phase extends from stabilization of the flight system under three-axis attitude control to 30 days after launch (L + 30 days). This is a period of initial operation, checkout, and calibration for the spacecraft and payload and includes an initial trajectory correction maneuver (TCM) to correct for launch vehicle injection errors.

Calibration of the payload and testing of the autonomous navigation system with some initial observations of the Moon are important objectives early in the commissioning phase. For the January 2004 launch of Deep Impact, the 1-year cruise between launch and the Earth flyby provided substantial time for checkout of the spacecraft subsystems, and the Earth flyby provided a unique opportunity for testing and calibration of the instruments and the AutoNav algorithms. With the 1-year launch delay, these activities must be prioritized and selectively scheduled during the commissioning or cruise phases, in order that the flight system and ground system reach full readiness at E−60 days for the approach and encounter phases of the mission.

Some initial science calibration measurements using the Moon are important to determine some characteristics of the payload and have been scheduled as early as 3 days after launch for the HRI and MRI to use the Moon as an extended and fairly uniform object. Similar measurements with the ITS are conducted as early as 13 days after launch after the impactor has been powered on and checked out. Tests of the AutoNav algorithms on the flyby spacecraft are also conducted using the Moon as a target. Subsequent tests of the 2-h encounter AutoNav encounter sequence use Jupiter as a target.

Because the HRI telescope structure is made with composite materials, it is subject to expansion if water is absorbed at the launch site in a brief period when the instrument is not purged with dry nitrogen. To provide for correction of the instrument focus due to this expansion, bakeout heaters provide the capability to warm the structure for up to 40 days and drive out the absorbed water. A focus test shortly after launch determines if this procedure is required.

6.3. CRUISE PHASE

The cruise phase begins 30 days after launch and ends 60 days before encounter with a duration of about 2–3 months, depending on launch date. Although a somewhat quieter period of transit between the busier commissioning and approach phases, this phase includes important science calibrations, an encounter demonstration test to run an encounter-like simulation on the flyby spacecraft, ground operational readiness tests, and a second TCM.

Monthly calibrations of the science instruments begin in this mission phase, with the three instruments observing a set of stars and nebula to further characterize the instrument performance and any time variation. To characterize the payload and attitude control system performance for the optical navigation imaging on approach to the comet, additional observations are collected.

Some initial long-range observations of Tempel 1 are attempted beginning in March 2005 for navigation purposes. The comet is at opposition from the spacecraft in early April, so the initial observations are at high sun angles and thermal control capabilities limit the observations to just a few minutes until the Sun–Spacecraft–Comet angle drops below 120° and the payload is fully shaded. The comet is also at opposition from Earth in early April (1.76 AU from the Sun), and ground-based observations are collected to update the comet's ephemeris for spacecraft targeting maneuvers in the approach phase.

With the fixed encounter date, the dates for cruise phase events beginning in late February 2005 are fixed on the calendar wherever possible to simplify scheduling. The date of TCM-2 is set for Thursday 10 March 2005, and fixed dates are planned for all subsequent maneuvers up to encounter. Some events early in the cruise phase, such as the end of the HRI bakeout, still vary with the actual launch date.

6.4. APPROACH PHASE

The approach phase extends from 60 days before encounter until 5 days before encounter (E−60 days to E−5 days, 5 May to 29 June 2005). This is the period of intensive observations to detect the nucleus of Comet Tempel 1 using the HRI and then refine the spacecraft and comet ephemerides. It also includes regular science observations of the comet, payload testing, and two maneuvers (TCM-3A and 3B) to target the exact encounter time. The start of the approach phase at E−60 days is the approximate time at which it should be possible to clearly identify the comet nucleus in the coma background using the HRI. Daily images and spectra are collected to further characterize the activity and light curve of the comet. One goal of this study is to better define the rotational pole and phasing at encounter and understand the implications for the autonomous targeting of the impactor.

Two trajectory correction maneuvers in this phase correct the targeting of the flight system toward the comet and make final adjustments to the impact time, which can be adjusted within a 55-min window on 4 July 2005 to optimize observations by the Hubble Space Telescope (HST). HST is in a 96-min orbit around Earth that is periodically occulted from seeing the comet, and only about 50 min of observations are available on each orbit.

For most of this mission phase the sun-spacecraft-comet angle is greater than 120°, and periods of observation are limited to approximately 15 min every 4 h by thermal constraints. About 10 days before encounter, this angle falls below 120° and the HRI can observe the comet continuously while shaded from the sun. This allows a high volume of optical navigation images to be collected to support the final targeting maneuvers (TCM-4 and 5) in the encounter phase.

6.5. ENCOUNTER PHASE

The encounter phase begins 5 days before and ends 1 day after the impact with Comet Tempel 1. This period includes the two final targeting maneuvers before release of the impactor, the 24-h impactor mission after release to achieve an impact on the lighted side of the comet nucleus, the deflection maneuver and subsequent imaging sequence for the flyby spacecraft during its close flyby of the nucleus with observations of the impact event and the resulting crater, and the playback of all the collected data.

Two final propulsive maneuvers complete the targeting of the impactor before it is released at E−24 h: TCM-4, executed at E−96 h, reduces the targeting error to well within the 30-km accuracy that is needed to provide a robust fuel margin for the impactor maneuver sequence after release. TCM-5, executed at E−30 h, provides a final, higher accuracy adjustment that has an accuracy of ∼2 km. This further reduces the propulsive requirements on the impactor and provides a good chance of achieving impact if the impactor is unable to execute any targeting maneuvers.

The impactor is released at E−24 h with a relative separation speed of ∼36 cm/s that is accounted for in the final targeting. Five minutes after release, the impactor first begins to fly on its own, establishing inertial pointing, turning the ITS toward the comet and initiating a 64-kbps S-band telemetry link with the flyby spacecraft, which is maintained until impact.

After releasing the impactor, the flyby spacecraft regains attitude control and turns to the proper attitude for the deflection maneuver of about 100 m/s, which targets a safe, 500-km flyby of the nucleus and slows the spacecraft down to provide a 850-s delay between the time of impact and the time of closest approach. An imaging attitude toward the nucleus is then reestablished, and images and spectra of the comet and the impactor data are broadcast to Earth on the 200-kbps X-band telemetry link. On Earth, the largest, 70-m, antennas of the DSN are focused on collecting the critical scientific and engineering data.

Beginning 2 h before the predicted impact, the AutoNav software on both spacecraft begin processing ITS images of the nucleus to update the ephemerides of the comet and spacecraft for pointing control and maneuver computations. On the impactor, the first targeting maneuver (ITM-1) is computed and executed at E−100 min. ITM-2 follows at E−35 min, and the final targeting adjustment is executed at E−7.5 min to place the impactor on a path toward collision on the lighted portion of the nucleus. Beginning at E−4 min, the impactor is pointed along the relative velocity vector and images are returned to characterize the nucleus topography at the impact site (Klaasen *et al.*, 2005). Imaging continues to impact, but dust impacts may degrade the image quality or impactor pointing in the last minute. Subframed images of only the CCD central pixels are used to most rapidly read out and transmit the highest resolution images of the impact site.

On the flyby spacecraft the AutoNav results are used to maintain pointing toward the nucleus and to update the predicted time of impact. With its deflected flight path, the flyby spacecraft can better resolve the time-of-flight uncertainty in the comet's ephemeris. Ground processing of the optical navigation images provides an improved impact time estimate that is relayed up through the flyby spacecraft and the S-band link to the impactor to better control the timing of the final imaging sequence. On the flyby spacecraft, an autonomous update is used to initiate a rapid imaging sequence shortly before impact to best capture the impact event and to characterize the ejecta and formation of the crater. Imaging and spectral data are collected on the flyby spacecraft, which turns to keep the instruments pointed at the nucleus until it reaches a distance of about 700 km from the nucleus, at which time the spacecraft is locked into a shielding attitude for safe passage through the inner coma at a closest distance of 500 km from the nucleus. Pointing of the high-gain antenna to Earth is maintained through the imaging sequence and shielded attitude to deliver the critical scientific data to the ground at the highest 200-kbps telemetry rate. This shielded attitude prevents further imaging until the flyby spacecraft is safely beyond the greatest dust hazard. At 22 min after closest approach, the flyby spacecraft is turned to collect lookback images and spectra of the opposite side of the nucleus at a higher phase angle of 117°. Redundant playbacks of the encounter data and periodic lookback imaging continue into the final, playback phase of the mission.

6.6. PLAYBACK PHASE

The playback phase begins one day after impact and continues to the end of mission at E+30 days (3 Aug 05). Completing the primary mission, this phase provides time to complete redundant playbacks of data stored during the encounter, to characterize the flyby spacecraft health after the encounter (including dust particle damage), and to leave the spacecraft in a known final configuration. Additionally, periodic lookback imaging of the comet is planned until 60 h after closest approach. A last calibration of the HRI and MRI is also conducted. Ground-based observations of Comet Tempel 1 continue indefinitely to allow monitoring of any changes in outgassing that would characterize evolution in the newly exposed surface layers.

7. Extended Mission Possibility

After the encounter, the flyby spacecraft is in a heliocentric orbit that has an orbital period of about 1.5 years. At this time, no follow-on mission is planned, but the orbital period means that the spacecraft re-encounters Earth in a distant flyby about 3 years after launch. If sufficient hydrazine remains in the fuel tank and other subsystems are healthy, this Earth flyby could be used as a gravity assist to another target. This possibility was first identified in early studies for the CONTOUR mission (Cornell University and Johns Hopkins University Applied Physics Lab, 1997).

Trajectory studies were conducted to identify small-body targets that could be accessible after the Earth flyby, under the following groundrules:

- no change to the baseline trajectory with a 4 July 2005 impact date,
- a minimum second Earth flyby altitude of 300 km, and
- reasonable expectations (100–150 m/s) of post-Tempel 1 maneuvering capability.

From several small-body targets identified in the initial search, five targets are possibly accessible within the spacecraft maneuvering capability and of great scientific interest. These include three comets (85P/Boethin, 103P/Hartley 2, and 10P/Tempel 2) and two asteroids (3200) Phaethon and (4015) 107P/Wilson-Harrington, which are considered possible extinct or dormant comet nuclei and which were considered as impact targets for the primary mission. Since the original studies, the discovery of numerous, very dark objects in cometary orbits has cast further doubt on brighter Phaethon being a dormant cometary nucleus, despite its association with a meteor stream. The Science Team selected Hartley 2 and Boethin as the targets of primary interest for more detailed study. Tempel 2 has a low probability of success, due to a higher maneuver requirement, and the asteroid targets have low approach phase angles that are actually bad for the Deep Impact spacecraft design, as the backside of the payload would be heated by the Sun. Observations of Hartley 2 have detected CO Cameron bands that have led to predictions of the abundance of CO_2, a prediction that could be tested with the Deep Impact instrumentation. Compared to Boethin, Hartley 2 has a longer flight time (encounter in October 2010), better Earth observing conditions (range 0.12 AU, Sun–Earth–Comet angle 121°), and a higher approach phase angle (108°). The Boethin encounter would occur earlier (December 2008), at a better approach phase angle (88°), but with less favorable Earth observing (range 0.89 AU, Sun-Earth-Comet angle 76°). Possible extended missions to these two targets are under study, but will depend on favorable interest from NASA, accurate launch conditions that conserve the propellant reserves, and the survival of the flyby spacecraft and payload during the passage through the inner coma at Tempel 1.

Acknowledgments

This research was carried out at the Jet Propulsion Laboratory, California Institute of Technology, under a contract with the National Aeronautics and Space Administration. The Deep Impact mission design evolved in a close partnership with the members of the Science Team and engineers at Ball Aerospace and Technologies Corp. and the Jet Propulsion Laboratory. At JPL, trajectory studies for the Deep Impact mission have been supported by Jennie Johannesen, Joan Pojman, Chen-wan Yen, Louis D'Amario, and Paul Penzo; and important launch vehicle studies have been supported by Greg Fruth and Jeff Tooley of the Aerospace Corp. Navigation studies were conducted by Ram Bhat, John Bordi, Raymond Frauenholz, Daniel Kubitschek, Nick Mastrodemos, George Null, Mark Ryne, and Stephen Synnott. Development of the overall mission strategy has been supported by John Aiello, Abraham Grindle, Grailing Jones, Jennifer Rocca, Calina Seybold, David Spencer, Ed Swenka, and Charles Wang.

References

Belton, M. J. S., *et al.*: 2005, *Space Sci. Rev.* **117**, XXX.

Bhaskaran, S., Riedel, J. E., Synnott, S. P., and Wang, T. C.: 2000, *The Deep Space 1 Autonomous Navigation System: A Post-flight Analysis*, Paper AIAA 2000-3935, AIAA/AAS Astrodynamics Specialist Conference, Denver, CO, August.

Bhaskaran, S., Riedel, J. E., and Synnott, S. P.: 2004, *Autonomous Target Tracking of Small Bodies During Flybys*, AAS Paper 04-236, AAS/AIAA Spaceflight Mechanics Conference, Maui, HI, February.

Cornell University and the Johns Hopkins University Applied Physics Laboratory: 1997, *CONTOUR Orals Briefing*, September 16, 1997.

Hampton, D. L., *et al.*: 2005, *Space Sci. Rev.* **117**, XXX.

Klaasen, K., Carcich, B., Carcich, G., Grayzeck, E., and McLaughlin, S.: 2005, *Space Sci. Rev.* **117**, XXX.

Mastrodemos, N., Kubitschek, D., and Synnott, S.: 2005, *Space Sci. Rev.* **117**, XXX.

Meech, K., A'Hearn, M., Fernandez, Y., Lisse, C., Weaver, H., Biver, N., and Woodney, L.: 2005, *Space Sci. Rev.* **117**, XXX.

Yeomans, D. K., Giorgini, J., and Chesley, S.: 2005, *Space Sci. Rev.* **117**, XXX.

AN OVERVIEW OF THE INSTRUMENT SUITE FOR THE DEEP IMPACT MISSION

DONALD L. HAMPTON[1,*], JAMES W. BAER[1], MARTIN A. HUISJEN[1],
CHRIS C. VARNER[1], ALAN DELAMERE[1], DENNIS D. WELLNITZ[2],
MICHAEL F. A'HEARN[2] and KENNETH P. KLAASEN[3]

[1]*Ball Aerospace & Technologies Corporation, P.O. Box 1062, Boulder, CO 80301, U.S.A.*
[2]*Department of Astronomy, University of Maryland, College Park, MD 20742-2421, U.S.A.*
[3]*Jet Propulsion Laboratory, California Institute of Technology, 4800 Oak Grove Drive,
Pasadena, CA 91109, U.S.A.*
(*Author for correspondence; E-mail: dhampton@ball.com)

(Received 21 August 2004; Accepted in final form 3 December 2004)

Abstract. A suite of three optical instruments has been developed to observe Comet 9P/Tempel 1, the impact of a dedicated impactor spacecraft, and the resulting crater formation for the Deep Impact mission. The high-resolution instrument (HRI) consists of an $f/35$ telescope with 10.5 m focal length, and a combined filtered CCD camera and IR spectrometer. The medium-resolution instrument (MRI) consists of an $f/17.5$ telescope with a 2.1 m focal length feeding a filtered CCD camera. The HRI and MRI are mounted on an instrument platform on the flyby spacecraft, along with the spacecraft star trackers and inertial reference unit. The third instrument is a simple unfiltered CCD camera with the same telescope as MRI, mounted within the impactor spacecraft. All three instruments use a Fairchild split-frame-transfer CCD with $1,024 \times 1,024$ active pixels. The IR spectrometer is a two-prism (CaF_2 and ZnSe) imaging spectrometer imaged on a Rockwell HAWAII-1R HgCdTe MWIR array. The CCDs and IR FPA are read out and digitized to 14 bits by a set of dedicated instrument electronics, one set per instrument. Each electronics box is controlled by a radiation-hard TSC695F microprocessor. Software running on the microprocessor executes imaging commands from a sequence engine on the spacecraft. Commands and telemetry are transmitted *via* a MIL-STD-1553 interface, while image data are transmitted to the spacecraft *via* a low-voltage differential signaling (LVDS) interface standard. The instruments are used as the science instruments and are used for the optical navigation of both spacecraft. This paper presents an overview of the instrument suite designs, functionality, calibration and operational considerations.

Keywords: comets, CCD cameras, IR spectrometer

Abbreviations: A/D – analog to digital converter; CCD – charge coupled device; DAC – digital to analog converter; DN – digital number (or data number); EEPROM – electrically erasable programmable read-only memory; FPA – focal plane array; FPGA – field programmable gate array; FWHM – full width at half maximum; HRI – high-resolution instrument; IC – instrument controller; ICB – instrument and impactor crosslink board; IP – instrument platform; IR – infrared; IRS – (Spitzer) infrared spectrograph; ITC – instrument time code; ITOC – Instrument Test and Operations Console; ITS – impactor targeting sensor; LVDS – low-voltage differential signaling; LVPS – low voltage power supply; MIPS – multiband imaging photometer for Spitzer; MRI – medium-resolution instrument; MSSRD – Mission, Science, and Systems Requirements Document; MTLM – mechanism and telemetry board; MWIR – mid-wave infrared; NVM – non-volatile memory; PROM – programmable read-only memory; PSF – point spread function; QE – quantum efficiency; S/C – spacecraft; SCU – spacecraft control unit; SIM – spectral imaging module; SNR – signal to noise ratio; SRAM – static

Space Science Reviews (2005) 117: 43–93
DOI: 10.1007/s11214-005-3390-8

random access memory; TPG – timing pattern generator; TVAC – thermal vacuum (test); VME – Versa Module Europa (IEEE 1014-1987 standard); VTC – vehicle time code

1. Introduction

The NASA Deep Impact mission is designed to deliver an impactor spacecraft with a mass of about 360 kg to collide with comet Tempel 1 in early July 2005, at a relative speed of approximately 10 km/s. The impactor and flyby spacecraft are launched attached to one another and separate about 24 h before impact. The flyby spacecraft diverts and slows down to fly by the nucleus at a range of 500 km, at about 14 min after the impact. The goal of the impact is to create an impact crater in the surface of the comet and to expose pristine material underneath the nuclear crust to be observed by the flyby. Minimal knowledge of the physical properties of the surfaces and interiors of cometary nuclei results in a large range of possible ejecta plumes and crater depths and diameters. This mission is truly an experiment and almost any observable outcome will significantly improve our knowledge of the properties of cometary nuclei. The challenge in the design of the instruments is to be able to get useful data for any outcome within the range of possibilities.

The Deep Impact scientific suite consists of three optical instruments: a high-resolution instrument (HRI) and a medium-resolution instrument (MRI) on the flyby spacecraft, and an impactor targeting sensor (ITS), mounted on the impactor. The HRI incorporates a 30-cm aperture, 10.5-m focal length telescope feeding both a filtered CCD camera covering the wavelength range from 0.32 (at 50% transmission) to 1.05 μm ("visible light") and a long-slit IR spectrometer with spectral range from 1.05 to 4.8 μm. The MRI incorporates a 12-cm aperture, 2.1-m focal length telescope that feeds a filtered visible-light CCD camera. The ITS telescope is identical to the MRI telescope, but feeds an unfiltered CCD camera. Table I summarizes the physical attributes of the instruments.

Figure 1 shows the block diagram of the HRI, and Figure 2 shows the block diagram of the MRI or ITS, which are identical, except for the inclusion or exclusion of the filter wheel.

The detectors and mechanisms are controlled with a set of dedicated instrument electronics – one for each instrument. These electronics are controlled by commands *via* a MIL-STD-1553 interface. Health and status telemetry from the instruments also are transmitted to the spacecraft *via* this interface. Image data are transmitted to the spacecraft *via* a dedicated low-voltage differential signaling (LVDS) interface, as shown in both Figures 1 and 2.

In this paper, we describe the mission objectives and how these translated into requirements on the instruments. We then describe the design of the instruments from optical and mechanical through electronics and software. We then describe the instrument imaging modes and the data flow from the instruments to the Deep Impact spacecraft. The integration and test of the instruments is described along

TABLE I

Instrument optical summary.

	HRI	MRI	ITS
Telescopes			
Diameter (cm)	30	12	12
Focal length (m)	10.5	2.1	2.1
$f/\#$	35	17.5	17.5
Visible			
Format	1,024 × 1,024 split frame transfer CCD		
Pixel size (μm)	21	21	21
IFOV (m-rad)	2.0	10.0	10.0
FOV (mrad)	2.0	10.0	10.0
FOV (°)	0.118	0.587	0.587
Scale per pixel (m)	1.4 at 700 km	7 at 700 km	0.2 at 20 km
IR			
Format	512 × 256 HgCdTe FPA[a]		
Pixel size (μm)	36		
Spatial			
IFOV (μ-rad)	10.0		
FOV (mrad)	2.5		
FOV (°)	0.15		
Spectral			
IFOV (μ-rad)	10.0		
FOV (μ-rad) (slit width)	10.0		
Spectral range	1.05–4.8 μm		
Minimum $\lambda/d\lambda$	216		

[a] 1,024 × 512 rebinned 2 × 2.

with the results of that testing. Finally, some operational considerations for the instruments are presented.

2. Requirements and Mission Objectives

The general mission objectives are discussed by A'Hearn *et al.* elsewhere in this issue. From the general mission objectives it was possible to derive a set of requirements for the capabilities of the instruments. However, the uncertainty about which physics is relevant to the impact and thus about what will be observable, led to necessary compromises in the requirements. The resulting requirements were thus

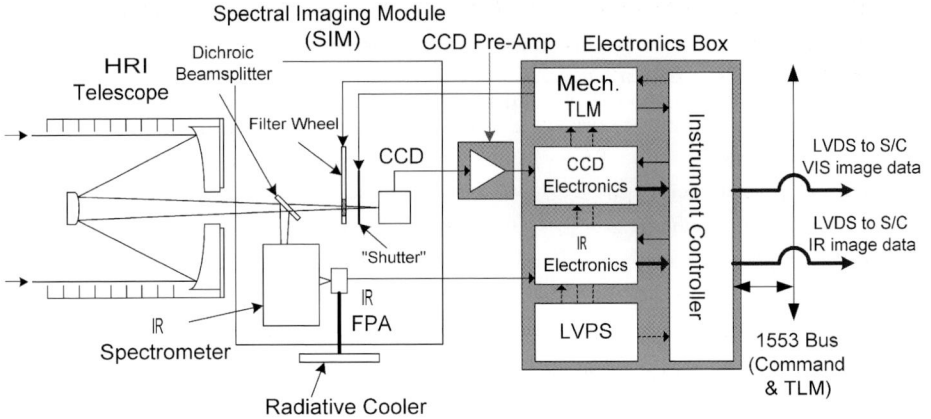

Figure 1. HRI block diagram.

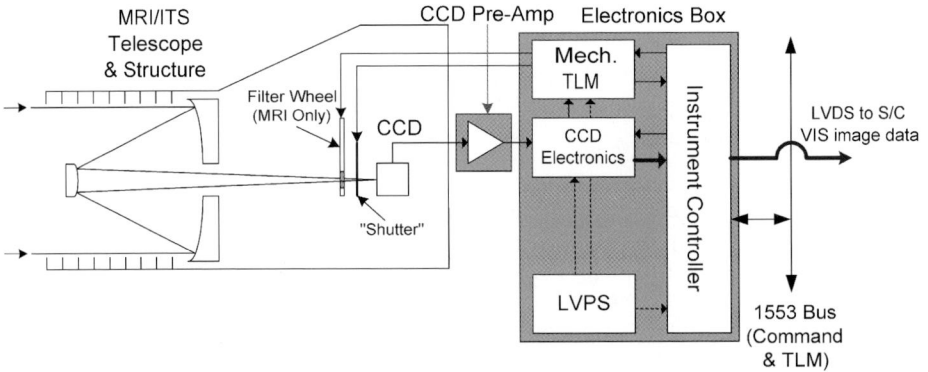

Figure 2. MRI/ITS block diagram.

targeted at the baseline scenario (see article by A'Hearn *et al.*, Schultz *et al.*, and Melosh *et al.*, in this issue) but intended to ensure that the mission would still return useful information in the event that the baseline assumptions are not correct. As examples, crater formation could be considerably slower than assumed, by as much as a factor of 3; the crater depth could vary from the baseline assumption by a factor of 2; and the diameter of the crater could be as little as 2 m or as large as 200 m according to various models. The requirements have been codified in the Mission, Science, and Systems Requirements Document (MSSRD). Cost restrictions have resulted in de-scopes from the originally proposed instruments. In this description, we will use the current version of the MSSRD (Revision C, 2004), which incorporates all de-scopes to date. The requirements on the instruments derive directly from Section 3.2 "Science Observational Requirements", which is reproduced, in part, later. The MSSRD document defines the albedo, phase function, phase angle and range from the sun that are the baseline values accepted by the science team. These values are implicit in the following requirements.

The following observational requirements assume that pointing accuracy is sufficient to avoid the need for (mosaic imaging maneuvers) and that pointing has recovered from any previous dust impacts.

1. Optical images shall be obtained showing growth of the impact crater and the ejecta cone, including, if possible, bright ejecta emanating from the impact.
2. IR spectra shall be obtained of the down-range ejecta near the crater.
3. At least 660 s (three times the baseline prediction for crater formation time) shall be used to observe the crater growth, to allow for uncertainties in the predicted phenomena.
4. The instruments shall perform visible and infrared observations of the comet, as follows: (a) visible imaging shall cover the spectral range from 0.3 to 0.95 μm; (b) infrared spectroscopic imaging shall cover the spectral range from 1.05 to 4.8 μm.
5. The instruments shall observe comet brightness variations due to rotation prior to its being spatially resolved, plus the natural comet shape and surface features before and after impact, and after orbit plane crossing.
6. A sequence of pre-impact images from the impactor of the surface, including a context image of the predicted crater location, shall be obtained with the following image scales and fields of view:

Observation	Image scale	Field of view (minimum)
Nucleus w/coma	30 m/pixel	15×15 km
Full nucleus	10 m/pixel	2.5×2.5 km
Impact target area	1.5 m/pixel	180×180 m
Target point	0.6 m/pixel	30×30 m

7. The ejecta cloud, including the inner coma up to two times the nucleus' radius above the surface, shall be observed, in the visible range, at a pixel scale of 25 m or less, with an SNR of $\geq$30.
8. The fully developed crater shall be observed in the visible range,

   (a) at <3.4 m total resolution (i.e. system point spread function (PSF), full width half max) with an SNR of $\geq$70 and
   (b) at <5.5 m total resolution with an SNR of $\geq$100, using no filter in either case, all with a field of view of >360 m (three times the predicted crater diameter).

9. The fully developed crater shall be observed in the visible range using geology filters at <6 m total resolution with an SNR of $\geq$50, with a >360 m FOV.
10. The crater shall be observed spectrally (across the IR wavelength range) at a spectral resolving power $\geq$200 and at spatial resolution of <25 m.
11. The entire comet shall be observed spectrally (across the IR wavelength range) at a spectral resolving power $\geq$200 and at spatial resolution of <130 m.

12. An IR spectrum of ejecta and outgassing in the area of the down-range limb shall be obtained in the 1.05–4.8 μm wavelength range after impact.
13. To provide area context, near-simultaneous visible observations shall be taken with five times the field of view of the high-resolution instrument.
14. To facilitate analysis, a complete set of pre-impact context observations shall be obtained as close to the time of impact as practical of the impact point (visible and IR), the down-range limb area (IR only), and the inner coma (visible only).
15. The SNR for visible global imaging of the comet for the expected geometric albedo (4%) and phase function (value = 0.0135) and a phase angle of <70° shall be ≥250 for broadband nucleus observations and ≥100 when observing the nucleus using geology filters.
16. The minimum SNR for IR observations of the comet nucleus (averaged over the illuminated portion of the nucleus) in the 1.05–4.8 μm range for the expected reflectivity (4%) and a phase angle of <70° shall be ≥50.
17. The visible and IR imaging shall have spatial, sensitivity, and spectral characteristics that are capable of achieving the science goals at the assumed comet albedo and the illumination angles planned for the mission.
18. The performance for visible imaging of the cometary coma, for expected brightness (of 0.3 ergs/cm^2/s/sr over a bandpass from 520 to 540 nm) and a phase angle <70° shall (a) provide an SNR greater than or equal to 100 when observing the coma using dust (or geology) filters and (b) ensure a digitized signal level that is at least 2 DN (data numbers) higher than background when observing comet pre-impact gas emissions with gas emission filters.

3. Instrument Design

Translating the scientific observation requirements into instrument designs was accomplished by engineers at Ball Aerospace & Technologies Corporation, working to the requirements as set forth in the MSSRD and as translated by the systems engineers into more detailed "Level B Specifications". The approach, as described in the initial proposal and as it exists today (less de-scopes described later) was to have one visible camera system, called the ITS, on the impactor spacecraft and two instruments on the flyby spacecraft. This array of instruments would provide the scientific observations and also be used for autonavigation of the two spacecrafts.

Having two instruments on the flyby spacecraft provides some redundancy as well as some capability for optimization of the two instruments for different types of observations. The HRI incorporates both an infrared imaging spectrometer and a visible imaging camera with a wheel of nine filters optimized for surface characterization, while the MRI incorporates only a visible imaging camera, also with nine filters, including some optimized for coma studies and some identical with those in the HRI. Both instruments are mounted to an instrument platform, which also holds two star trackers and the inertial reference unit, maintaining all of these in fixed alignment with respect to each other independent of distortion in the body of the flyby spacecraft.

A planned, second infrared spectrometer on the MRI was eliminated in a de-scope. Similarly, the planned ability to scan the infrared spectrometer slit over the field of view of the HRI visible imager was eliminated in another de-scope; the spectrometer slit currently has a fixed location in the HRI field of view, so scanning the slit over the nucleus of the comet is accomplished by rotation of the entire spacecraft.

3.1. Optical and Mechanical

The parameters of the mission (flyby velocity and distance, desired resolution and fields of view, expected coma brightness, nucleus albedo and phase function, desired time resolution for impact and crater formation, and so on) constrain the design of the instruments. Within these constraints, we attempted to minimize design effort and enhance manufacturability to keep costs acceptable. Therefore, the ITS and MRI were designed to use identical telescopes, all visible sensors were designed to use identical CCDs, both MRI and HRI were designed to use identical filter wheels (but with different filters in the wheels), and all visible imagers were designed to use identical light blockers.

3.1.1. *Telescopes*
Two Cassegrain telescopes were designed for use in the Deep Impact instruments. The HRI telescope is a 30-cm aperture 10.5-m focal length optic resulting in a relatively slow $f/35$ focal ratio. Both the MRI and ITS have 12-cm apertures and 2.1-m focal lengths resulting in faster $f/17.5$ telescopes. The apertures, focal lengths, and resultant f-ratios for the telescopes were derived directly from the science requirements for resolutions, fields of view, and throughput of the visible imagers and the infrared spectrometer. The f-ratios for the primary mirrors, separations of secondary mirrors from primary mirrors, and distances from the back of the primary to focus were all determined from packaging and mounting constraints including accommodation of filter wheels, light blockers and, in the case of the HRI, the dichroic beamsplitter described later. In fact, a major driver for the HRI telescope was that it must mount on the spacecraft but not violate the envelope of the launch vehicle fairing.

The mirrors are made from Zerodur® glass ceramic, and the reflective surfaces are coated with Al with a SiO_2 overcoat for surface protection.

Because the instruments are intended to operate at low temperatures, the telescope optics were mounted on flexures within carbon composite structures having slightly negative coefficients of expansion, so that the separations of primary and secondary mirrors could be designed to remain essentially constant over large variations in structure temperature, eliminating the need for focus mechanisms. Tests of these instruments over ranges of temperature from $+60\,°C$, high above the expected

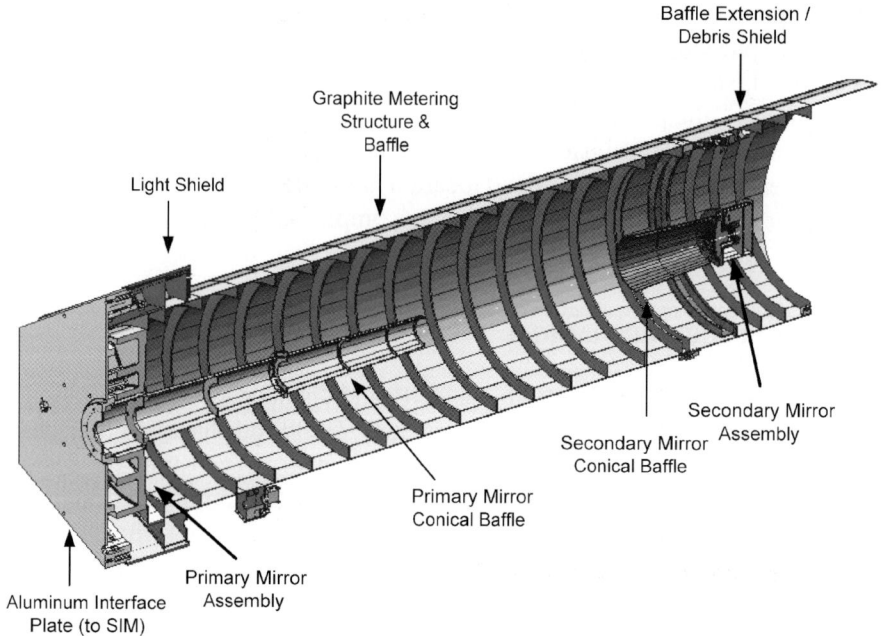

Figure 3. HRI telescope.

operating temperatures, to temperatures as low as −170 °C, close to the expected coldest operating temperature of the HRI telescope, showed that the telescopes would maintain correct focus over the ranges of their respective expected operating temperatures.

The telescopes were designed with several features to reduce stray light. In addition to including multiple internal baffles within the main telescope tube structure (see Figure 3), there are conical baffles around the outside of the secondary mirror, and within the central hole of the primary mirror. These two mirror baffles prevent any light outside of the instrument field of view from directly reaching the CCD, or the HRI IR spectrometer slit. The graphite structure that forms all of the internal telescope tube baffles typically has a normal incidence reflectivity of 15%.

3.1.2. *HRI Spectral Imaging Module*

The spectral imaging module (SIM) of the HRI incorporates both the visible detector with its associated filter wheel and light blocker and the infrared spectrometer, using a custom-designed dichroic beamsplitter to reflect the visible light and transmit the infrared. Figure 4 shows the layout of the SIM instrument.

The IR imaging spectrometer operates from 1.05 to 4.80 μm with a spectral resolving power ($\lambda/\Delta\lambda$) that varies from a minimum of 200 at middle wavelengths to higher values at shorter and longer wavelengths. The spectrometer uses prisms for dispersion and diamond-turned optics for collimating and imaging. Imaging along

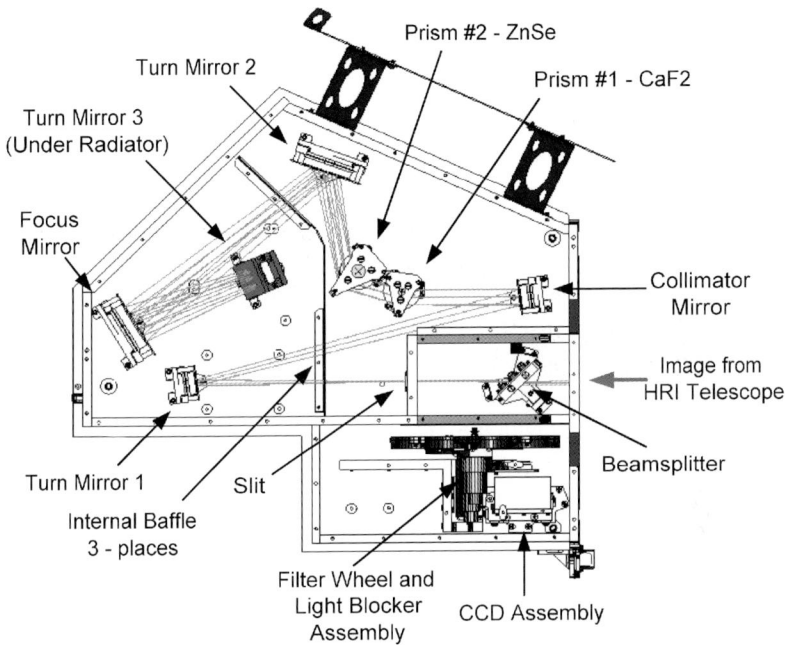

Figure 4. HRI spectral imaging module, without IR detector radiator. The structure at the top of the image is a debris shield that will protect the SIM from particle damage during the encounter.

the slit is maintained for the full 0.145° slit-length field of view. The spacecraft will rotate to scan a scene across the slit to obtain a spectral image cube.

A prism spectrometer was chosen over a more typical grating spectrometer for two reasons. First, the large spectral range (over two octaves) would require order sorting filters in a grating spectrometer, but not in a prism spectrometer. Eliminating such filters reduced the number of optics required to be mounted. Second, the large spectral range would results in a large range of spectral resolving power for a grating spectrometer. In the extreme of an ideal linear grating spectrometer, meeting the required resolving power of 200 at 1.05 μm would result in a resolving power of more than 25,000 at 4.8 μm, and require a detector with nearly 20,000 elements to cover the entire spectrum. Choosing materials with two complementing indices of refraction resulted in a relatively flat resolving power across the large wavelength range, and thus kept the spectrum at a manageable size to fit on an existing detector. The measured resolving power is shown in Figure 27, and discussed in Section 5.3.6.

After the infrared light passes through the beamsplitter and slit, a fold mirror reflects the rays onto a spherical collimator mirror. The collimated beam is refracted through a calcium fluoride (CaF_2) prism and through a zinc selenide (ZnSe) prism, reflected off another fold mirror, and then reflected by a spherical camera mirror that images the spectrum. After the camera mirror another fold mirror, "Turn Mirror #3" in the figure, reflects the light out of the plane of the figure and onto the IR detector.

The reason for the out-of-plane reflection is to permit the use of a radiative cooler, attached to the back of the IR detector, viewing radiatively cold open space. An impressive part of the optical design is the incorporation of an aspherical surface on the exit face of the ZnSe prism, which acts as a Schmidt corrector for the spherical camera mirror providing, without the use of additional optical surfaces, the large field of view required for accurate imaging of the entire length of the slit.

Other than the prisms, beamsplitter, and detector all components of the IR spectrometer are aluminum. The optical bench is made of honeycomb aluminum with bonded aluminum face sheets. Mounting points in the optical bench for optics and side walls are aluminum inserts. The sides and top cover are all machined aluminum. The reflective optics are diamond-turned aluminum, mated to aluminum optics mounts. The prisms and beamsplitter are mounted within aluminum frames *via* spring-loaded plastic pads that maintain tension at room temperature so that the prisms survive the launch environment, and take up tension as the aluminum frames contract. The mounting of the IR detector to its two-stage radiative cooler is described in Section 3.2.3. Without the prisms, the all-aluminum optical design would be largely a-thermal. The prisms have different thermal expansion characteristics, and the index of refraction of the CaF_2 and ZnSe also change with temperature (Tropf, 1995)

The requirements to observe both the comet nucleus and the coma (see Section 2, numbers 11, 12, 13, and 15), resulted in a dynamic range that could not be accommodated by the single IR detector. The sub-solar point on the nucleus could be as warm as 300 K (Lisse, personal communication, 2003), while CO emission lines at 4.6 and 4.7 μm could be as dim as 100 kilo-Rayleighs. This results in a ratio of 1:10,000 in brightness. The problem is due mostly to the warm nucleus producing a high flux at the long-wavelength end of the spectrometer that will saturate the detector in the minimum full-frame exposure (2.9 s). Therefore, a spectral attenuator was added at the entrance slit that reduces the long wavelength flux over half the slit length. The filter is a made of polished, 1.0-mm thick, Schott WG 295 glass. The spectral characteristics are shown in Figure 5. The filter is arranged to cover the middle third of the slit, so that the top and bottom thirds are un-filtered. Since the imaging sequence will place the nucleus in the center of the IR detector, the filter arrangement will attenuate the warm nucleus signal while not filtering the low signal coma gas and dust. During the last minutes of the encounter, the nucleus may extend beyond the filter, and the long wavelength measurements may saturate. However, unlike the CCD (see Section 3.3.1), the saturated signal does not bleed across the array, and the spectrum below 3.5 μm should not be affected.

The beamsplitter reflects light in the visible band to the HRI CCD, and transmits light in the IR band to the slit of the IR spectrometer. The requirements on the beamsplitter were quite stringent with a precise 1.05-μm crossover and providing high-efficiency transmission over more than two octaves of IR wavelength, and high reflectivity from 0.34 to 1.05 μm. Figure 6 shows the resulting reflectivity and transmission. The crossover point is well positioned at 1.05 μm.

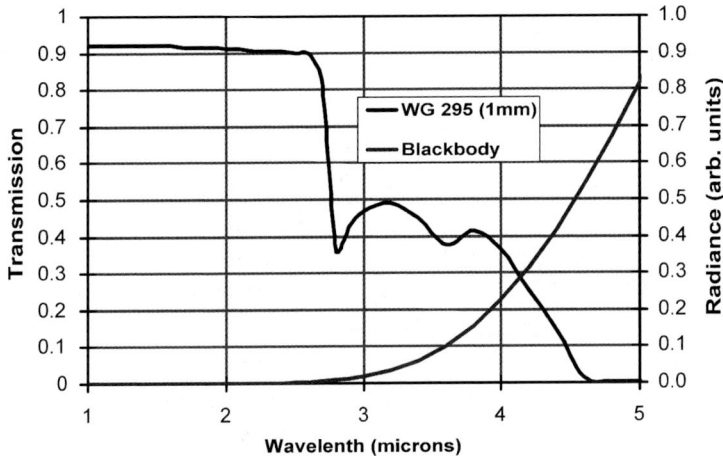

Figure 5. Transmission of the IR attenuator filter. The filter is made of 1-mm thick Schott WG 295 glass. Also shown is the radiance of a 300 K blackbody showing how quickly the brightness changes with wavelength at 3–5 μm. Portions of the comet nucleus surface may be as warm as 290 K.

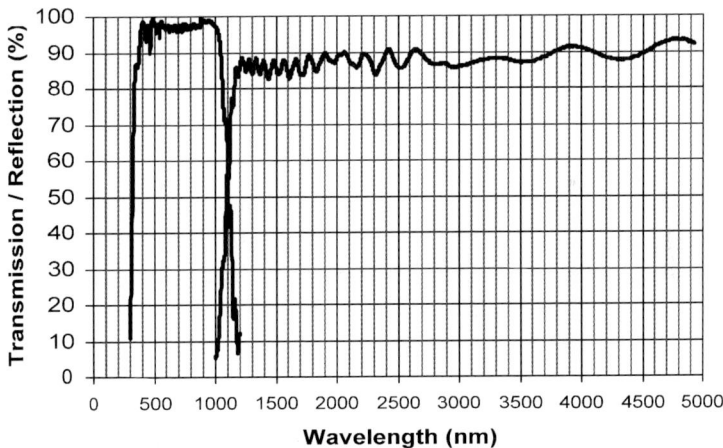

Figure 6. HRI beamsplitter optical properties. The darker profile on the left is visible reflectivity, the lighter profile is the IR transmission.

3.1.3. *Mechanisms*

3.1.3.1. Filter Wheel. Figure 7 shows the 9-in. diameter HRI filter wheel with its nine optical filters spaced at 40° intervals. The MRI filter wheel design is identical except for the mounting brackets. The filter cells are separate and complete assemblies that mount into the magnesium wheel. Both filter wheels are required to move between adjacent filter positions within 2.2 s. The filter wheel mounts directly to the output shaft of the stepper motor gearhead. All lubricated and moving components are contained within the motor and gearhead assembly.

Figure 7. HRI filter wheel and light blocker.

The motor is a 30° two-phase stepper motor which, with a 96:1 gearhead ratio. It rotates the filter wheel 0.3125° per motor step so that 128 steps are required to rotate the 40° between adjacent filter positions. There are mechanical stops at filter positions 1 and 9 that limit the wheel rotation to a total of 320°. A command that causes the filter wheel to bump against a stop causes no harm to the motor or mechanism.

During normal operations, the motor is driven at 120 steps per second with 60% of the maximum possible drive current to the motor, which provides sufficient torque and inertia margin to move the wheel at the required rate. The instrument controller (IC) can directly set the mechanism Field Programmable Gate Array (FPGA) registers that control the motor pulse rate and the current drive level to the motor (0, 19, 60, and 100% of maximum).

The wheel is moved by an absolute filter position command from the spacecraft to the IC. The IC in turn sends a command to the motor control FPGA (on the mechanism and telemetry (MTLM) board) to move the desired number of steps in the correct direction. A homing command to the IC will move the wheel from any position to the clear filter in position 1. Four Hall-effects sensors provide 4-bit absolute position indication to FPGA registers that are read by the IC. Note that the wheel is moved by open-loop step commands, and the Hall sensors are used only for position confirmation. The main reason for providing absolute position indication is to allow relatively quick resumption of critical sequence imaging following an unplanned reboot of the IC.

On initialization of the instruments, the software will check the 4-bit Hall sensor reading. For a known reading, the software will store the current measured position and consider the filter wheel initialized and proceed with executing commands. If the 4-bit Hall-effects sensors do not match a known code, then the software will begin stepping the filter wheel toward the home position (filter 1). At each step, the software will read the Hall-effects sensors and compare it to the known filter position codes. When a Hall sensor pattern that matches a known code is received, the software then sweeps several steps on either side of that position and records the Hall-effects sensor pattern at each step. The software then uses an optimization routine to determine the most central position for that code and moves the filter wheel to that position. It then stores the current position and considers the filter wheel initialized, and will proceed with executing commands. This initialization sequence takes 3 s or less, depending on where the filter wheel starts. This initialization sequence can also be commanded at any time the instrument software is in the operate state.

Table II shows the filter parameters for each filter in the HRI and MRI filter wheels, and Figure 8 shows the measured filter transmission for each filter.

Other than filter positions 1 and 6, each filter cell is made of two pieces of glass. One is a fused silica substrate on which the interference filter is deposited, while the other is a longpass or shortpass blocking filter to reduce the throughput of resonance transmission bands in the bandpass filter. The thickness of both pieces

TABLE II

Filter wheel filter characteristics filter.

Filter wheel position	MRI center (nm)	MRI filter width (nm)	MRI filter target measurement	HRI center (nm)	HRI filter width (nm)
1	650	>700[a]	Context	650	>700[a]
2	514	11.8	C_2 in coma	450	100
3	526	5.6	Dust in coma	550	100
4	750	100	Context	350	100[b]
5	950	100[c]	Context	950	100[c]
6	650	>700[a]	Context	650	>700[a]
7	387	6.2	CN in coma	750	100
8	345	6.8	Dust in coma	850	100
9	309	6.2	OH in coma	650	100

[a]Filters in positions 1 and 6 are uncoated and not band limited.
[b]The coating on the 350-nm filter is shortpass, the substrate limits the short wavelength throughput.
[c]The 950-nm filter is longpass.

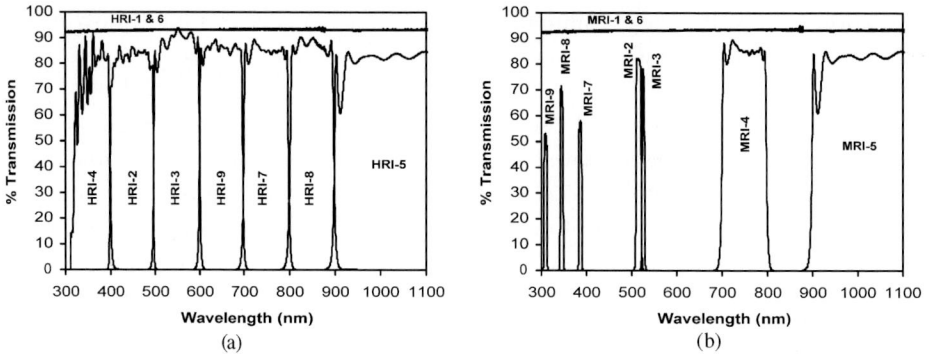

Figure 8. HRI and MRI filter transmissions. The labels refer to the filter position in the filter wheel (see Table II).

of glass in each filter cell was designed to maintain a constant optical thickness for all filter positions, and therefore maintain the same focus at all filter wheel positions.

3.1.3.2. Light Blocker. The light blocker is used in all three instruments, and acts to cover the CCD aperture between exposures. Note that we refer to this mechanism as a light blocker rather than a shutter. The length of a visible exposure is set by the start of the CCD frame transfer (see Section 3.3.1), not by the closing of a light blocker. The light blocker is used to limit the amount of light falling on the light-sensitive parts of the CCD during readout of an image; if the light were not limited during the relatively long readout times of full-frame or near-full-frame images, the light-sensitive part of the CCD could be saturated and bleed signal charge into the frame-storage area, corrupting the image.

As such, the light blocker is required to cover the aperture within 0.1 s of the end of an exposure and open within 0.2 s of the beginning of an exposure command. The light blocker mechanism is illustrated in Figure 7.

A common mounting interface is used for the light blocker in all three instruments. The blade mounts directly to the output shaft of the stepper motor gearhead. All lubricated and moving components are contained within the motor and gearhead assembly. The motor is a 30° two-phase stepper motor. The gearhead rotates the light blocker 3° per motor step, so that 12 steps are required to rotate the 36° between the "open" and "closed" positions. Mechanical stops at both ends limit the rotation of the light blocker. Single Hall-effects sensors at both ends of travel are used to sense permanent magnets mounted in the shutter, to give an indication of light blocker position. The light blocker may be moved by open-loop commands to move a number of motor steps, and the Hall sensors are used only for position confirmation. No damage to the motor or mechanism is caused by driving the light blocker against a stop.

In normal operation, the light blocker is opened and closed under control of the CCD timing pattern generator (TPG) FPGA that also controls the clock signals that drive the CCD readout. The IC can directly set the mechanism FPGA registers that control the motor pulse rate, current drive level to the motor (0, 19, 60, and 100% of maximum), and the number of steps to move between the open and closed positions. During normal operations, the motor is driven at 120 steps per second with 60% of the maximum possible drive current to the motor, which provides sufficient torque and inertia margin to open or close the light blocker. For emergency and test situations, the IC can override TPG control of the light blocker operation by writing to a TPG register. In those situations, the IC can send direct commands to move the light blocker a desired number of steps in a given direction.

3.1.3.3. Visible Stimulator. Although not a mechanism, the visible stimulator is described in this section, since it is packaged in the light blocker. The visible stimulator is used as an on-board diagnostic for the CCD systems. A light emitting diode (LED) is mounted inside the light blocker mask, aimed at approximately the center of the light blocker blade in front of the CCD. The LED is powered by one of the instrument heater circuits, supplied by the flight system thermal control subsystem. The LED intensity is a function of both the LED temperature and the current supplied to the LED. None of the voltages on the heater circuits that supply the LEDs is regulated, and the light blocker temperature may vary during the mission, so the stimulator is not an absolute intensity measurement. Instead, it is meant to be a method of checking for high spatial frequency changes in the CCD sensitivity – i.e. dust or other obscurations on the CCD itself – and checking for gross changes of gain or offset across quadrant boundaries.

3.2. THERMAL

The thermal design of the instruments is highly integrated with the spacecraft thermal and mechanical design. The instruments sit on the opposite side of the large ($7 \, m^2$) solar panels and so have excellent views of cold space for almost all of the mission, including the encounter with comet Tempel 1.

The IR spectrometer is the major driver in the instrument thermal design. With a cutoff wavelength of 4.8 μm, the IR detector will be saturated in its normal readout time (3 s) if it looks at a room temperature background. Further, the IR detector dark current will also saturate the detector full-well in a readout time, if the detector temperature is >110 K. The thermal design made use of the clear view of cold space to cool the detector and the SIM bench passively. The SIM is cooled by its exposed upper surface, which has been extended about 8 cm beyond the SIM structure to increase the cooling efficiency. The detector is cooled by a two-stage passive radiator mounted to the SIM optical bench. The second (warmer) stage is isolated from the SIM bench by three fiberglass composite flexures. The primary radiator is isolated from the second stage by similar fiberglass composite flexures. The flexures are thin to reduce thermal conductivity, but are stiff enough to maintain the detector spacing and survive the mechanical loads at launch. The effective surface emissivity of the radiator surfaces is enhanced by adding aluminum honeycomb in surface pockets. Thermal-vacuum tests along with correlated thermal models predict that the bench will operate at ≈136 K, and the IR detector will operate at <85 K at the encounter, which will produce <1,000 DN/s of dark signal.

The CCDs benefit from cold operation, since their dark current is also reduced. The HRI CCD, which is mounted to the SIM bench, is actually warmed to 160 K by integrated heaters. This reduces the risk of mechanical stress to the detector, which has been tested as low as 155 K. The MRI CCD is mounted directly to the MRI telescope structure. Most of the MRI telescope is covered with multi-layer insulation, but part of the top surface of the telescope cover is left uncovered to aid in cooling of the structure and CCD to around −85 °C. The ITS structure is largely internal to the impactor spacecraft, so it cannot cool the CCD effectively. Therefore, a dedicated radiator and copper heat pipe were added with a direct link to the ITS CCD. The radiator looks out the front of the impactor spacecraft so that it will cool the CCD during the cruise, as well as the 24 h of the impactor spacecraft free flight before impact. During this period, the ITS CCD will be maintained at less than −30 °C.

All of the critical optical components are fitted with heaters so that during the first few days after the launch their temperatures can be maintained above those of the rest of the structure. These heaters are intended to reduce the risk of contamination in the critical time period after launch.

Ground tests determined that the focus of the telescopes was within specifications when the water content of their graphite epoxy structure was reduced below 10% of saturation. In ambient conditions in the launch vehicle fairing, the composite structure will absorb moisture. Therefore, to reproduce the proper focus, the water in the structure must be reduced to less than 10% of saturation in flight. A set of heaters was added to the structure of each telescope so that it could be raised to temperatures above 0 °C to bake out the moisture. The heaters are expected to raise the temperature to about 5 °C, and the HRI telescope structure will require 30–40 days to reach 10% of saturated moisture content or less.

Since instrument thermal control must be maintained even during periods when the instruments are not powered on, all of the heaters are controlled by the spacecraft thermal subsystem. The spacecraft thermal subsystem also maintains a set of thermal monitors for the key components. Each instrument electronics subsystem has 10 platinum resistor thermal monitors distributed among the structure and optics on that instrument, as well as five integrated circuit temperature transducers (Analog Devices AD590) to monitor the instrument electronics board temperature (there are six on HRI). These temperature sensors only supply telemetry to the spacecraft and ground when the instrument electronics are powered on and analog data collection is commanded on.

3.3. DETECTORS

There are two types of detectors on the Deep Impact instruments. A CCD detector, with response from 0.3 to 1.1 μm, is used on all three instruments. An infrared focal plane array (IR FPA) with response from less than 1.0 to a 4.8 μm cutoff is used on the HRI IR spectrometer only.

3.3.1. *CCD*

The CCDs are Fairchild Imaging (formerly Lockheed Martin Fairchild) custom design split frame transfer CCDs (ID CCD424) with four independent quadrants, each with an associated output amplifier. The image area has 1,024 × 1,024 pixels, while both storage regions have 1,024 × 530 pixels. The storage areas are over-sized to allow for image mask mis-alignments. The pixels are 21 μm^2, producing an image area that is 2.15 cm on a side. Figure 9 shows the CCD architecture.

Image transfer and readout are controlled by three clocking phases in both the parallel and serial directions. Each quadrant is fully independent in the parallel direction (in fact, the image and storage regions have independent gates), while one phase is shared on either end in the serial direction. Charge can be transferred across the center line of the CCD in the parallel direction. This means that any pixel may be read out of any of the four amplifiers on the CCD – although this option is not implemented on DI.

With the large pixels, detector full-well is measured at about 400,000 electrons. This high capacity was implemented in response to an early requirement to be able to image the comet and guide stars simultaneously within the dynamic range of the CCD system. While subsequent navigation design has moved away from this requirement, the large full-well does help guard against saturation due to uncertainties in the surface brightness of the nucleus and coma. The image area is thinned to about 17 μm thickness and the image is formed on the back of the CCD which increases quantum efficiency, especially in the near-UV and blue. The quantum efficiency of most of the CCDs characterized for DI peaks at a value of 0.7 at a wavelength of about 600 nm as seen in Figure 10. The data in the figure were extrapolated above 900 nm, since the CCD characterization equipment was limited in wavelength coverage to 900 nm. The CCDs were tested at the flight readout rate of 187,500 pixels/s. The CCD test set was not designed to separate the test set noise from the CCD noise, but the read noise of the combined test set system was 10–12 electrons. This is a small contribution (noise combines in quadrature) to the flight system read noise which is slightly less than 28 electrons (about 1 DN).

3.3.2. *IR FPA*

The IR detector used for the DI IR spectrometer is a hybrid array with a mercury cadmium telluride (HgCdTe) infrared sensor indium bump-bonded to a Rockwell Science Center HAWAII-1R multiplexer. The readout circuit was originally developed for the HST Wide-Field Planetary Camera III instrument. For Deep Impact,

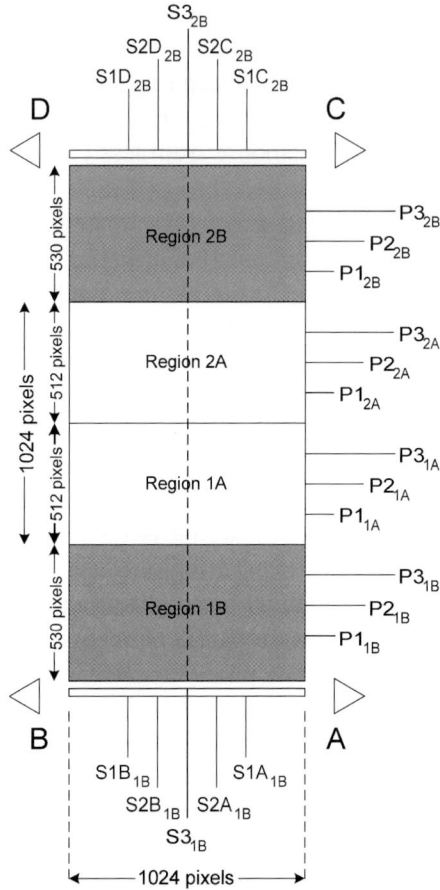

Figure 9. Deep Impact CCD architecture. Regions 1A and 2A are the imaging area. Regions 1B and 2B are the storage regions.

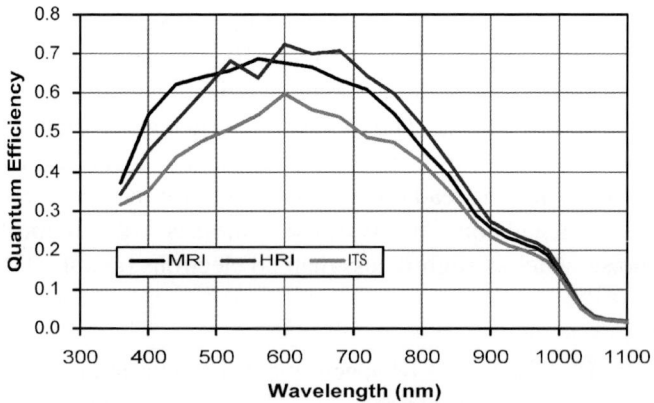

Figure 10. CCD quantum efficiency. The quantum efficiency for all three flight detectors is shown as measured in the CCD characterization test set.

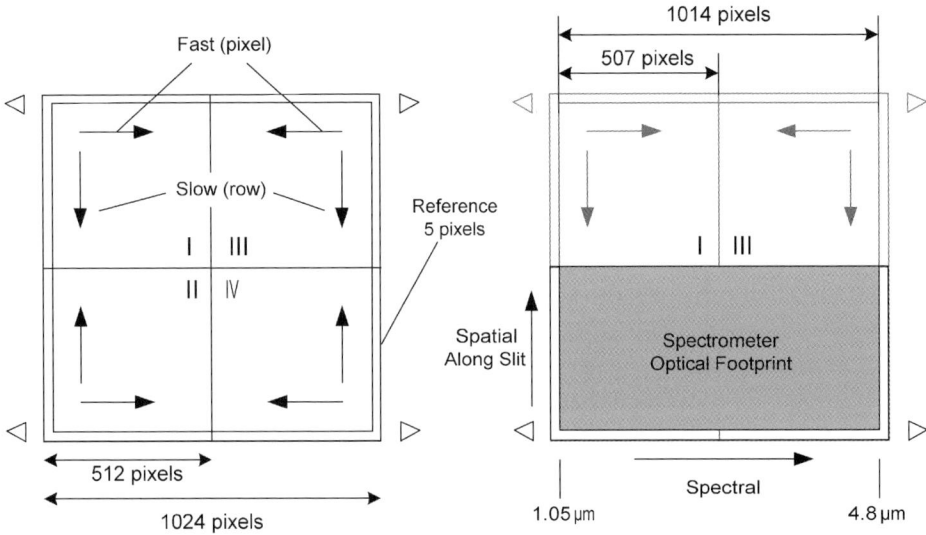

Figure 11. IR FPA architecture. The figure on the left shows the architecture of the full HAWAII-1R array. The figure on the left shows how the spectrometer is mapped to one half of the array.

a custom MWIR HgCdTe substrate with a cutoff wavelength at 4.8 μm was developed. The CMOS readout array has 1,024 × 1,024 pixels that are 18 μm^2, and has four electrically independent quadrants with separate readout amplifiers. The optical system was designed such that only two of the quadrants are illuminated. (Thus, while each IR image only has two halves, these halves are still called quadrants.) The detector orientation was chosen such that the fast readout direction is along the spectral direction, while the slow readout is along the slit, in the spatial direction, as seen in Figure 11. Therefore, the spectrum of each spatial sample has the same time history. Because the detector is read in a ripple mode, one spatial sample will have a slightly different time history than the another. Section 4.1.2 describes the timing from line to line.

The detector has five reference rows on each edge. Four of the rows have a fixed capacitance, while the fifth (outer) row has a series of four capacitance values, cycling across the detector. The detector QE is shown in Figure 12. The cutoff wavelength is 4.8 μm. The QE drops rapidly below 2 μm due to a passivation layer added to the HgCdTe surface. This layer reduces significantly the detector readout noise, which somewhat compensates for the low sensitivity at short wavelengths. Further, surface spectra are expected to look much like solar spectra, so that the greatest fluxes will be at short wavelengths. As seen in Section 5.3.5, the low QE at the short ends flattens out the response for surface spectra.

In the DI flight operating configuration the detector full-well is measured at 120,000 electrons. With the 4.8-μm cutoff, the detector dark current exceeds the full-well for the minimum 2.9-s full-frame readout time, unless the detector

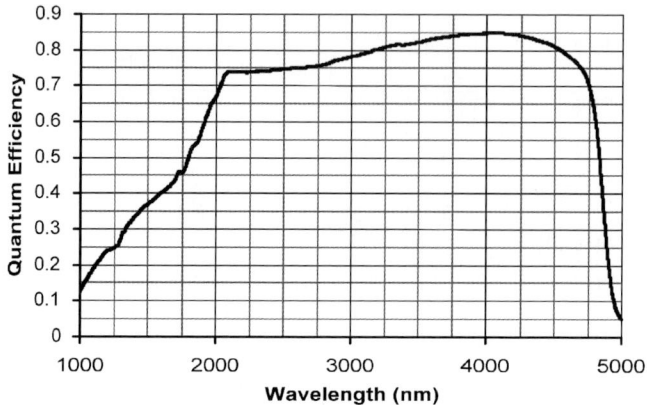

Figure 12. IR FPA quantum efficiency. The steep change below 2 μm is due to a passivation layer on the HgCdTe detector layer to reduce detector readout noise.

temperature is less than 110 K. The radiative cooler described in Section 3.2 is expected to cool the detector to 85 K or less. At this temperature, the detector dark current produces about 3,000 electrons in a nominal readout time, less than 3% of the detector full-well.

To achieve this temperature for the IRFPA, the number of linkages between the detector and the radiator was reduced as much as possible, while still protecting the mechanical integrity of the FPA. The cooler, made of aluminum, has a much greater coefficient of thermal expansion than the detector. The detector is bonded directly to a molybdenum substrate , which is mounted to a flexured BeCu support, which is mounted directly to the aluminum cooler. This stepping of materials with gradual changes in expansion coefficient results in a low mechanical stress on the detector substrate, which protects the relatively delicate indium bump bonds. The detector has survived six thermal cycles from room temperature to operating temperature with no detectable increase in non-operable pixels. Only the thermal cycle in space remains at the time of writing.

A mask is mounted immediately in front of the IR FPA to shield additional responsive areas of the array from out-of-spectrum light.

3.4. ELECTRONICS AND SOFTWARE

Each instrument has two electronics boxes associated with it: a main electronics box, mounted internal to the spacecraft, and a CCD pre-amp/clock driver box. Figure 13 shows the placement of the main HRI and MRI electronics boxes relative to their respective instruments. Because the boxes are mounted internally, they require less energy from the spacecraft to maintain the desired operating temperature. With these boxes mounted on the spacecraft, as shown in the figure, the cable lengths between the electronics and main instrument structure are 1 m or less.

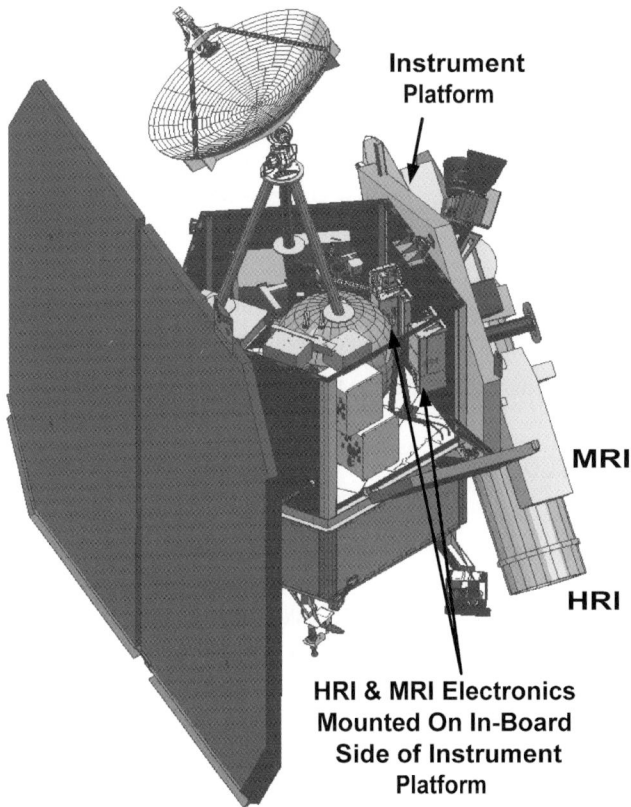

Figure 13. Instrument electronics box placement in Deep Impact spacecraft. The HRI and MRI main electronics boxes are mounted internal to the spacecraft structure. They are mounted on the panel adjacent to the instruments to reduce the cable length to less than 1 m.

3.4.1. *Instrument Electronics*

An early decision in the development of the DI instrument electronics was to have a modular electronics design, with one main function for each board, and one main electronics box for each instrument. Not only did this allow for parallel development, test and final production of each board and box, it also allowed for packaging that reduced the electrical noise on the sensitive detector readout boards. The main electronics box contains a motherboard, which provides interconnects between the specific subsystem boards. There are four daughter boards in the MRI and ITS boxes and the five daughter boards in the HRI electronics box. Physically, the boards conform to a 120 mm tall 6U VME form factor. Electrically, the connector pinouts are not VME standard, but are custom to DI. The boards are shown in the block diagrams (Figures 1 and 2), and their main functions are described in Table III. There is no image storage or manipulation in the instrument electronics. In fact, other than adding an image header, the IC acts only as a data conduit to the spacecraft

TABLE III

Boards in the main instrument electronics boxes.

Board	Functions
Instrument controller (IC)	Command and Control
	Telemetry collection and broadcast to SCU
	Error handling and notification
	Image data flow from other boards to SCU
Low-voltage power supply (LVPS)	Power conditioning from S/C bus
	Power distribution to other boards
Mechanism and telemetry (MTLM)	Mechanism control and monitoring
	Analog telemetry collection
Visible system (CCD)	CCD exposure command execution
	CCD timing pattern generator
	CCD analog data conditioning and digitization
Infrared system (IR)[a]	IR exposure command execution
	IR analog data conditioning and digitization
	IR FPA bias generation and adjustment
	IR FPA timing pattern generator
	IR FPA clock driver

[a]HRI only.

control units (SCUs) for the science images from the detectors. The IR and CCD boards are shielded from the other boards by placing aluminum spacers between those boards and adjacent boards. This reduces electrical cross-talk between the detector boards, and the IC board that handles high-frequency data transmission, and results in detector systems that meet their noise performance requirements. The following sections describe the functionality and design of each board in more detail. With the exception of the IR board, which is only in the HRI electronics box, the electronics box for each instrument contains one of each of these boards.

3.4.1.1. Low-Voltage Power Supply. The LVPS receives unregulated spacecraft bus power and supplies regulated voltages to all other boards in the electronics box. The LVPS will operate as required for an input bus voltage between 24 and 35 V, which are the required limits on the spacecraft bus. To reduce detector system cross-talk, the LVPS supplies each board with its own set of isolated regulated voltages, even those boards that use common voltages (e.g. 5 V). The LVPS switching frequencies are derived from the operating pixel frequencies of the CCD and IR detector signal processing boards. This ensures that LVPS switching noise is synchronized with the detector clocking, making sure switching does not happen when the detector board is performing a critical measurement.

The LVPS board also supplies the system reset to the entire electronics system, to ensure that the electronics are in the reset condition during either power-up or power-down transition.

3.4.1.2. Mechanism and Telemetry (MTLM) Board. The MTLM board contains the pulse drivers for stepper motors that drive the two mechanisms on each instrument – the light blocker and the filter wheel (ITS has no filter wheel). These drivers can run at 0% (off), 19, 60 and 100% of their current capacity and at an arbitrary step rate. Both the current level and step rate can be set by external command. On instrument power on, the current level is set to 60%, and the rate to 120 steps/s, which supplies sufficient torque to the motors. Commanding the drivers to 100% current capacity and reducing the step rate will result in greater torque on each motor. This can be used as a contingency to overcome any unforeseen mechanism "stiction" that may develop during the mission.

The MTLM board is also the focal point of all analog telemetry collected for that instrument. The MTLM FPGA cycles through its set of analog telemetry points every 2 s, and transmits the data to the IC on the same schedule. This analog telemetry is stored in software registers and is sent out as instrument telemetry when requested by the spacecraft. Some of the analog telemetry is placed in the 100-byte image header when the image is transferred to the SCU.

3.4.1.3. Instrument Controller Board. Figure 14 shows a block diagram of the IC board. The design is centered around a radiation-tolerant TSC695F single-chip SPARC 32-bit microprocessor. The processor hosts a real-time operating system that receives commands from the spacecraft *via* the 1553 bus. The IC communicates with the other boards *via* 3-Mbit/s low-voltage differential signaling (LVDS) lines that are controlled by the IC FPGA. Data from the detectors are routed through the IC FPGA and to the SCU by means of a 12-Mbit/s LVDS interface.

The instrument flight software image is stored redundantly in 2 MB of non-volatile EEPROM. On instrument power-up, the EEPROM image is transferred to 4 MB of SRAM, and the processor operates from the software image in SRAM. The choice of which EEPROM image to transfer is controlled by a bi-level relay, which is toggled by the remote interface unit on the spacecraft. To accommodate quick turn-on during the encounter, another bi-level relay controls whether the boot process includes a memory check or not. Normal bootup (including the memory check) takes 28 s, while the fast bootup takes 21 s, a 7-s savings.

The IC is also responsible for generating a 100-byte image header placed in each image that is transferred from the instrument electronics to the spacecraft control unit. The header contains routing information from the exposure command, software and hardware states from the flight software, analog telemetry and a time stamp based on the spacecraft vehicle time code (VTC). The timing of the header generation is important because that determines the time that is associated with

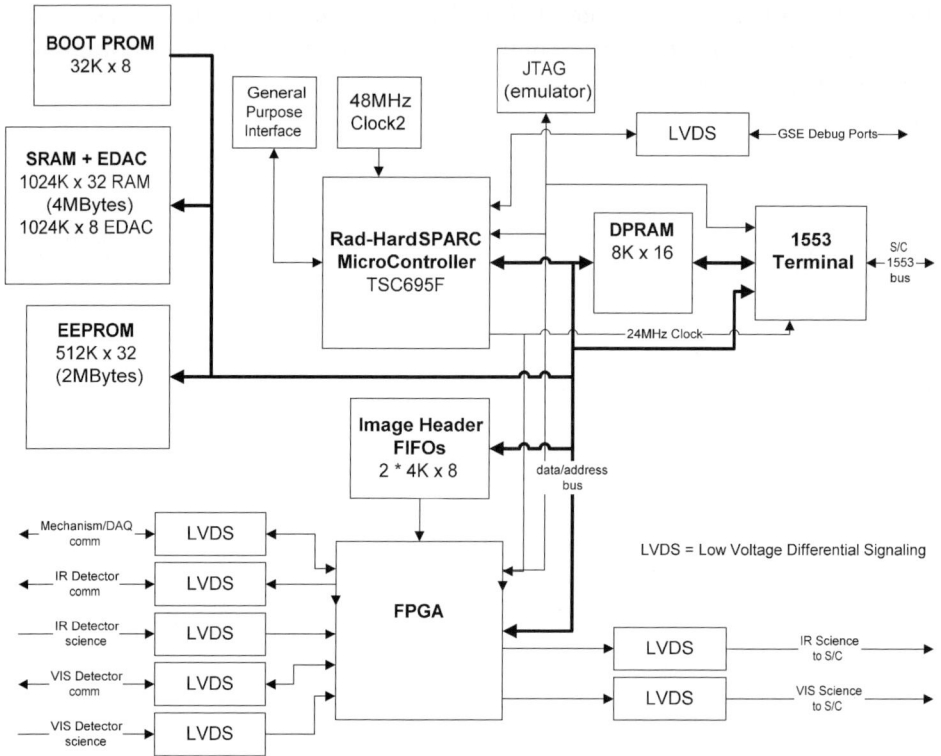

Figure 14. Instrument controller board block diagram.

that exposure. The detailed relationship between the header time stamp and the exposure time of the images is discussed in Sections 4.1.1 and 4.1.2.

The VTC is collected from the spacecraft *via* the 1553 command channel. The sequence is basically "at the tone the time will be...", followed by a tone. The spacecraft sends out a message that includes an upcoming VTC. The IC stores that VTC in a register. The spacecraft then sends a message that tells the instrument to set its internal time to the VTC that was just transmitted. The accuracy of the message and the latching maintain the spacecraft to instrument VTC to within 64 μs. The VTC is updated at a 1 Hz rate.

3.4.1.4. IR Signal Processing Board. The IR board FPGA contains the TPG that drives the readout of the IR FPA. Since the FPA is a digital multiplexer that is operated by TTL signal levels, and because the FPA output amplifiers are capable of driving a differential signal over a cable 1 m in length, there was no need for an IR FPA pre-amp close to the detector. Thus, the detector biases are also set on the IR signal processing board. Since the performance of the IR FPA is sensitive to these biases, three of them are controlled by means of 8-bit DACs that are controlled by external commands. The output of the IR FPA produces a zero signal level at a positive voltage (around +5 V). Two similar 8-bit DACs, one for each quadrant, are used to adjust these offsets to be on-scale for the 14-bit A/D converter. The 14-bit A/D converter and 120,000-electron full-well led to selecting a conversion gain of 15 electrons per DN.

The IR FPA is fundamentally different from the CCD in that it essentially has a portion of an amplifier in each pixel. In many applications this is desired, since it allows multiple non-destructive readouts of the array. The best noise performance for detectors of this type comes from doing correlated double sampling, i.e. each integrated signal is correlated with the offset value from the beginning of that integration. To accomplish this, an offset value must be read from each pixel before the charge is collected in that pixel during the integration period. For many IR systems this is done by collection of both a reset frame (the offset levels) and a read frame (the integrated levels) with a subsequent subtraction during image analysis; however, this results in two images per measurement. We reduce the number of images required to be returned by incorporating memory on the IR signal processing board in which the reset frame is stored. The subtraction that results in a correlated image is then calculated on the IR board during the readout of the integrated frame. The memory and storage of the offset frame also allows for 2×2 binning of the images on the IR signal processing board.

The details of the readout modes for the IR detector are discussed in Section 4.1.2.

3.4.1.5. CCD Signal Processing Board. The FPGA on the CCD board contains the TPG that drives the readout of the CCD. The clocking lines from the FPGA go to the clock driver board, which is part of the CCD pre-amp clock and bias box (see the following section) where the timing patterns are amplified to their voltage rail values. The CCD board also receives differential signals from each quadrant of the CCD, after they have been conditioned by the CCD pre-amp. The CCD board performs a clamp and sample operation during each pixel readout, based on TPG signals, producing a correlated double-sampled analog signal. The board then converts the analog signal into a 14-bit digital number (DN) by a 14-bit A/D converter. The digital signal is transferred first to the IC and finally to the SCU. With the 14-bit resolution and the requirement to span the data over the originally expected 450,000-electron full-well of the CCD, the conversion gain of the CCD system was selected to be 30 electrons per DN.

The details of the readout modes for the CCD system are described in Section 4.1.1.

3.4.1.6. CCD Pre-Amp, Clock and Bias Box. The separate CCD pre-amp, clock and bias box (pre-amp box for short) contains two printed circuit boards. One board has four 1:1 amplifier circuits that convert the CCD output video signal from single ended to differential signal, allowing for low noise transmission over a 1 m cable length. The DC bias voltages required by the CCD are also set on this board. The other board contains a set of voltage converters that set the upper and lower rails of the clock lines that transfer the charge on the CCD. These convert a TTL signal from the TPG on the CCD signal processing board into the clocks that drive the CCD charge transfer. The two boards are separated by a solid metal partition so that the amplifier does not pick up noise from the clocks. The pre-amp boxes are located within 18 in. of the CCDs on each instrument. This means that for the flyby spacecraft they reside on the instrument platform, and therefore have heaters and temperature sensors to maintain them above their allowable flight temperatures and maintain a stable operating temperature at encounter. For the ITS, the pre-amp boxes reside within the cover of the impactor spacecraft and will not be at risk of cooling below acceptable operating temperatures.

3.4.2. Software
The software for the DI instruments leveraged heavily off the Spitzer (formerly SIRTF) IRS and MIPS instrument software development. Figure 15 shows the major components of the instrument flight software. All code is written in ANSI-C.

The DI instrument software performs the following main functions:

1) Receive and process command data.

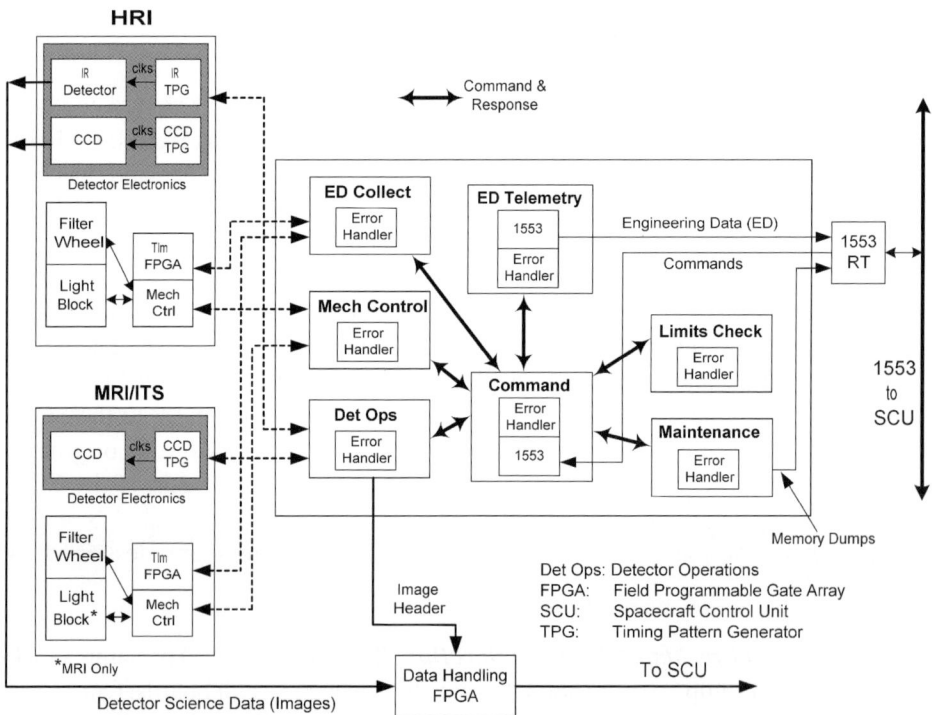

Figure 15. Instrument flight software architecture diagram.

2) Control and coordinate the HRI/MRI/ITS instrument activities based on defined operational parameters within the accepted commands.
3) Configure the visible detectors and infrared detector, and control the collection of the science image data *via* the detector electronics interface.
4) Configure the instrument mechanisms (filter wheel, light blocker) for mechanism calibrations and for observations *via* the mechanism electronics interface.
5) Gather, format and output engineering telemetry data.
6) Initiate science data-taking and the flow of science data images to the spacecraft computer (SCU) *via* the science data channel.
7) Monitor instrument safety parameters and maintain the instrument in a safe configuration.
8) Perform memory self tests.
9) Initialize and maintain the Instrument Payload electronics and software systems of the instrument.
10) Gather calibration data. For mechanism calibration, the calibration data are a subset of the engineering data routinely collected by the Instrument Payload Flight Software and forwarded to the SCU.

3.4.2.1. Boot State. When power is applied to the main electronics of any instrument, or when a reset occurs, the IC processor will begin execution of the boot state software stored in a radiation-hardened PROM non-volatile memory area. The primary purpose of the boot state software is to perform basic memory testing, followed automatically by the copying of one of two software images from EEPROM to RAM, followed automatically by the transfer of control to that software image in RAM. Since the boot-code will exist in PROM, it will not be possible to modify the boot-code after launch.

3.4.2.2. Operate State. Once the instrument has completed the steps described in the boot state, it transitions to the operate state. In this state, it will perform on demand the following activities:

1. Accept and execute commands received over the 1553 interface.
2. Collect and monitor engineering data.
3. Configure and control the mechanisms.
4. Configure the detectors.
5. Initiate science data-taking and the flow of science images to the SCU.
6. Format and output engineering telemetry data to the 1553 interface.
7. Provide the ability transfer data between instrument memory and the ground system *via* the spacecraft SCUs.

In addition to these functionalities, the software also performs some rudimentary fault protection, as well as interfacing to the spacecraft fault protection system. This includes error reporting from software and hardware monitors. There are 498 errors that can be reported. The software also monitors high and low limits on analog telemetry points – voltages and temperatures – reporting out of limits conditions as an error. The limits reside in an internal software table that includes a persistence count for each telemetry point – e.g. temperatures must be out of limits for 30 consecutive queries before an error is posted.

If an error is deemed sufficiently bad, the instrument will post a request for action to the spacecraft fault protection system. The possible actions are to reset the IC, power cycle the instrument, or turn off the instrument and alert the ground. The severity of the action can also be determined by the criticality of the current spacecraft state. Errors that would result in a request for power-off during the 6-month cruise may result in simply posting the error when the spacecraft is engaged in the encounter with Tempel 1.

4. Data Format and Pipeline

This section describes the different imaging modes, including details on the timing, and the data pipeline. The TPGs on the IR and CCD electronics boards are able to produce several different imaging modes each to accommodate requirements on the imaging sequence. As such, not only do the instrument electronics designs need to be dynamic in accepting imaging command parameters to produce these modes, the rest of the data pipeline, all the way to the spacecraft control unit non-volatile memory, must react correctly to the different imaging modes.

4.1. IMAGING MODES

4.1.1. *Visible (CCD) Imaging Modes*
One mission requirement is to observe the impact at imaging rates significantly greater than 1 Hz, which can only be accomplished with image sub-framing. Further, at times when the nucleus is not resolved, or does not fill up the field of view of the instrument, a sub-frame image will produce the same information on the comet nucleus surface while taking up much less of the limited storage space on the spacecraft computer. With four output amplifiers, symmetric sub-framing of the CCD results in approximately a factor of two reduction in readout time for each factor of two reduction in linear image size. Thus, the 1.4-s readout of the full $1,024 \times 1,024$ pixels is reduced to about 0.7 s for reading out 512×512 pixels. The CCD imaging modes are shown in Table IV. All the sub-frame modes are symmetric about the center of the CCD, i.e. each quadrant makes up one-quarter of the stored image.

A second consideration for the imaging modes is whether the light blocker is used for each image. The light blocker can be controlled by either a direct command to the mechanism FPGA, or by the CCD TPG. The light blocker was added so that during the readout of a full-frame image, the scene in the imaging area does not saturate

TABLE IV

Visible (CCD) imaging modes.

#	Mode	Stored size $(X$ and $Y)$	Serial O'clocked size (X)	Serial O'clocked co-add	Parallel O'clocked size (Y)	Parallel O'clocked co-add	Light blocker used on each image	Dither?
1	FF	1,024	8	0	8	4	Yes	No
2	SF1	512	4	0	4	4	Yes	No
3	SF2S	256	4	0	4	4	Yes	No
4	SF2N	256	4	0	4	4	No	No
5	SF3S	128	2	0	2	4	Yes	No
6	SF3N	128	2	0	2	4	No	No
7	SF4O	64	0	0	1	2	No	No
8	SF4NO	64	0	0	0	0	No	No
9	FFD	1,024	8	0	8	4	Yes	Yes

TABLE V

Visible mode frame-to-frame times.

#	Mode	Minimum commanded t_{INT} (ms)	F-T-F time for minimum t_{INT} (s)	F-T-F time for $t_{INT} = 100$ ms (s)
1	FF	0	1.634	1.735
2	SF1	0	0.737	0.838
3	SF2S	0	0.430	0.531
4	SF2NS	4	0.232	0.328
5	SF3S	0	0.312	0.413
6	SF3NS	4	0.113	0.209
7	SF4O	4	0.062	0.158
8	SF4NO	4	0.062	0.158
9	FFD	0	1.634	1.735

the CCD and bleed charge into the storage area, corrupting the image currently being processed. For the smallest sub-frames, the readout times are significantly shorter and the light blocker should not be required to prevent charge bleeding. For modes 1 and 2, the light blocker is used for each image, whereas for modes 7 and 8 (64 × 64 pixel sub-frame size) the readout time is 0.06 s as shown in Table V, and the light blocker is not required. For intermediate image sizes (128 and 256 pixels square), there is a choice of whether the light blocker is used or not. For modes where the light blocker is not used, the ratio of exposure time to frame-to-frame time becomes very efficient, with only a 10- to 15-ms delay between the end of a previous exposure and the start of the next.

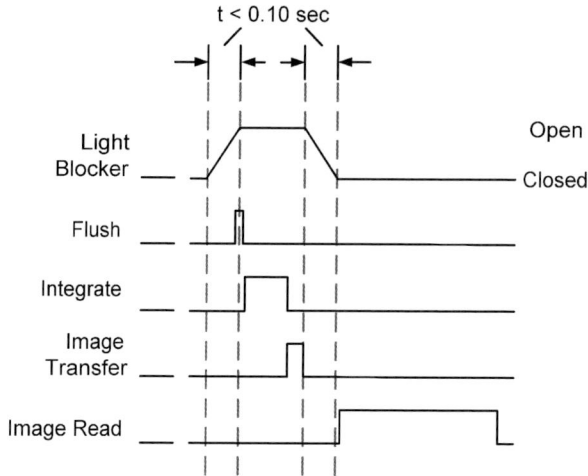

Figure 16. CCD image exposure sequence for modes where the light blocker is used for each image. The timing is not to scale. Generally, a trace shows inactive low, active high – e.g. image transfer is active just before the light blocker starts to close. Note that the image is not read out, while the light blocker is moving to avoid electrical noise from the stepper motor contaminating the image data.

The imaging cycle when the light blocker is used for every image is shown in Figure 16. The overall frame time is increased by the opening and closing of the light blocker. We chose not to set the timing so that the CCD electronics are reading the CCD while the shutter is moving to avoid inducing noise in the CCD data. Upon receiving the exposure command, the CCD TPG begins by opening the light blocker. This takes slightly less than 100 ms. At about 50 ms into the light blocker motion, the TPG begins flushing the CCD. This is the same pattern as transferring the charge from the image area to the storage area, but in this case the pattern continues until the image and storage area have been completely transferred five times. The end of this flush happens after the light blocker is completely open, and designates the beginning of the integration period. The TPG waits for the commanded integration delay time and then begins the frame transfer. The transfer takes 5.7 ms. Once the transfer is complete, the TPG closes the light blocker. This also takes the same time as the opening process, 100 ms. When the light blocker is closed, the TPG either begins the readout of the CCD for the full-frame modes (1 and 9) or performs a partial transfer of the storage area to get to the first row saved for a sub-frame image. When the last row of pixels is read out, the TPG FPGA either begins the entire cycle again if the command requested more than one image, or on the final image it closes out its registers, sends a message back to instrument flight software, and is ready to receive the next exposure command.

For modes that do not use the light blocker between images, the cycle is shown in Figure 17. As in the light blocker mode, the first action that the TPG takes is to open the light blocker. It also performs the similar five-image flushes starting

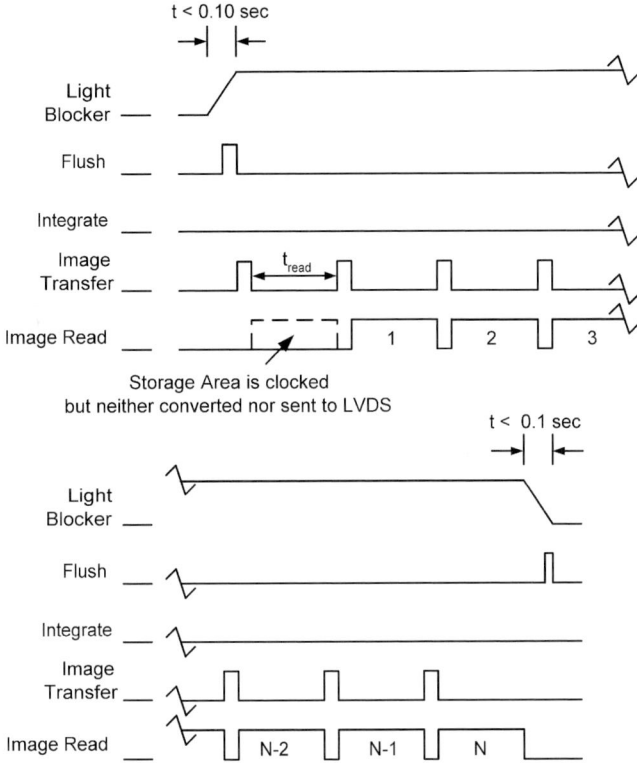

Figure 17. CCD multi-image exposure sequence for modes where the light blocker is not used for each image. In this case, N images were requested in the image command. The format is the same as Figure 16. The timing is not to scale.

50 ms into the light blocker move. In this case, however, the TPG begins the cycle by immediately starting a frame transfer at the end of the flush. The image that was transferred was not integrated for the proper time (it does not include the readout time from a previous frame), so while the TPG does read the image out of the CCD to set the proper timing, it does not command conversion of the charge into digital signals and therefore does not transfer an image to the SCU. As soon as the TPG finishes the readout, and waits for the commanded integration time, it begins another frame transfer and immediately begins the readout of the CCD. This image, the first of the sequence, is converted to digital signals and transmitted to the SCU. This cycle is repeated for the number of images commanded (1–255). After the last image is read out, the TPG then closes the light blocker, clears the FPGA registers and signals the software that it is ready to receive the next image command. Note that in this mode, the true integration time is actually the minimum frame-to-frame time minus the frame transfer time plus the commanded integration time.

An important parameter to understand for interpretation of the data for both science and the navigation, is the relationship of the time stamp in the image

TABLE VI

IR imaging modes.

#	Mode mnemonic	Mode	Stored image size	Minimum exposure time (s)	Frame-to-frame time for minimum exposure (s)
1	BINFF	Binned full frame	512 × 256	2.86	2.862
2	BINSF1	Binned sub-frame 1	512 × 128	1.43	1.432
3	BINSF2	Binned sub-frame 2	512 × 64	0.71	0.717
4	UBFF	Unbinned full frame	1,024 × 512	2.86	2.862
5	ALTFF	Alternating binned full frame	512 × 256	1.43	2.868
6	DIAG	Diagnostic	1,024 × 512	1.43	1.432
7	MEMCK	Memory check	1,024 × 512	N/A	2.862

header to the center of the integration for that image. For the CCD the timing is straightforward: the time stamp is latched into the header when the TPG sends the signal to transfer the charge from the image area to the storage area. This means that whether or not the image mode uses the light blocker for each image or not, the center of integration is the time stamp minus half the true clock-stopped integration time for that image.

4.1.2. *IR Imaging Modes*
The IR imaging modes are shown in Table VI.

The IR detector modes are divided into two different readout methods; interleaved readout and alternating readout.

The Hawaii 1-R allows an entire row to be reset at one time and permits a very efficient interleaved readout method. A row of integrated values is read out (on the first pass, though these values are discarded). That same row is then reset, and the offset values are read from that row, and immediately stored in memory in preparation for the next exposure. The TPG continues this process through each row of the detector. When it reaches the end of the detector it returns to the beginning and reads out the integrated values from the first row, subtracting off the corresponding offset value stored in the IR board memory on the previous cycle. With this interleaved readout method, the only integration time lost is the time to reset and read a single row, provided that multiple exposures are taken one after another by a single command.

In the alternating readout method, a row of the detector is reset and the offset values are read into memory. The TPG continues with only this operation until it reaches the end of the detector. The TPG then returns to the beginning of the detector and reads through each row, reading the integrated values and subtracting their corresponding offset values. The frame-to-frame time for alternating readout is the same (for a similar sized frame) as for the interleaved readout, but only half of the time is dedicated to integrating signal in the pixels. This method is useful if the approximately 3 s for integration in the interleaved readout will result in saturation. But from a time series standpoint, half of the sampling time is missing. Thus, a fast event that happens during the offset readout of the alternating mode will be missed. The interleaved sampling will not resolve the same fast event, but it will capture the average signal during the event, while the alternating readout has a 50% chance of missing it entirely. Modes 1–4 use the interleaved readout method, while modes 5–7 use the alternating readout method.

As with any multiplexer readout, each row of the DI IR detector has slightly different start and stop times for integration, while the total integration time for each row is the same. This can be problematic when trying to interpret a dynamic scene (as is expected at the time of impact), but the timing for the DI detector readout is systematic and repeatable, and is well correlated to the time stamp that accompanies each image.

Modes 1–3, and 5 make use of the ability of the IR signal processing board to bin the output 2×2. Modes 4 and 6 leave the pixel data unbinned rather than using the 2×2 binning option. In mode 6, the offset image and the integrated image are not subtracted in the instrument, but are transferred to the spacecraft individually. This mode allows a check to verify that both images are within the range of the A/D converter. Mode 7 allows a check of the memory on the IR signal processing board that is used for correlated double sampling and binning. The FPGA loads either a single value or a double ramp pattern into memory and then reads the memory out into the image data stream. This test is intended only to check the integrity of the memory and does not test the analog portion of the IR electronics.

As mentioned in Section 3.3.2, the IR detector is read out in a ripple mode, and each spectral line has a slightly different time history. The frame-to-frame time for a full-frame interleaved image is 2.862 s. This is the time, ignoring any overhead, taken to read 512 physical rows of the detector. Using this as a quick measurement shows that each row is read in about 5.5 ms for interleaved modes. This is actually the time taken to read through the row twice, once for the integrated values and once for the reset values. For mode 4 (unbinned, interleaved), this is the amount of time between the beginning of integration for one row and the beginning for the next row. Two rows that are 20 rows apart would have integrations that begin approximately 110 ms apart.

For binned interleaved rows, the row-to-row time is doubled, since it takes four reads of each row to produce a single binned row. Thus, for modes 1–3, the integration time offsets are approximately 11 ms times the difference in row number apart. For the alternating readout modes, the integration time offset for binned and unbinned modes are a factor of 2 shorter, approximately 2.75 ms per row for unbinned, and 5.5 ms per row for binned.

This shows the timing between rows, but an important parameter is to know how this relates to the instrument time code (ITC), which is synchronized to the VTC at a 1 Hz rate, that is placed in an IR image header. Like the CCD time stamp, the IR

TABLE VII

Parameters to determine line time relative
to the image header time stamp.

IR mode	A (ms)	B	C (ms)
1	11.01	1	0
2	11.01	1	0.341
3	11.01	1	1.02
4	5.50	0	0
5	11.01	1	0
6	2.75	0	0

time stamp is latched into the header buffer on the IC by timing signals from the IR TPG. The IR TPG sends a signal to the IC 200 μs before it is ready to begin reading a new set of integrated values. The IC then loads exposure command parameters and telemetry into the header FIFO buffer and sends that header over the LVDS interface to the spacecraft control unit (SCU). When the IR TPG begins the new readout of the detector it sends a second signal to the IC, which then latches in the current ITC from its register, into the header buffer and sends it immediately to the SCU. It can do this in 4–5 μs, but the latching of the time is accurate to within two 12-MHz clock cycles, or ±166 ns. The limiting knowledge of the image time to the spacecraft time is the 64 μs synchronization of the 1553 discussed in Section 3.4.1.

Now the IR TPG begins the sampling of the detector. Even in unbinned modes (4 and 6) the TPG will take two of its clock cycles, 5.33 μs each, in housekeeping and overhead before it will sample the first pixel. For the binned modes, it will take one full row of readout and calculation before it starts producing data that it will send to the SCU. For the sub-frame modes (which are all binned – see Table VI), there is added overhead as the TPG must step down the number of rows needed to get to the first row of that sub-frame, and then there will be a full row period before the data begin to flow.

To determine the center of the exposure for an IR image requires two steps. First, find the time that the readout of a row begins by using the following formula: Tstart $=$ ITC $+ A$(row $- B) + C$, where A, B and C are mode dependant and are defined in the Table VII, and ITC is the time stamp in the image header.

When this offset time is calculated, the mid-point of that line's exposure is calculated by subtracting half of the combined minimum exposure time and integration delay for that image mode.

4.1.3. Data Flow

The control of the instruments begins and ends with the spacecraft control unit (SCU). When power is provided to the instruments, they will initialize autonomously and within 30 s will be ready to receive commands. Commands will

TABLE VIII

Visible exposure command parameters.

Parameter	Range	Description
IMAGE_MODE	1–9	What CCD readout mode to execute – see Table IV
EXP_ID	1–99,999,999	What exposure ID to give this set of images. Must be unique or you risk overwriting a previous image
PRIORITY	1–XX	The downlink priority for images that are automatically telemetered. Lower goes earlier
DATAPATH_HW	Several	Which SCU should receive the data, and whether the images is compressed, uncompressed or both
DATAPATH_COMP	0–3	What to do with files on the compressed channel: Do not save, save to file only, save to file and send to the downlink queue
DATAPATH_UNCOMP	0–3	What to do with files on the uncompressed channel: Do not save, save to file only, save to file and send to the downlink queue
ROUTE_TO_NAV	0–1	Whether the uncompressed image will be routed to be processed by either Optical Navigation or AutoNavigation
ANAV	Several	Five parameters that contain parameters for AutoNavigation processing
LUT_SELECT	0–3	For images routed to be saved as compressed, which LUT to use.
IMAGE_COUNT	1–255	The number of images to be taken with this single image command. All images have the same parameters
INTEG_TIME	0–10,48,576, in ms	Commanded integration time
DELAY_TIME	0–16,383, in ms	Commanded time between images for multiple image exposure command[a]

[a]If only one image is commanded, this time is added to the end of the command execution and will delay the execution of the next exposure command by that much time.

come from any one of a set of 10 sequence engines in the SCU. The most common commands will be exposure commands and filter wheel commands. Commands for visible images (common to all three CCD imagers) have the parameters shown in Table VIII.

Data flow through each detector signal processing board and the IC is real time (other than the IR board offset frame storage) in that there is no buffering of the data in the instrument electronics. Once the instrument electronics have generated

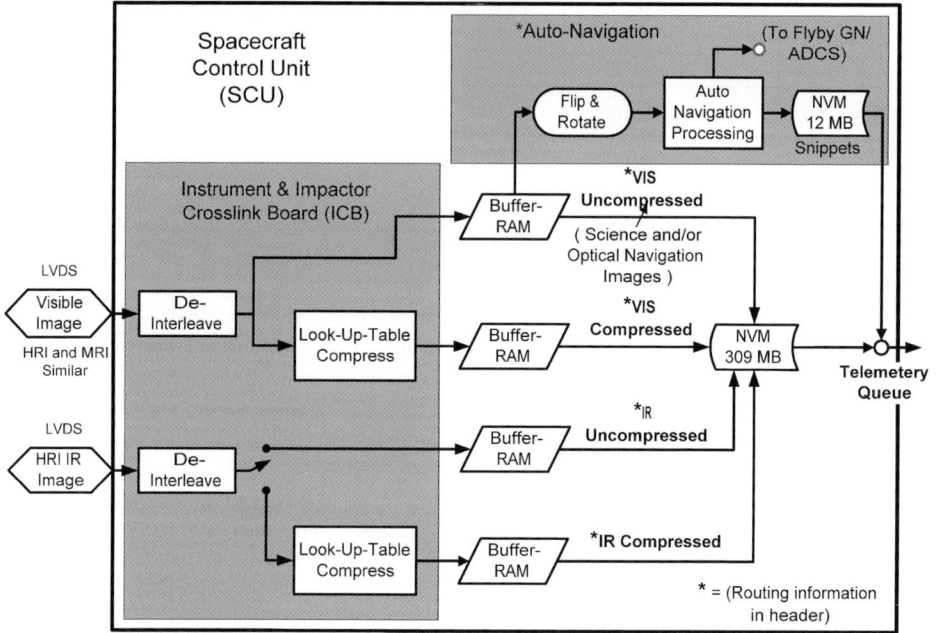

Figure 18. Data routing from the instruments to storage in the SCU, and to the downlink telemetry stream.

the image data they are transferred to the SCU *via* 12-Mbit/s LVDS interfaces. One LVDS interface is dedicated to each detector. Prior to the start of the detector readout, the same handshaking signal sent to the IC by the detector signal processing board to create the image header is sent, *via* the LVDS interface, to the Instrument and Impactor Crosslink Board (ICB) in the SCU. This enables the channel on the ICB so that it is ready to receive the forthcoming image. The IC then transmits the header, with the ITC time stamp, followed by the image data for that image. When the handshaking signal is disabled, it indicates the end of the image to the ICB.

Figure 18 shows the data path options once an image reaches the ICB. The first operation is to pull off the first 100 bytes of header data and put it aside in memory. The ICB and the instrument images software running on the SCU then snoop the header for routing and compression information to determine where the image should go. For visible images there are 27 possible permutations of routing, including to which SCU it is saved, whether it is saved as compressed, uncompressed or both, and whether it is only stored or stored and telemetered. In addition, CCD images may or may not be routed to be used by onboard AutoNavigation software. IR images can be saved only as compressed or uncompressed, not both, resulting in 18 permutations for IR images.

The first operation on the actual image data is to de-interleave the data. As the images come out of the instruments, there are four pixels converted for one CCD at the same time. Thus, the first four pixels from the instrument to the ICB are the pixels from the outer edge of each quadrant, the next four are the next pixel from each quadrant and so forth. The ICB separates the four pixels from the CCD images into essentially four different data paths (two paths for the two IR quadrants) to reconstruct the quadrants individually in memory.

Images to be compressed are converted by means of a compression lookup table. These tables reside in EEPROM on the ICB and are simply registers with 16,384 entries, which corresponds to the maximum value of a 14-bit digital value. Each entry contains an 8-bit value that will be written to the compressed image, if the uncompressed image contains that 14-bit signal value. There are four lookup tables on each channel of the ICB – i.e. four for the HRI visible channel, four for the HRI IR channel and four for the MRI visible channel on the flyby spacecraft. A parameter in the exposure command determines which lookup table will be used for that image. The lookup table value is also captured in the image header. The lookup tables are programmable in flight to both fix any bit errors that may occur in flight and to be able to accommodate shifts in detector outputs.

The choice of lookup tables was driven largely by simplicity and timing, and also by the fact that the size of the compressed image is deterministic. The compressed image size from most compression algorithms depends strongly on the details of the image itself. With limited storage capability and requirement for high-speed imaging, larger-than-expected images could fill the DI memory before the final highest-resolution images were taken, thereby compromising the science. Visible images can be saved as both compressed and uncompressed from the same exposure command; this is to support science imaging and navigation more efficiently. The IR channel allows only compressed or uncompressed images to be stored for one imaging command, as can be seen in Figure 18.

Lookup table compression from 14 to 8 bits is not a lossless process. The tables that are currently baselined, however, make use of the fact that the photon shot noise increases as the square root of the signal. The lookup tables follow a square-root function so that at higher signal levels, the step in uncompressed signal between successive compressed values is larger. With encoding steps proportional to the square root of the signal acceptable signal-to-noise ratio can be maintained. With a lossy compression, there is always a concern that for low signal levels significant science can be compromised. In the case of Deep Impact, the concern is loss of detail in jets or ejecta from the impact. The mission team has taken several steps to collect data that reduces the impact of the loss. Since there is already uncertainty in the scene brightness of the nucleus, the image sequence includes images using several exposure times. The four tables are used to highlight different science. For images of the nucleus, two of the tables have a range that cover the full A/D converter range – 16,384 DN. For images planned to look specifically at the coma, two tables are built that span only a portion of the range of the A/D converter. With the smaller range, the low end of the table produces compressed step sizes that are one-to-one with the uncompressed step sizes. Since Tempel 1 is a low-activity comet, the expected brightness of the coma is at maximum 1/10th of the expected surface brightness of the nucleus and 1/30th is expected. Surface brightnesses at 1/30th of nuclear signal will result in signal levels within the one-to-one range of the coma compression tables. Further, the imaging sequence includes several images that are stored on the spacecraft uncompressed.

After the ICB determines the compression routing, images are collected in software-controlled RAM buffers. When the full image has been collected in a RAM buffer, it is transferred to SCU non-volatile memory by means of direct memory access transfer. The non-volatile memory on the SCU is set up with a file system, so each image is given an unique filename. The filenames are codified with information about the instrument and detector of origin, the exposure ID and image count for that exposure ID, whether the image is compressed, and which SCU the file was collected on. Those files that are tagged for near real-time downlink are immediately sent to the telemetry queue. Each file is given a telemetry priority, and those with the highest priority will be downlinked first.

Figure 18 shows the data path to one of the two redundant SCUs. During flight, the second SCU will be powered on and will accept images to be stored in its non-volatile memory. Since the exposure commands allow routing to a single, the effective storage volume can be increased by storing images singly to one or the other SCU. However, in some cases this is not allowed, as described in Section 5.4.

5. Integration and Test

5.1. MECHANICAL INTEGRATION AND TEST

Designing a 30-cm aperture, 10.5-m focal length HRI telescope with a secondary axial magnification of 61, to operate successfully at a temperature of 140 K without a focus mechanism was quite a challenge. Therefore, we designed our test program to build confidence that the instruments would ultimately meet their performance requirements, by beginning the testing at the lowest possible component or assembly level, rather than waiting until the entire system was built, at which point fixing any problems would be costly.

One of the first tests was a cryogenic test of the figure of the HRI primary mirror. The test was conducted with an engineering model mirror with a spherical surface. In fact, the test showed that the cryogenic figure of the mirror distorted at the predicted HRI telescope flight temperature and would not meet performance requirements. A second test of a different (flight model) primary mirror showed that the EM mirror was from a flawed batch of Zerodur®, but the preliminary result illustrates the utility of early testing during a hardware development program with challenging requirements.

Next, the telescope mirrors were assembled with their structure to form the telescopes. The telescopes were tested first at room temperature for alignment and focus. A point source, formed by a Zygo interferometer with a transmission sphere, was placed at various points across the focal plane of the HRI telescope. When this light exited the front of the telescope it was returned by a full aperture reference flat. This formed a double-pass measurement, with the beam returning to the interferometer when the reference flat was adjusted properly. The entire telescope was mounted on a stage that rotated about its long axis. With this, the alignment and focus were measured at two orientations to produce an averaged "zero-g" performance.

The MRI and ITS telescopes were aligned using a collimated beam from a 4-D Technology PhaseCam instantaneous interferometer. Five high-precision ball bearings were placed at the mechanical location of the telescope focal plane. When the collimated light was aimed into the aperture correctly, it reflected from one of the balls, returning along the same path, again forming a double-pass measurement. Focus offsets were introduced by moving the PhaseCam source toward and away from the focus of the collimator, and comparing the position of zero-power with that of a reference flat. Alignment was sensed by comparing the input angle of each ball-bearing field point to the angle of maximum return from an alignment cube fixed to the structure. The MRI and ITS alignment fixtures were also capable of rotating the telescope to determine the "zero-g" alignment and focus.

After adjustments and characterization at room temperature, the telescopes were placed in a thermal-vacuum chamber and baked at $+60\,°C$ to eliminate water in the structure. The telescopes were then measured at room temperature and at their expected operating temperatures using a collimating test system similar to the one used for the ITS and MRI room temperature alignment, and a high precision fused silica window. With this we were able to determine the effects of both moisture content and temperature on the focus and alignment of the telescopes, with only one orientation with respect to gravity.

The components of the SIM were manufactured and assembled at the same time as the telescopes. The placement of the SIM optical components was accomplished by observing well-characterized fiducial marks on the components with high-precision theodolites. The IR FPA and its radiative cooler were installed, and the SIM was placed in another thermal-vacuum chamber with the top of the SIM and radiative cooler staring at a liquid helium reservoir. The SIM was tested at its expected operating temperature of 135–137 K, as well as at 128 and 142 K. The multiple temperatures allowed us to determine the change in spectral dispersion with change in prism temperature, as well as the change in dark signal with change in the overall SIM temperature. The alignment and focus of the SIM were checked using an external test telescope that simulated the $f/35$ beam of the HRI telescope. The focus was found to be within tolerance, but the alignment was shifted slightly to the red, which could risk failure to capture the CO emission lines at 4.6 and 4.7 μm. This was corrected after the SIM was returned to room temperature by moving the final focus mirror so that the spectrum was moved toward shorter wavelengths. A second thermal-vacuum test of the SIM showed that the alignment and focus were within requirements, and a first calibration of the spectrometer throughput was performed. During this test, a flaw in the IR FPA light shielding was discovered that we were able to fix with a simple shield near the IR FPA (also described in Section 3.3.2). The change resulted in low cost and schedule impacts to the program, again demonstrating the advantages of early testing.

After the SIM was tested, the HRI and MRI were assembled entirely and were mounted on the instrument platform. The platform was mounted on a test fixture that allowed tests with the normal to the platform up or down with respect to gravity. In this condition, alignment and focus tests were conducted on the two instruments at room temperature with the added test of checking the alignment of the two instruments relative to one another in order to meet the derived co-alignment requirement of 1 mrad. After the room temperature tests, the platform was placed in a thermal-vacuum chamber. The focus of both telescopes was tested at elevated temperature, during composite bake-out, at expected operating temperature, and the MRI focus was tested at $-30\,°C$, the expected operating temperature of the ITS at encounter. This last test was used as a surrogate for testing the focus of ITS at its operating temperature. The radiometric throughput performance of the instruments was also tested during the thermal-vacuum tests, as described in Section 5.3.

5.2. ELECTRONICS AND SOFTWARE INTEGRATION AND TEST

The testing of the instrument electronics and software also proceeded under the philosophy of testing as early as possible. Prototype microprocessors were purchased early in the program to test functionality and the software development tools. (We even found that multiplication did not work correctly in the purchased real-time operating system and wrote code to fix the problem.) Each instrument electronics board design (other than the motherboard) had a dedicated test station with custom FPGAs and LabVIEW® interfaces. These were used for board-level testing and board-level certification before integrating the flight boards in the instrument electronics. We used prototype detectors, integrated to these board-level test sets, with engineering model detector boards, to finalize the proper detector bias settings and timing patterns before the flight model electronics boards were produced. Two engineering model electronics boxes were also produced that served as safe surrogates for troubleshooting several interface and timing problems.

An Instrument Test and Operations Console (ITOC) was developed to simulate interfaces between the instrument electronics box and the spacecraft. The interfaces that were tested included commanding, telemetry, and image data flow *via* LVDS. The ITOCs were used to certify the electronics boxes, perform the flight software qualification testing, and then used during instrument level testing – taking images for calibration and alignment testing and collecting analog telemetry that allowed us to verify the conversion coefficients from raw telemetry data to engineering units.

5.3. MEASURED AND PREDICTED PERFORMANCE

5.3.1. *Focus – Over Temperature – and Expected Point Spread Function*

Focus is a concern because it directly affects the resolution that can be achieved at closest approach. The combination of telescope and instrument tests during thermal vacuum and at room temperature resulted in a good understanding of the major environmental factors that affect the alignment and focus of the instruments. For HRI the environmental factor that has the greatest effect on the focus is the telescope temperature. Figure 19 shows the relative focus offset for several telescope temperatures. The two thick horizontal lines for each temperature show the focus tolerance range over which the telescope will meet the PSF requirement. The results show that the focus (under one-g conditions) at the predicted telescope temperature is close to the optimal focus. As part of the thermal-vacuum tests, the telescopes were baked 48 h at 60 °C to remove moisture before being measured at operating temperature. The effects of moisture at room temperature are also shown in the figure. This shows that elimination of water is critical to achieving the precise focus adjustment necessary to meet the system requirements.

Similar measurements made on MRI, as seen in Figure 20, show that temperature is the dominant effect, and that moisture may not need to be removed in flight to meet requirements because the effect is so small in the shorter telescopes.

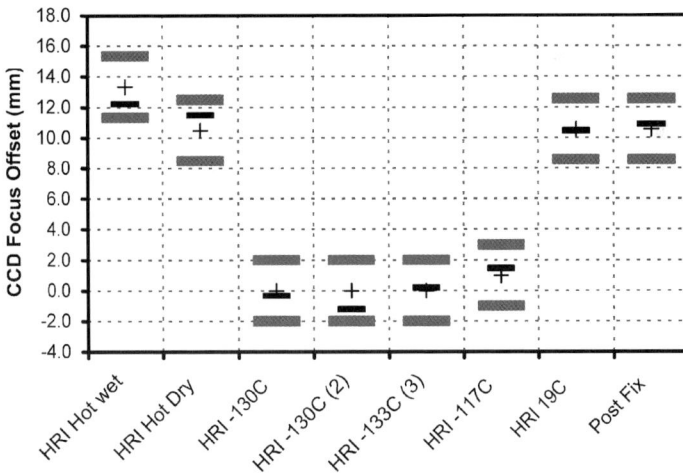

Figure 19. History of the CCD focus position during the HRI environmental test program. The format is as follows: the cross is the expected position, based on previous telescope-only tests, and the heavy gray bars are the tolerance to meet wavefront requirements. The black bar is the measured focus position. The first two measurements show the effects of moisture content on the focus an effective shift of the CCD position of 2–3 mm, which could jeopardize the focus at operating temperatures (−130 °C). All four cold measurements were within tolerance. The SIM cover was removed after the thermal-vacuum test, and a room temperature measurement shows that the focus did not shift due to this activity.

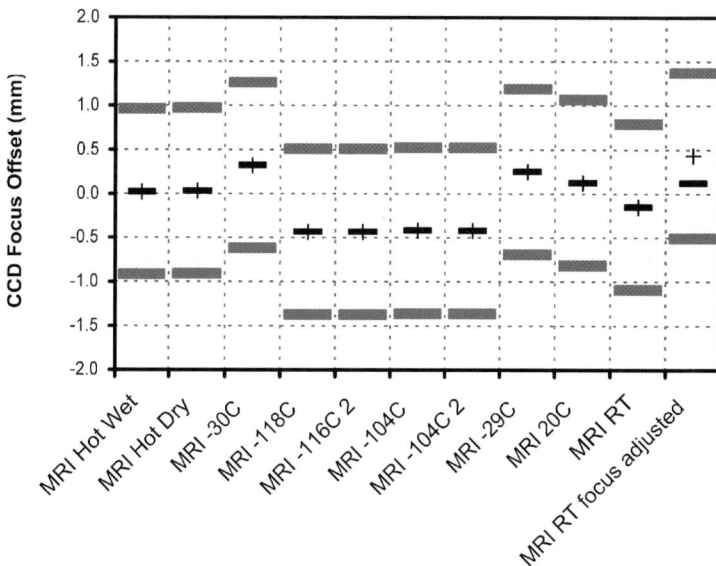

Figure 20. History of the MRI CCD telescope focus over the MRI environmental test program. The figure has the same format as Figure 19. In all cases the MRI CCD focus was in its range of focus tolerance to meet the mission requirements. A final adjustment was made to try to correct the slightly negative offset measured at cold temperatures.

The measured MRI PSF is about 1.5 pixels full width at half maximum (FWHM), which is right at the edge of that required at encounter. The HRI PSF shows a FWHM less than 2 pixels, which does meet its derived requirement of 2.5 pixels or less. The PSF was measured with a back illuminated pinhole at the focus of the collimating test set described earlier. The illumination source was a xenon flash lamp that had a few microsecond flash. This was used to reduce the effects of air turbulence and relative collimator to instrument motion. However, analysis showed that the pinhole image did jump by a few pixels from image to image. This indicates that due to the large apertures and narrow fields of view of the telescopes, these distortion effects could not be eliminated entirely, and they affected the fidelity of the measurements. Especially in the case of the HRI, the in-flight performance is expected to match or exceed what was measured. In the case of the MRI, this PSF is probably approaching the limit that can be expected due to charge diffusion in the CCD. The CCD characterization tests did not measure charge diffusion, since it was beyond the scope of the project.

5.3.2. *Alignment*

A derived requirement from the observational requirements (see Section 2) is that the boresights of the two instruments on the flyby spacecraft be co-aligned to within 1 mrad – or one-half the size of the HRI FOV, with a desire to be within 250 μrad. To measure and track this alignment, a set of high-precision alignment

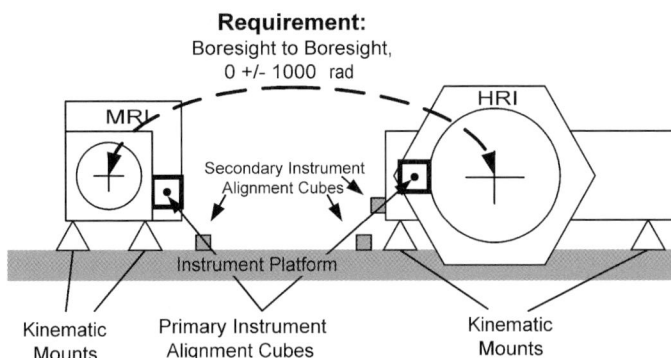

Figure 21. Placement of the instrument alignment cubes used for alignment measurements. The two alignment cubes on the instrument platform are near the kinematic mounts of the respective instruments. The second alignment cube on the HRI is on the SIM bench, whereas the primary cube on HRI is on the telescope structure.

cubes were attached to each major component on the instrument platform. Figure 21 shows their positions and the method for tracking alignment. The cubes on the MRI and HRI telescopes became surrogates for the internal boresight, and at each major I&T event the cube-to-cube alignment was measured using theodolites. There were also three main tests where this surrogate cube was measured to the actual instrument boresight; before environmental tests (vibration, and thermal vacuum), thermal-vacuum, and after environmental tests. From these measurements we have a good understanding of the effects of gravity and temperature on the alignment, and confidence that the predictions for flight are accurate.

Using two fundamental assumptions – that gravity and temperature effects are independent, and that the zero-g position is the average of those measured at the two telescope orientations that we measured at room temperature – we predicted an in-flight alignment. This result is shown in Figure 22, and ranges from exactly aligned to an offset of 60 μrad. If this is the actual performance in flight, it would allow a re-thinking of the image modes selected at the time of impact (see Klaasen), though using a smaller sub-frame does put greater reliance on the spacecraft pointing knowledge at the time of impact. The instrument-to-instrument alignment will be verified during flight calibration.

5.3.3. *Plate Scale and Targeting*
Tests using the steerable flat mirror on the collimator test station and a theodolite to measure the mirror position with high accuracy resulted in a measurement of the telescope plate scales, and a first check of telescope distortion. The effective focal length of the HRI telescope was measured, at operating temperature, to be 10.50 m, with no significant distortion in the field ($\pm 1.6\%$). The MRI telescope, also measured at operating temperature, was found to have an effective focal length

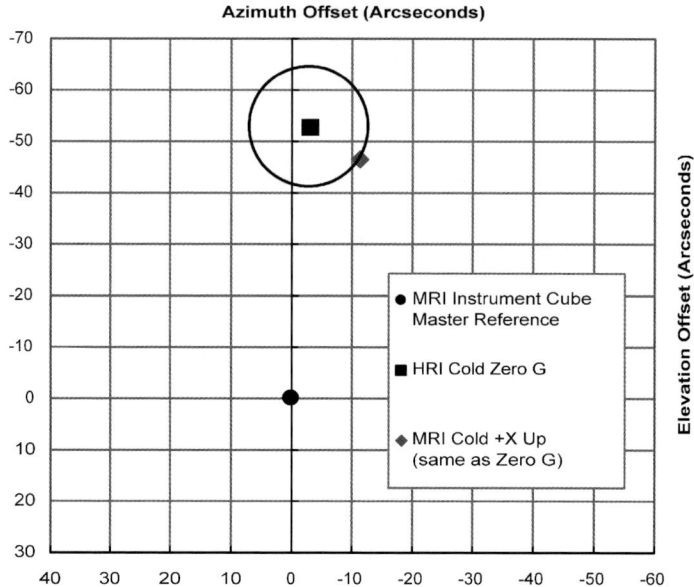

Figure 22. Results of co-alignment performance prediction, based on ground alignment measurements. All measurements are relative to the MRI Instrument Cube – the circle at (0,0). The two remaining measurements are the boresight of HRI and MRI relative to that measurement. The MRI boresight measurement is stable for both orientation of the instrument with respect to gravity. The error circle is 60 μrad in radius based on the scatter in the HRI measurements.

of 2.10 m, again with no significant distortion. The ITS focal length was not measured at operating temperature, but its room temperature focal length was also 2.10 m with no significant distortion. Data were taken that imaged a test target rastered across the field of view of each instruments, but has yet to be analyzed. This test and planned observations of open clusters during the mission cruise will further define any distortions in the focal plane and aid in the interpretation of both the navigation data and the scientific interpretation of the effects of the impact event.

5.3.4. *System Noise and Offset*

Images taken during thermal-vacuum testing of the instrument platform and flight system show that the CCD read noise is 1 DN, which allows the system to easily meet science imaging requirements. The LUT compression system applies the same selected LUT to all quadrants on a CCD image, or to both halves of an IR image. Thus, if one quadrant has an offset value that is significantly different than the others, it can adversely affect the possible SNR of other quadrants of compressed images. For example, if one quadrant of a CCD system has an offset that is 100 DN lower than the three others, over one-third of the LUT would be used up accommodating the one low quadrant. The three other quadrants would only have 155 levels above

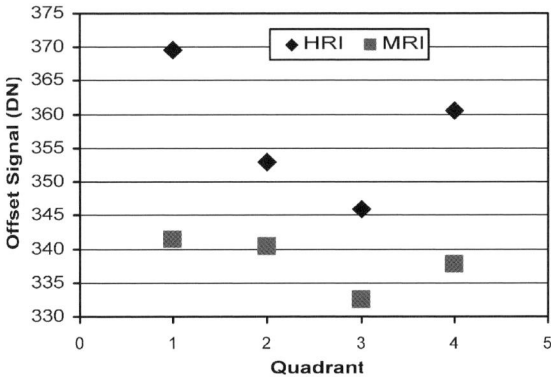

Figure 23. HRI and MRI CCD serial overclocked average signal levels. There is a nearly 23 DN difference between quadrants 1 and 3 in HRI. The spread in MRI is less than 10 DN.

their offset, and therefore would have their dynamic range significantly reduced. Therefore, a derived goal for the system was to have the offset of the quadrants within 15 DN of one another on all systems but the ITS.

Figure 23 shows the CCD offsets for HRI and MRI with the CCDs near their operating temperatures. The HRI shows nearly 20 DN difference from the highest to the lowest offset level. The MRI shows less than 10 DN from the highest to lowest level.

The scientific modes of the IR detector perform a subtraction of the reset levels of each pixel from the integrated values. Thus the signal left in the imaging area of the detector is the dark signal, which is a combination of the detector dark current, and the background IR radiation generated by components on the SIM bench. The detector does have five rows of reference pixels around the edge of the detector that are not connected to the HgCdTe substrate. In the DI three of these edges are available in full-frame images. of the images that are collected on DI. While the expected performance is to all be at a signal level of zero after the subtraction, the reality is that there is a slight variation across columns. While the source of this profile is not explained, the condition is stable over a wide range of conditions, and can be used as a quick diagnostic.

The two tests of the IR spectrometer at the SIM level were by far the best tests of the IR detector during the entire development program. In this case, the detector reached an operating temperature of 85 K, close to the predicted flight temperature. These tests were used to determine both a bad pixel map of the detector (see Section 5.3.7) and the dark signal performance of the spectrometer. Figure 24 shows the results of an analysis of dark signal over a wide range of temperature conditions. The figure shows a good Arrhenius curve of dark current versus temperature, after the prism temperature has been subtracted. This indicates that a large portion of the dark signal results from the prisms which are in the light path, and have a higher emissivity than does the bare aluminum surfaces of the mirror.

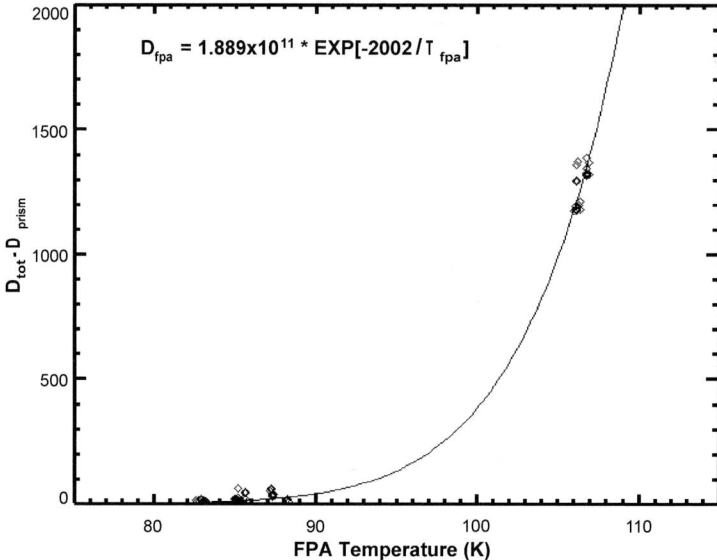

Figure 24. Dark signal for the HRI IR FPA. This plot shows the relationship between dark signal and the FPA temperature. The effect of prism temperature has been removed. The signal is that which is measured during a 2.86-s exposure time, the minimum for IR imaging modes 1 and 4.

5.3.5. *System Throughput in the Visible and IR, and Expected SNR at Encounter*

A major driver in the instrument resolution at the encounter is the length of the exposure time. A shorter exposure reduces the time that the imager is subjected to jitter of the spacecraft platform. A measurement of the instrument throughput will allow for a better prediction of the needed exposure times and thus an estimate of the resolution at closest approach.

Part of the calibration was to image a calibrated integrating sphere for both flatfield and throughput measurements. A forward model of the instruments, using all the measured components, was used to predict the throughput from each instrument. The model and the measured signal compared favorably. With the good comparison, the forward model can then be used to produce a calibration factor for all filters on both HRI and MRI, as shown in Figure 25. The calibration factors are calculated for a flat spectrum with unit flux of units μ W/(cm^2 sr^2 nm^1). Also shown in the figure are the same calibration factors for the Multi-Spectral Imager used on the Near Earth Asteroid Rendezvous (NEAR) mission as reported by Hawkins *et al.* (1997). The model can also be used to predict the signals that are expected at the encounter. The predicted signal rates for each filter of the HRI imager are shown in Table X. This model includes the effects of coma, and the calculation only include the SNR of the surface features. In this case, the coma was assumed to be one-third of the nucleus signal, which is conservative for Tempel 1. The mission requirement is that the crater be imaged with an SNR of 70 at closest approach. The table, along with the noise measurements of the CCD system and an estimate of the coma signal,

TABLE IX

Predicted signal rates for HRI at comet Tempel 1, and exposure times to meet SNR.

Filter	Rate (DN/s)	Rate (electrons/s)	SNR						
			10	30	50	70	100	150	250
1	3114.5	87,829	0.004	0.019	0.044	**0.080**	0.158	0.347	0.953
2	398.9	11,249	0.032	0.145	0.342	0.628	1.233	2.712	7.443
3	688.5	19,415	0.018	0.084	0.198	0.364	0.715	1.572	4.312
4	78.1	2,203	0.162	0.742	1.744	3.208	6.296	13.85	38.00
5	214.4	6,045	0.059	0.270	0.636	1.169	2.295	5.047	13.85
6	3110.2	87,707	0.004	0.019	0.044	0.081	0.158	0.348	0.955
7	505.2	14,247	0.025	0.115	0.270	0.496	0.974	2.142	5.876
8	303.7	8,564	0.042	0.191	0.449	0.825	1.620	3.563	9.775
9	666.2	18,787	0.019	0.087	0.205	0.376	0.738	1.624	4.456

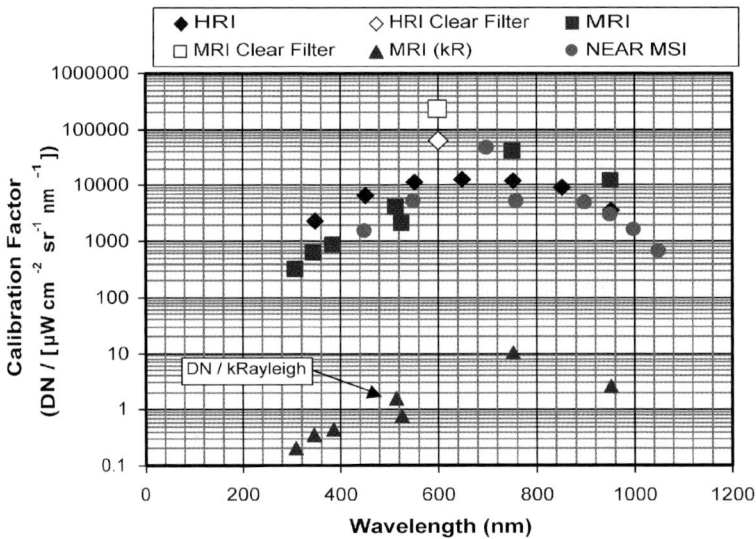

Figure 25. Calibration factors calculated for HRI and MRI for a 1-s exposure. Also shown are calibration factors for the NEAR MSI camera.

shows that exposure times of 0.080 s will produce a surface SNR of 70 or greater for images using a clear filter (e.g. Filter 1).

For MRI and ITS, similar tables are produced and these are used to set exposure times for the image sequence. The gas-filter observation of the coma take significant exposure times, and may be accumulated from a number of shorter exposures. With low relative readout noise this is possible.

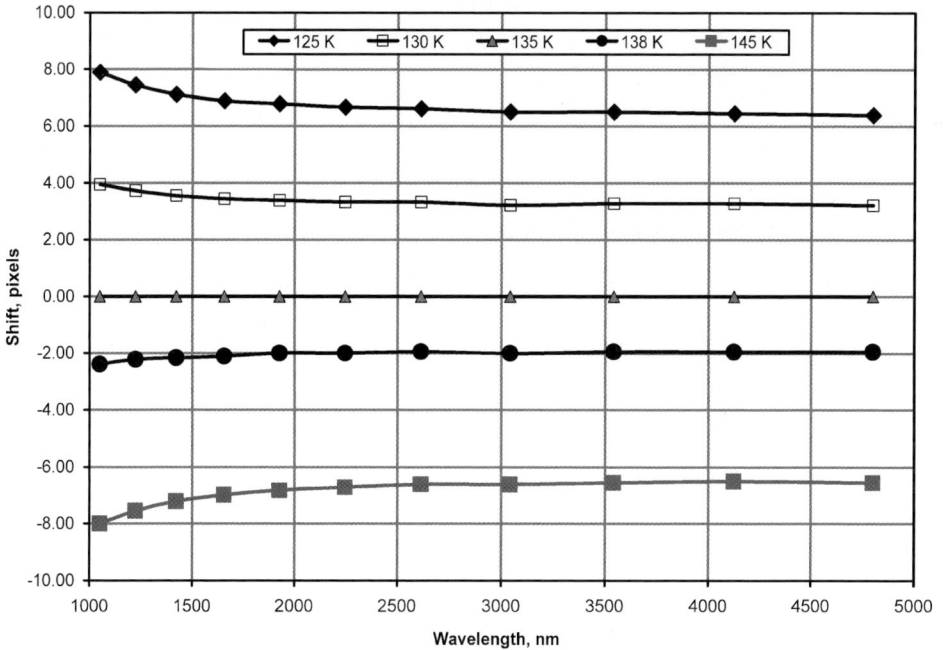

Figure 26. Wavelength shift of the IR spectrometer vs. prism temperature. The shift in center wavelength from the nominal center wavelength at a prism temperature of 135 K is plotted for each column for several cases. The wavelength shift with prism temperature is largely the same, 0.6 pixels/degree, over most of the spectrum. Below 2 μm, however, the rate of shift is wavelength dependant.

5.3.6. *IR Spectrometer Alignment, Wavelength Map and Spectral Resolution*

Section 5.1 describes the testing of the IR spectrometer at the SIM level. During these tests, the SIM optical bench was adjusted to three temperatures, along with the prisms. This allowed a measurement of the change in dispersion of the prisms with temperature. Using spectral lines of Kr, Ar, CO, and NH_4, as well as the ever-present CO_2 and H_2O when observing sources outside the chamber, a wavelength map was produced for each bench temperature. Figure 26 shows the wavelength maps at the three test temperatures. The wavelength shifts were compared to empirical formulae produced by Tropf (1995) and found to match closely.

With the wavelength mapped, the geometric spectral resolution is easily calculated. Figure 27 shows the spectral resolution at 137 K, the nominal flight temperature. The minimum geometric resolution is 216, which meets the required minimum of 200. Measurements of the line width for spectral lines of Kr and Ar that are known to be singlets shows that line widths are 1–1.1 binned pixels wide. At worst case the spectral resolution is 196.

As with any imaging spectrometer, the spectral lines show some curvature, or "smile" on the focal plane. The slit misalignment to the detector is less than one pixel across the 512 rows.

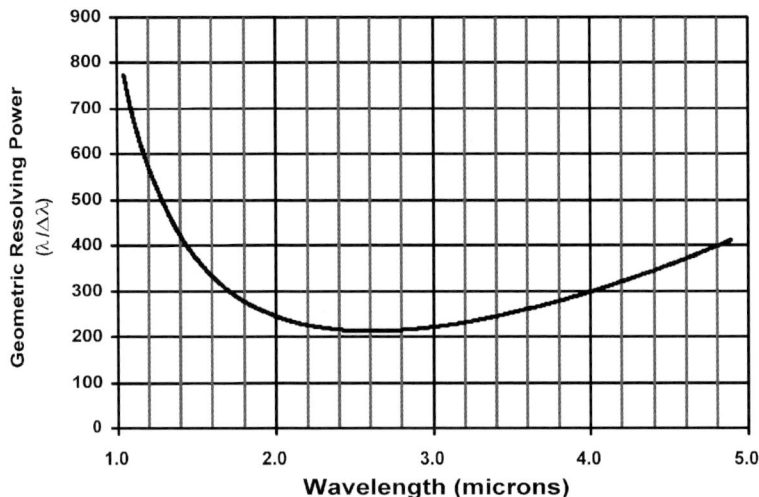

Figure 27. Geometric resolving power for the HRI IR spectrometer. The resolution is shown for the optical bench at an operating temperature of 137 K.

5.3.7. *Detector Bad Pixel Maps*

Interpreting the data requires an understanding of which pixels are inoperable, or have performance significantly less than the average pixel. Data from several tests could be used to make this determination.

The IR detector showed about 3.5% inoperable pixels on an initial analysis. The criterion used to determine an inoperable pixel was that it was greater than four standard deviations from the mean signal. Many of the pixels that have been initially categorized as inoperable may still have reduced performance, but information may be recovered. This is the subject of future work in analyzing the ground calibration data.

The CCDs showed impressively low numbers of bad pixels. The CCD in the HRI has one pixel that at room temperature is considered a "hot" pixel, and leaves an obvious trail of charge when operated. However, when the CCD is cooled below $-40\,°C$, the pixel seems to behave normally.

Figure 28 shows the boundary between the two image areas (see figure CCD-1 in Section 3.3.1). Because the readout circuits of the two halves are electrically isolated, there is additional traces at this boundary that reduce the effective collection area of the edge pixels. The figure was taken while the HRI was looking at an integrating sphere for flat-field testing. The signal in the two edge rows is consistently 15.5% less than that of the main region. There is a slight color variation on the order of 2%.

5.4. OPERATIONAL CONSIDERATIONS

There are necessarily limitations on how the instruments are commanded and controlled, and how they interface with the spacecraft. Some of the limitations were

Region 2A

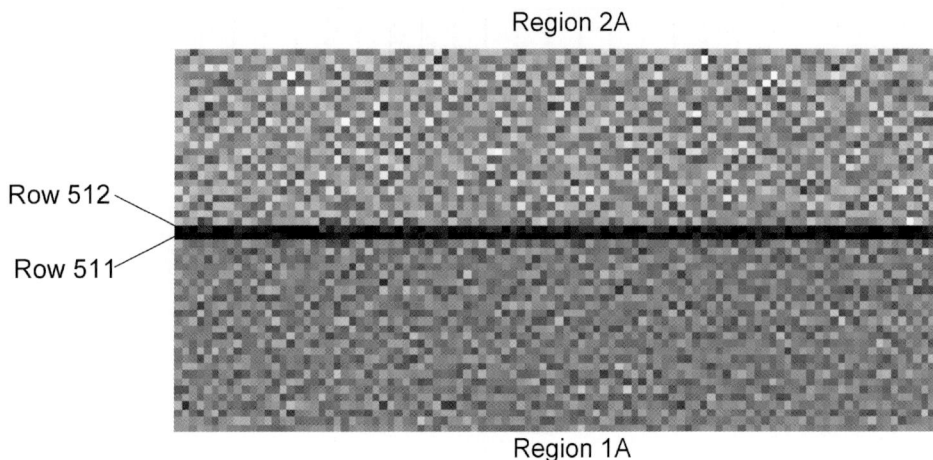

Row 512

Row 511

Region 1A

Figure 28. Close-up of a CCD flat field image at the boundary between the two imaging regions. The two rows that are reduced are due to extra traces in the CCD readout required to maintain independence between the two halves of the CCD. Regions 1A and 2A refer to the regions shown in Figure 9.

planned to reduce complexity and maintain cost, while others were discovered during interface testing with the spacecraft.

5.4.1. *IR Exposure Commands*

Two limitations, that were not planned, have to do with how the IR detector exposure commands are issued. Both have to do with interface issues with the ICB. First, there is a minimum commanded exposure time of 250 ms for IR mode 4 (interleaved, unbinned) images. This is required to ensure that the IR channel of the ICB is re-initialized. Second, IR images commanded to only one SCU will lock up the IR channel of the ICB in the other SCU. In other words, commanding IR images to be stored only on SCU-A will put the IR channel of the ICB in SCU-B into an indefinite "busy" state. This limitation is avoided by sending IR images to both SCUs in each command, which is the new baseline for the encounter imaging sequence.

5.4.2. *Data Rates and Limitations*

To support the unique mission scenario, the detector readout rates were set at 187,500 pixels per second per quadrant, and all three detectors from the flyby spacecraft may be transmitting data simultaneously. This scenario – such a high data rate from three instruments – is unique to planetary missions, to the knowledge of the authors. The ICB can receive the data at this rate from all three detectors as well, but once the data are being written to the SCU, all three detectors are attempting to access the same non-volatile memory simultaneously. Effectively, the data storage rate has been found to be 1.4 MB/s. During the 13 min from impact to closest approach, running at this rate would result in over 1 GB of image data being produced. That much storage is not available, so on average, the storage rate to non-volatile memory is not a limitation. The last few images are the highest resolution and therefore of the highest importance to the surface science goal of the mission and are taken in rapid succession. Early testing showed that short bursts of imaging at rates slightly above 1.4 MB/s resulted in one to six images being lost. Further adjustments to the imaging sequence and the management of the file system have resulted in subsequent tests storing all commanded images.

6. Conclusions

The Deep Impact instruments will collect the images from the first active experiment on a comet nucleus. They were designed based on the needs and constraints of the Deep Impact mission. Other than the de-scopes early in the program, required to meet budget constraints, the instrument team has produced instruments that match or exceed the desired performance envisioned at the beginning of the project. The testing showed that the ground performance will meet all the requirements. The ability to recover from anomalies during integration and test was enhanced by testing at the lowest assembly level as possible.

Acknowledgements

The authors wish to express their deep gratitude and thanks to the entire Deep Impact instrument team. A more dedicated and enthusiastic team would be hard to find. Of special note are Tom Yarnell whose leadership and dedication to the mechanical design were unmatched, Sheree Burcar who kept an energetic software development team on course, and Lyle Hunter for managing an aggressive test schedule. A special thanks to Bill Smythe at JPL for his oversight and sharing his past experience. This work was supported by NASA *via* the Deep Impact project.

References

A'Hearn, M. F.: 2005, *Space Sci. Rev.*, this issue.
Ball Aerospace & Technologies Corporation: 2002, *561177 Instrument B-Specification*, Revision D.
Hawkins, S. E., Darlington, E. H., Murchie, S. L., Peacock, K., Harris, T. J., Hersman, C. B., *et al.*: 1997, *Space Sci. Rev.* **82**, 31.
Klaasen, K. P., Carcich, B., Carcich, G., and Grayzeck, E. J.: 2005, *Space Sci. Rev.*, this issue.
Richardson, J., Melosh, H. J., Artemeiva, N. A., and Pierazzo, E.: 2005, *Space Sci. Rev.*, this issue.
Schultz, *et al.*: 2005, *Space Sci. Rev.*, this issue.
Tropf, W. J.: 1995, *Optical Eng.* **34**, 1369.
Jet Propulsion Laboratory: 2004, *Mission, Science, and Systems Requirements Document*, JPL D-16496, Revision C.

AUTONOMOUS NAVIGATION FOR THE DEEP IMPACT MISSION ENCOUNTER WITH COMET TEMPEL 1

NIKOS MASTRODEMOS*, DANIEL G. KUBITSCHEK and STEPHEN P. SYNNOTT

Optical Navigation Group, Navigation and Mission Design Section, Jet Propulsion Laboratory, California Institute of Technology, 4800 Oak Grove Drive, M/S 301-150, Pasadena, CA 91109, U.S.A.

(*Author for correspondence; e-mail: Nickolaos.Mastrodemos@jpl.nasa.gov)

(Received 19 August, 2004; Accepted in final form 8 December, 2004)

Abstract. The engineering goal of the Deep Impact mission is to impact comet Tempel 1 on July 4, 2005, with a 370 kg active Impactor spacecraft (s/c). The impact velocity will be just over 10 km/s and is expected to excavate a crater approximately 20 m deep and 100 m wide. The Impactor s/c will be delivered to the vicinity of Tempel 1 by the Flyby s/c, which is also the key observing platform for the event. Following Impactor release, the Flyby will change course to pass the nucleus at an altitude of 500 km and at the same time slow down in order to allow approximately 800 s of observation of the impact event, ejecta plume expansion, and crater formation. Deep Impact will use the autonomous optical navigation (AutoNav) software system to guide the Impactor s/c to intercept the nucleus of Tempel 1 at a location that is illuminated and viewable from the Flyby. The Flyby s/c uses identical software to determine its comet-relative trajectory and provide the attitude determination and control system (ADCS) with the relative position information necessary to point the High Resolution Imager (HRI) and Medium Resolution Imager (MRI) instruments at the impact site during the encounter. This paper describes the Impactor s/c autonomous targeting design and the Flyby s/c autonomous tracking design, including image processing and navigation (trajectory estimation and maneuver computation). We also discuss the analysis that led to the current design, the expected system performance as compared to the key mission requirements and the sensitivity to various s/c subsystems and Tempel 1 environmental factors.

Keywords: autonomous navigation, Tempel 1, simulations

Introduction

MISSION OVERVIEW

Deep Impact is a dual spacecraft mission planned for launch in January 2005 with the engineering goals of impacting comet Tempel 1 on July 4, 2005, observing the impact event and ejecta plume expansion, and obtaining high-resolution images of the fully developed crater using the Medium Resolution Imager (MRI) and the High Resolution Imager (HRI) on the Flyby spacecraft (s/c) for the scientific purpose of exposing and understanding the interior composition of a comet nucleus.

After a 6-month cruise, the two spacecraft will separate 24 h prior to the expected time of impact (TOI). The encounter geometry will result in an illumination phase angle of approximately 64° for the Tempel 1 nucleus. The Flyby s/c will perform a slowing maneuver with a ΔV of approximately 102 m/s to provide 800 ± 20 s of post-impact event imaging and control the flyby minimum altitude to 500 ± 50 km. During the first 22 h following release, the Impactor s/c will acquire and telemeter science and navigation reconstruction images to the ground using the Flyby s/c as a data relay. The Flyby s/c will acquire and telemeter MRI and HRI visible and HRI infrared (IR) images of the nucleus and coma.

The fundamental requirements for the autonomous navigation (AutoNav) system are as follows:

1. Guide the Impactor to impact in an illuminated area on the nucleus surface that is viewable from the Flyby s/c for 800 s after TOI;
2. Provide comet-relative position information at TOI that will result in a Flyby HRI pointing accuracy of better than 100 μrad (3σ) to capture the impact event in the HRI 256×256 pixel subframe and observe the ejecta plume expansion dynamics;
3. Provide comet-relative position information and a crater-pointing offset that will result in a Flyby HRI pointing accuracy that is better than 1 mrad (3σ) at E+800 s to capture the fully developed crater.

The autonomous phase of the encounter will begin at 2 h before TOI. The type of autonomy used for Deep Impact can be classified as scripted autonomy (Zimpfer, 2003). A critical sequence running on-board the Impactor s/c will spawn science and navigation subsequences that issue Impactor Targeting Sensor (ITS) commands to produce navigation images at a 15-s interval. The Autonomous Navigation (AutoNav) software (Bhaskaran et al., 1996, 1998; Riedel et al., 2000), originally developed and demonstrated during the Deep Space 1 (DS1) mission, processes images to form observations for the purpose of trajectory determination (OD). OD updates are used to support computation of trajectory correction maneuvers (TCM) and provide relative position information for the purpose of pointing the navigation and science instruments. Three (3) Impactor targeting maneuvers (ITM) will be computed by AutoNav on the Impactor s/c and executed by the Attitude Determination and Control System (ADCS): ITM-1 at E$-$100 min (E$-$designates time before impact), ITM-2 at E$-$35 min, and ITM-3 at E$-$7.5 min. At E$-$4 min, the Impactor ADCS will point the ITS along the AutoNav estimated comet-relative velocity vector to capture and telemeter high-resolution (between 3 m and 20 cm) ITS images of the impact site prior to impact. Meanwhile, the AutoNav software on the Flyby s/c will process MRI images of the comet every 15 s and update the trajectory model of the Flyby s/c every minute to continuously point the MRI and HRI instruments at the nucleus. Figure 1 shows a schematic diagram of the encounter activities.

This paper describes the expected performance of the AutoNav terminal guidance on the Impactor s/c and the expected performance of the AutoNav tracking

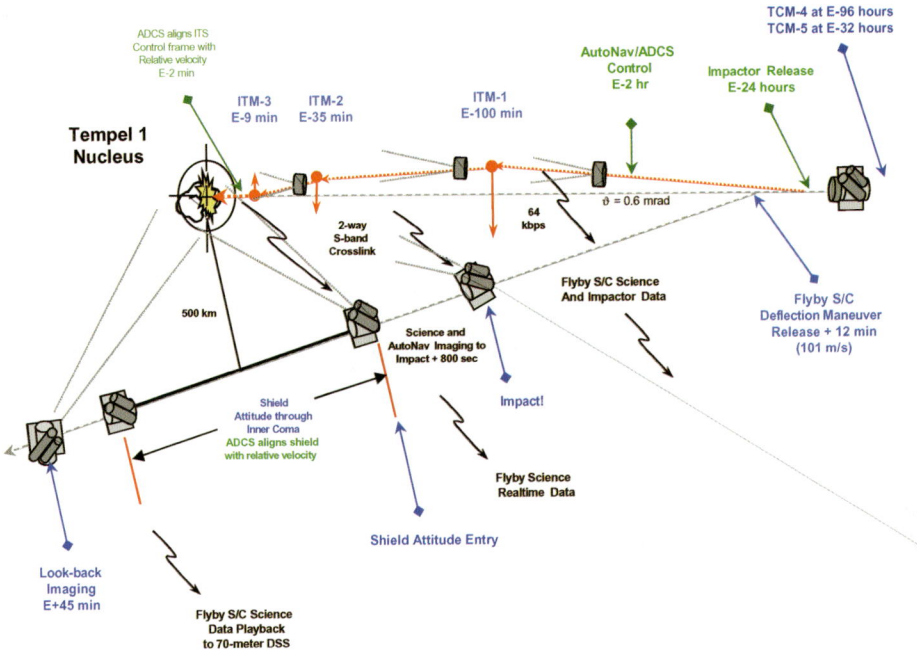

Figure 1. Tempel 1 encounter schematic for the Deep Impact mission.

process on the Flyby s/c at the two key science epochs: 1) TOI and 2) Time of highest-resolution imaging following the impact event.

FLYBY SPACECRAFT FLIGHT SYSTEM

The Flyby s/c, shown in Figure 2, was designed and built at Ball Aerospace Technologies Corporation. The Flyby s/c consists of a solar array to provide power to the combined Flyby and Impactor flight system, while the Impactor is mated to the Flyby s/c during cruise checkout and testing of the Impactor s/c; two redundant RAD 750 computers for processing; the MRI instrument that will be used for autonomous navigation during encounter; the HRI instrument with visible and IR detectors, which will be used for approach phase optical navigation (OpNav), Science imaging, and as the backup navigation camera during AutoNav operations; a high gain antenna (HGA) for real-time data return during encounter; a S-band antenna for communication with the Impactor s/c following release; a three-axis stabilized momentum wheel-based control system with a four divert/four RCS thruster hydrazine propulsion system for TCMs and momentum dumps; and an ADCS system that estimates the attitude in the ICRF inertial reference frame based on optical measurements from two StarTrackers and rates and linear accelerations from an Inertial Reference Unit (SSIRU). The mass of the Flyby s/c will be approximately 660 kg.

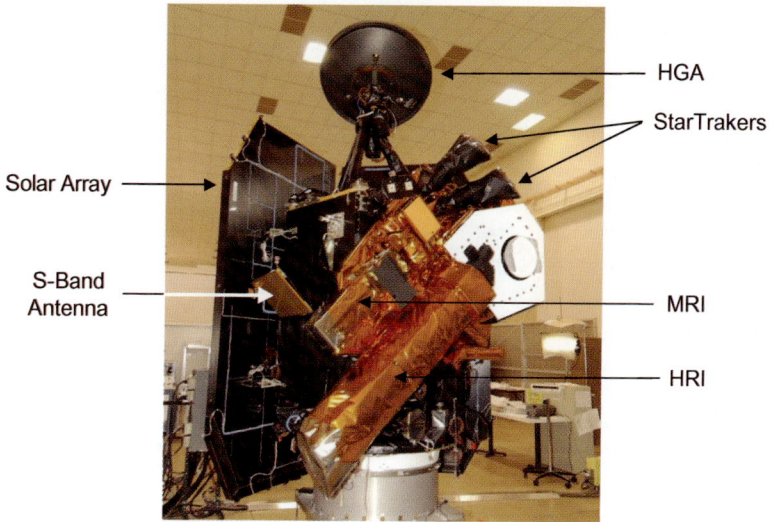

Figure 2. Flyby spacecraft in the clean room at Ball Aerospace prior to environmental testing.

The MRI camera has a 12 cm aperture (73.5 cm^2 collecting area with 35% obscuration), a focal length of 2.1 m and a 10 mrad field-of-view (FOV). The 1,024 × 1,024 pixel CCD is a split-frame transfer device with electronics that provide 14-bit digitization (16,384 DN full-well). The HRI instrument has a 30 cm aperture, a focal length of 10.5 m and a 2 mrad FOV with CCD electronics that are the same as the MRI.

IMPACTOR SPACECRAFT FLIGHT SYSTEM

The Impactor spacecraft, shown in Figure 3, was also designed and built at Ball Aerospace and consists of a battery for power during the 24 h free-flight, a single RAD 750 SCU for processing, an ITS consisting of a simple inverting telescope

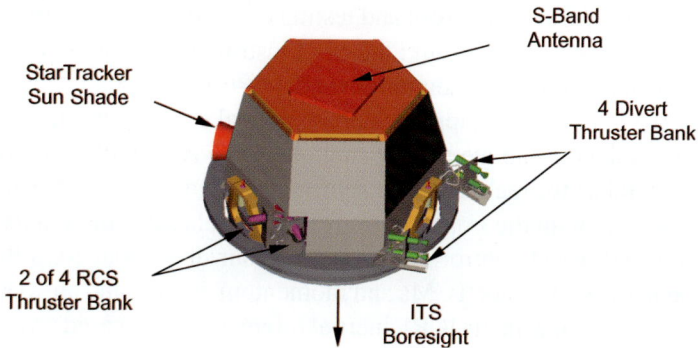

Figure 3. Impactor spacecraft flight system configuration.

and CCD detector, a S-band communications link to the Flyby s/c, a three-axis stabilized rate control system (RCS), a four divert/four RCS thruster hydrazine propulsion system with a ΔV capability of 25–30 m/s, and an ADCS subsystem that estimates the attitude in the ICRF reference frame based on measurements from a single StarTracker as well as rates and linear acceleration from an Inertial Reference Unit (SSIRU). The mass of the Impactor s/c will be approximately 370 kg with an all-copper fore-body cratering mass.

The ITS camera has a 12 cm, a focal length of 2.1 m and a 10 mrad FOV like that of the MRI. The 1,024 × 1,024 pixel CCD is also a split-frame transfer device with electronics that provides 14-bit digitization. The ITS serves a dual purpose: 1) provide navigation images and 2) provide pre-impact high-resolution (20 cm) science images.

SELECTION OF THE IMPACTOR TARGETING STRATEGY

General and specific guidance and control laws have been developed for several interceptor designs (Zarchan, 1997). There are two basic categories: 1) The use of proportional navigation or augmented proportional navigation techniques, and 2) the use of predictive guidance techniques.

The proportional navigation strategy is driven by measurement of the closing velocity and line of sight angular rates to control the acceleration via thrust vectoring. On the other hand, predictive guidance makes use of the dynamics (equations of motion) of both the target body and interceptor via state estimation using available measurements that can be related back to the state of the interceptor based on observations of the target. The former is what can be considered a non-dynamic or reduced-dynamic approach; the latter is a dynamic approach, which has the advantage of requiring fewer observations and being less susceptible to large, random errors in the measurements or observations.

If we consider the Impactor measurement system, which consists of ITS optical observations of the target body center of brightness; the Impactor targeting (maneuver) system characteristics; the well-known target body dynamics; the updating of the Impactor state based on the optical observations of comet Tempel 1; and the use of a few, discrete, lateral burns (ITMs) based on the predicted s/c and target body locations at the time of intercept, the Impactor targeting system for Deep Impact can be classified as an interceptor that uses predictive, pulsed guidance (few discrete burns) to achieve impact at the desired location.

It should be noted that the best quality optical observations are obtained during non-thrust periods, which suggests that a pulsed guidance system be used, as was selected for the Impactor s/c. Even though the *a-priori* position of the target body relative to the s/c can have a large uncertainty, which is reduced using optical navigation techniques, the dynamics are well known, except that (an important exception) the nucleus rotational dynamics and solar phase angle combine to induce motion (acceleration) of the center of brightness (CB) with time, which can cause targeting errors and influence the impact location on the surface of the nucleus *via* over-estimation of the lateral velocity required to intercept at the desired location. These are mitigated, to some extent, in the batch filtering process by having some knowledge of the nucleus rotation period and by selecting the appropriate arc length over which to perform an orbit solution. This suggests that a predictive guidance strategy be used for Deep Impact and we have selected a 20 min OD arc length. In general, the OD arc length is based on a number of considerations such as rotation period, time from last OD to impact, redundancy against image loss, etc. Detailed simulations have shown that for a rotation period of 40 h or longer, impact statistics are statistically similar for OD arc lengths in the 10–35 min range but begin to increase for longer arcs.

In summary, the Deep Impact Impactor s/c uses a predictive guidance strategy and pulsed guidance system consisting of three (3) lateral, discrete magnitude burns (ITMs) based on the integrated equations of motion of the Impactor s/c and the evaluated position of the target (apriori comet Tempel 1 ephemeris) at the TOI to compute the "zero effort" miss distance, which is then used to compute the magnitude and direction of each ITM to achieve impact.

Autonomous Navigation

The autonomous navigation system for Flyby s/c tracking of the nucleus and terminal guidance of the Impactor s/c relies on both the performance and interaction of the AutoNav and ADCS flight software and the MRI and ITS navigation cameras. AutoNav consists of three (3) distinct modules: 1) Image processing; 2) Orbit determination; and 3) Maneuver computation (Impactor only). AutoNav was originally developed to operate in two different modes: Star-relative mode, which uses images that contain both the target body (beacon) and two or more stars for determining the orientation of the camera at the time of each image exposure; and Starless mode, which uses the ADCS estimated s/c attitude and camera alignment information to determine the orientation of the camera at the time of each image exposure. For Deep Impact, the Starless AutoNav mode is used based on the expected quality of the ADCS estimated attitudes. The combination of the CT-633 StarTracker(s) and SSIRU rate sensor provides an estimated attitude bias of no more that $150\,\mu$rad (3σ) on the Flyby s/c and $225\,\mu$rad (3σ) on the Impactor s/c, a bias stability of $50\,\mu$rad/h (3σ), and estimated attitude noise of $60\,\mu$rad (3σ) (Trochman, 2001).

The steps involved in the Flyby tracking and Impactor autonomous guidance process are as follows:

1. Acquire images of the comet nucleus, every 15 s, starting 2 h before the expected TOI.

2. Process the images to compute pixel/line location of the nucleus center of brightness (CB).
3. Use observed CB pixel/line locations to compute measurement residuals for comet-relative trajectory estimation.
4. Perform OD updates, every 1 min, starting 110 min before the expected TOI.
5. Perform three (3) primary ITMs at 100 min (ITM-1), 35 min (ITM-2), and 7.5 min (ITM-3) during the terminal guidance phase.
6. Acquire ITS images on the Impactor s/c for computing a Scene Analysis-based offset, relative to observed CB, just prior to ITM-3 maneuver computation and use the offset in the maneuver computation for ITM-3.
7. Acquire HRI images on the Flyby s/c at 23 min before the expected TOI for computing a Scene Analysis-based offset, relative to the observed CB, which is applied as a pointing correction to observe the impact site and track the crater during formation.
8. Align the ITS boresight with the AutoNav estimated comet-relative velocity vector starting 4 min prior to predicted TOI to capture and transmit high-resolution images of the nucleus surface to the Flyby s/c.
9. Track the impact site with the Flyby imaging instruments for 800 s following the actual TOI.

PROCESSING IMAGES

Image processing for the AutoNav system serves the purpose of providing observations of the s/c–comet relative trajectory over time. All images received by AutoNav are used to compute the target body's brightness centroid. A few selected images are also used in order to establish the most suitable impact site. AutoNav uses two different methods in order to extract the centroid of the image: 1) Brightness centroiding of all pixels above a brightness threshold and within a predetermined pixel subregion (Centroid Box); 2) Finding each separate "blob" in the image, i.e. the one or more contiguous regions of pixel brightness above a brightness threshold, computing the centroid of each blob and providing the centroid of the largest blob (Blobber). For either method, the brightness centroid of the nucleus in pixel/line coordinates is computed *via* a moment algorithm; or 3) Scene Analysis, which is used to compute the most suitable impact site and provide this information as a pixel/line offset, relative to the CB location, as determined from either 1) or 2).

The first step in image processing consists of removing the image background, most of which consists of a fixed bias value that is different for each quadrant, by subtracting a dark frame from each image. Based on the best estimate of the s/c trajectory relative to the nucleus and the navigation camera attitude at the time the image was taken, a predicted pixel/line location is computed. For processing with the Centroid Box algorithm, all pixels above the brightness threshold in a $N \times N$ pixel region surrounding this predicted location are used and the CB is determined by computing the first moment values in the pixel and line directions. The brightness threshold can be a fixed value specified in the parameters file or dynamically computed based on a percentage of the "average peak" brightness. For Deep Impact we will use a 400×400 pixel region to cover uncertainties in the ADCS-estimated attitude and we will take 35% of the peak pixel brightness as the brightness cutoff. The brightness cutoff is used mainly to remove signal from the dust or coma surrounding the nucleus, so that the computed CB is entirely due to the nucleus.

The Blobber algorithm (Russ, 1999) has seen extensive in-flight use and differs from the Centroid Box in that it scans the full frame as it searches for each separate "blob" in the image. This removes the dependence on AutoNav's ability to predict where the nucleus will be located in the image array. The natural breakdown of the image into separate blobs, and subsequent selection of CB on the largest blob, offers a robust way to discriminate cosmic rays and other blemishes by means of minimum and maximum size criteria. This approach has many advantages over the Centroid Box approach, but has proven to be less stable for certain nucleus orientations where the illumination is such that the nucleus appears to have the shape of two disjointed large blobs of comparable size; in this case, a small change in the relative size of the two blobs can shift the CB abruptly and result in large targeting errors.

Scene Analysis images are processed using either Centroid Box or the Blobber algorithm to provide a reference CB location. In the first step, the image is scanned from the edge of the image towards the center to delineate the outer boundary of the nucleus by comparing each pixel with the brightness cutoff. Each on-nucleus pixel is further characterized as interior dark or bright by a similar comparison to the cutoff value. In the next step, make each on-nucleus pixel the center of a circle of radius equal to the 3σ error footprint, or control error. The control error itself is established *via* Monte Carlo simulations. For each candidate pixel, we compute a number of quantities: total number of on-nucleus bright pixels; total number of off-nucleus pixels and proximity to the direction of the Flyby s/c point of closest approach to the nucleus. In order for a pixel to qualify as a candidate for the impact site, it has to contain a certain portion of on-nucleus pixels within its control error. A typical such number is 95%. Qualifying pixels are then compared against each other with criteria considered in order of decreasing priority. The first criterion is maximizing the number of on-nucleus bright pixels within the control error (maximizing the lit area). When more than two sites meet the first criterion, the second criterion is minimizing the number of off-nucleus pixels. The third priority is proximity to the Flyby s/c's point of closest approach, which we call "biasing". Everything else being equal, biasing tends to move the impact site toward a location on the nucleus that is more likely to be visible from the Flyby s/c near its closest approach point where the view angle increases to $45°$, thus enhancing the ability of the Flyby s/c to image the fully developed crater at the time of highest-resolution imaging. The difference between the CB reference location and the selected site is computed and converted to an inertial correction vector that is stored and used by the AutoNav maneuver computation software. If no suitable site is found, then no correction vector is returned, which results in a default to targeting the CB. Note that the period of rotation does enter into the computations of the best impact site and there is no modeling of the pole or the orientation of the nucleus at encounter, since this information is not assumed to be available at any time prior to encounter.

TRAJECTORY DETERMINATION

For the Flyby and Impactor s/c, the trajectory is estimated and updated every minute during the last 2 h of encounter. The trajectory determination software supports both ADCS attitude control and AutoNav maneuver computations.

After images are processed, the important information needed to relate the observations back to the state of the spacecraft are stored in an on-board optical navigation (OpNav) file. This information consists of the time the image was exposed, the camera inertial orientation (right ascension, declination, and twist), the pixel and line location of the observed CB, and the data weight associated with a given observation and whether or not the observation was declared useful. When AutoNav receives the command to perform orbit determination, the best estimate of the s/c position and velocity is read from a second on-board file called the orbit determination file. The trajectory is integrated, making use of on-board accelerometer data that is stored in a non-gravitational history file, to the time of each observation and the predicted pixel and line location of the nucleus center of mass is computed. The difference between the computed pixel/line location and the observed pixel/line location corresponding to the observations contained in the OpNav file represents the residual, which is minimized in a least-square sense using AutoNav's batch-sequential processing algorithm to estimate position, velocity and two navigation instrument cross-line-of-sight attitude bias drift parameters. The on-board OD file is then updated with the best estimate of the s/c state vector. The OD arc length was selected to be 20 min for Deep Impact and with image processing every 15 s, each OD arc contains 80 observations.

Following completion of each orbit determination, the trajectory is updated and a representation of the estimated trajectory is generated in the form of Chebyshev polynomial coefficients. This polynomial represents a time series of the predicted Impactor position relative to the nucleus that is passed to the ADCS flight software. ADCS evaluates the polynomial, computes the relative position vector and aligns the ITS camera boresight on the Impactor and the HRI and MRI camera boresights on the Flyby s/c with the comet-relative position vector to center the nucleus in the instrument FOV. Figure 4 shows simulated MRI AutoNav images at E−2 h, E−0 min (Impact), and E+750 s, and the HRI AutoNav Scene Analysis image at E−23 min.

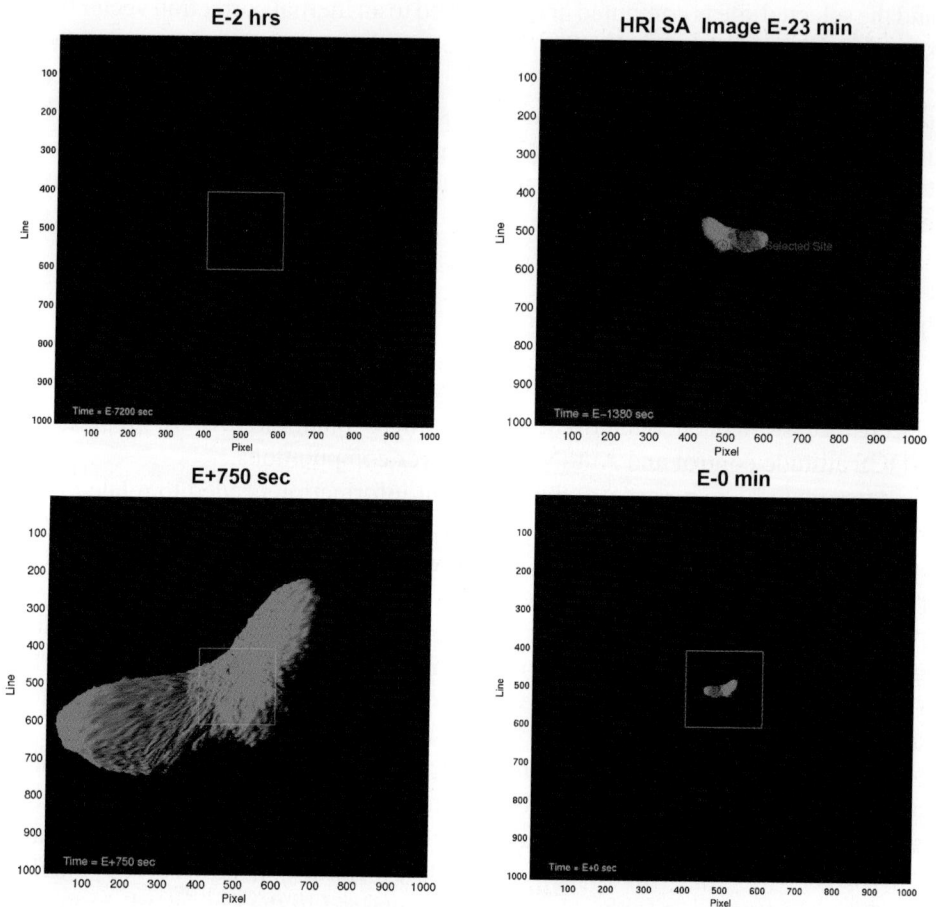

Figure 4. Flyby MRI AutoNav images with HRI FOV superimposed at E−2 h, E−0 min, and E+750 s. The upper right image is the Flyby HRI AutoNav Scene Analysis image at E−23 min.

IMPACTOR TARGETING MANEUVERS

Impactor targeting maneuvers are initiated *via* an AutoNav sequence command. These commands are issued from the engineering sequence that will be running during encounter. When AutoNav receives the command, an impulsive maneuver (magnitude and direction) relative to the ICRF frame is computed for the time contained in the command packet and passed to ADCS in the form of a command issued by AutoNav. ADCS receives the maneuver ΔV information in the command packet, computes the finite duration burn start time, and populates the necessary flight software current value table. When the current value table is populated, ADCS issues a command to spawn a trajectory correction maneuver (TCM) transition sequence consisting of the following commands:

1. Disable AutoNav image processing.
2. Command ADCS to transition (turn) to burn attitude.
3. Reset accelerometer accumulated ΔV.
4. Command ADCS to delta-V mode in preparation for ITM execution.

When the burn start time is reached, the burn is initiated. The ADCS flight software continually monitors the accumulated ΔV information based on incremental ΔV values measured by the SSIRU accelerometers and continuously adjusts the pointing and thruster duty cycle until the accumulated ΔV matches the desired maneuver ΔV, which results in burn termination. When the burn is complete, ADCS returns to instrument point mode and issues another command to spawn a second TCM transition sequence to re-enable AutoNav image processing. The need for TCM transition sequences and for disabling AutoNav image processing arises from the non-deterministic burn duration of each ITM. This allows instrument commands for AutoNav images to be issued every 15 s without regard to when a particular maneuver will occur or how long the burn will last. Since the quality of the images acquired during the burn may be degraded, this also allows AutoNav to simply ignore the navigation images received during these periods.

ITM-1 (ENCOUNTER $-$ 100 MIN)

The terminal guidance phase of the mission begins 120 min before impact. ITM-1 is the first targeting maneuver based on ITS observations and an updated trajectory. The primary purpose of ITM-1 is to remove the Flyby pre-release delivery errors. ITS observations have to be good enough to provide improvement over the pre-release orbit determination, but waiting too long for ITM-1 increases the amount of ΔV required to remove the delivery errors that are expected to be 6 km (3σ) due to the quality of the Flyby approach phase optical navigation solution. The delivery requirement is 30 km (3σ). Correcting for a 6 km pre-release delivery error requires $\sim$1 m/s ΔV for a maneuver placed at E$-$100 min. The maneuver execution error will be on the order of 4 cm/s for a 1 m/s burn. ITM-1 targets the nucleus CB (dot in Figure 5). Figure 5 shows a simulated image of the nucleus at a range of approximately 60,000 km. The optical signal from the nucleus spans approximately 10 pixels at the time of ITM-1.

ITM-2 (ENCOUNTER-35 MIN)

The second ITM, ITM-2, was placed at E-35 min to provide redundancy in the form of improved targeting over ITM-1 in the event ITM-3 fails to execute. ITM-2 will require approximately 11 cm/s of ΔV to remove the ITM-1 maneuver execution errors. The trajectory solution is based on ever improving ITS observations of the nucleus. As seen in Figure 6, the nucleus spans more than 25 pixels. ITM-2 also targets the nucleus CB (dot at tip of arrow).

N. MASTRODEMOS ET AL.

ITS Instrument: Image #1

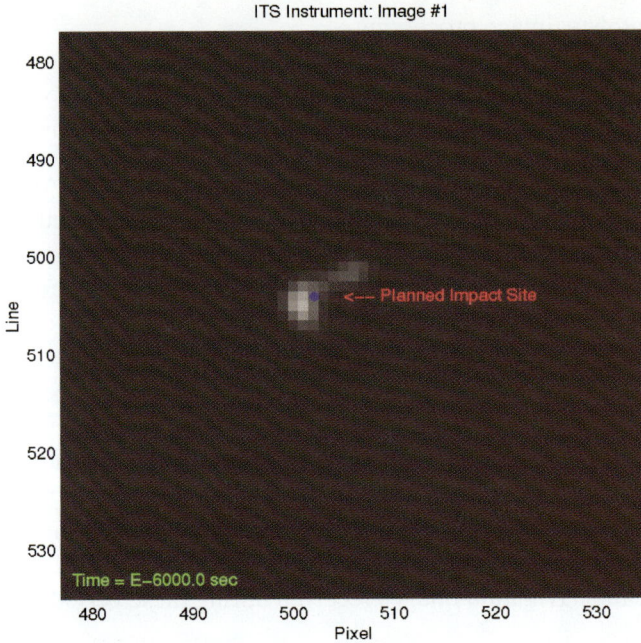

Figure 5. Simulated image of the comet nucleus with a 65° illumination phase angle at the time of ITM-1 (E−100 min).

ITS Instrument: Image #1

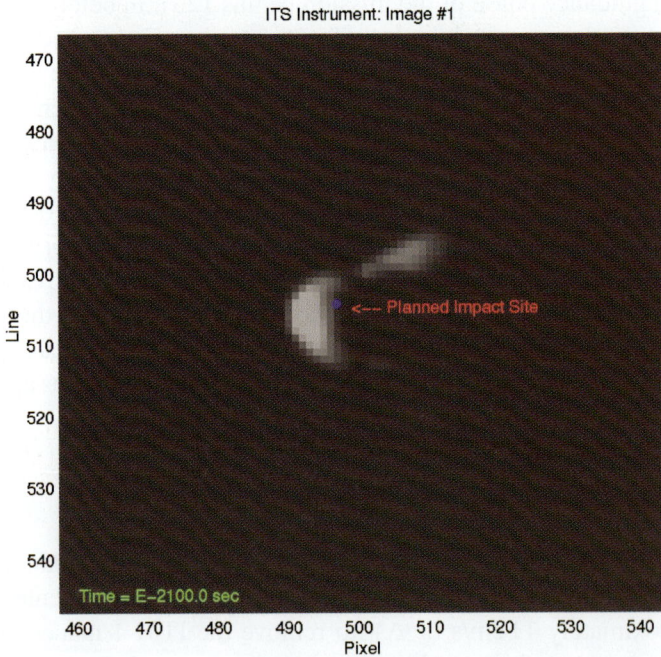

Figure 6. Simulated image of the comet nucleus with a 65° illumination phase angle at the time of ITM-2 (E−35 min).

ITM-3 (ENCOUNTER -7.5 MIN)

The third and final ITM, ITM-3, is the most important and provides the fine targeting for an illuminated impact. The trajectory determination for ITM-3 is based on CB observations of the nucleus. However, at E-11 min a Scene Analysis image is sequenced to provide a targeting offset from the CB. This offset, as previously mentioned, provides an increased probability of an illuminated impact and increases the probability that the crater will be seen from the Flyby s/c at the time of highest-resolution imaging. As seen in Figure 7, the nucleus spans nearly 100 pixels at E-11 min. The blue dot represents the computed CB. The red "+" symbol and surrounding circle represent the Scene Analysis selected impact site. It is important to note that for this nucleus orientation (particularly challenging), the CB lies in a shadowed region on the nucleus surface and would likely result in a dark impact had that been the target.

ITM-3 will nominally be centered at 7.5 min before impact and could require a B-plane correction of as much as 4 km to take the Impactor from intercept at the CB to intercept at the selected site. A 4 km correction at E-7.5 min requires approximately 7 m/s of ΔV. The associated maneuver execution error is expected to be no more than 10 cm/s, which results in a B-plane error of only 54 m. Placing

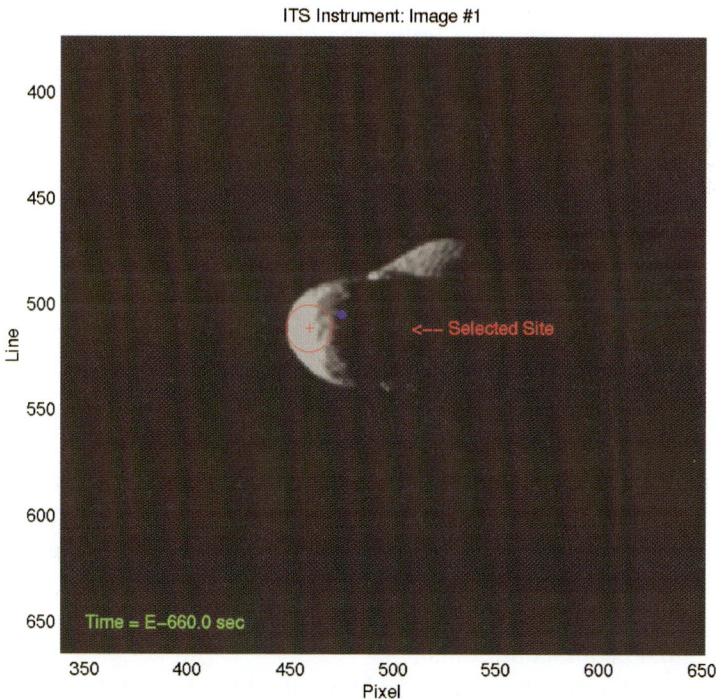

Figure 7. Simulated image of the comet nucleus with a 65° illumination phase angle at the time of scene analysis for ITM-3 (E-11 min).

ITM-3 as late as possible is ideal for minimizing maneuver computation errors and maneuver execution errors, however, the Impactor s/c is constrained by the amount of propellant (25 m/s ΔV capability) available for ITMs and the worst-case coma environment could result in ADCS control upsets due to particle impacts. To ensure a high-probability of successful maneuver completion it was determined that ITM-3 be completed prior to E−6 min, which represents a range of approximately 3,600 km from the nucleus.

Expected Autonav Performance

The performance of AutoNav is examined by simulations and testing at increasing levels of fidelity:

a) Monte Carlo simulations, which replicate the basic image processing, orbit determination and maneuver computation functions including error sources such as those from ADCS attitude estimation and maneuver execution. These simulations form the bulk of the analysis and are used not only for insight into the sensitivity of the various parameters, but also to establish important encounter parameters.
b) Monte Carlo simulations, which trigger the actual flight code and are used to confirm that the flight software will meet its expected performance.
c) Software test bench runs that utilize the closed-loop interaction between AutoNav and ADCS in the flight system environment which are used to verify the encounter sequence design and the AutoNav/ADCS interface. All three levels of testing are essential for performance validation at encounter.

This paper discusses the results from the Monte Carlo simulations from category a) described earlier.

NUCLEUS TEST-MODEL DEVELOPMENT

One of the unique technical challenges of the Deep Impact mission is the lack of detailed knowledge of the Tempel 1 nucleus prior to the actual encounter and the need for robust performance within a broad range of cometary characteristics such as different shape models, topographic relief maps, average size, rotation period, pole direction and dust environment. The most recent data from ground- and space-based observations, presented in detail elsewhere in this issue, indicate that the Tempel 1 nucleus has an average radius just over 3 km, a mean albedo in the range 0.03–0.04, an axial ratio of ∼3:1, a rotation period in the 38–41 h and a shape that may be irregular and not ellipsoidal.

Over the past few years, the Optical Navigation Group at JPL, in collaboration with the Deep Impact Science Team, has constructed a number of shape models driven by past observations of Solar System small bodies or theoretical predictions

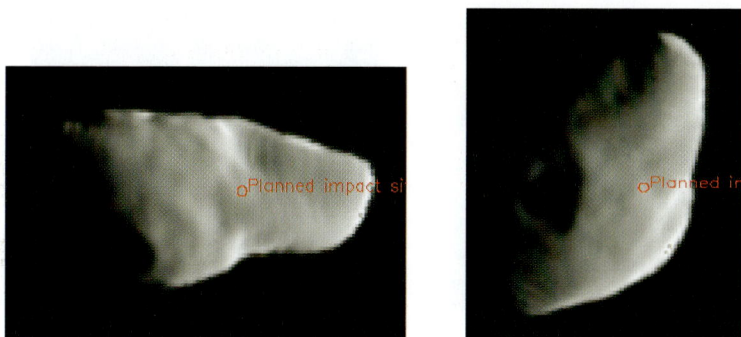

Figure 8. Examples of a Halley-like nucleus with an ellipsoidal shape and few topographic variations.

that span a range of possibilities that bound what is currently predicted for comet Tempel 1. Three such categories of models are highlighted as follows:

1. Halley-like models (Figure 8) derived from the unpublished shape models developed by P. Stooke and A. Abergel (Halley Nucleus Shape Model, personal communication via M. Belton, 2000). These models have a general ellipsoidal shape with an axial ratio of 2:1 and a relatively smooth surface;
2. Borrelly-like models (Figure 9) derived from the DS1 observations of comet Borrelly (Kirk *et al.*, 2004). They have axial ratios in the range of 2.5:1 (for the baseline model) to 3.5:1 (for the worst-case model) and a double-lobed structure that results in significant shadowing. These models cause the largest targeting errors and their center of brightness can be very near the limb; and
3. Accretion models (Figure 10) that result from theoretical considerations (Weidenschilling, 1997). These models have extreme topographic variations on every length-scale with heights up to half the size of the horizontal length-scale of the surface fluctuations. They tend to result in a higher frequency of dark impacts.

KEY AUTONAV ASSUMPTIONS AND ERROR SOURCES

The Monte Carlo analysis described in this paper relies on a number of key assumptions and sources of error that directly affect the performance of the AutoNav targeting and tracking algorithms: 1) Cometary characteristics; 2) ADCS attitude estimation errors; and 3) Maneuver execution errors.

Cometary Characteristics
In addition to the unknown shape, axial ratio and topography of Tempel 1, which is addressed by the variety of nucleus models discussed earlier (Figures 8–10), other comet model parameters that affect navigation performance include: the coma brightness, average size, rotation period, pole direction, nucleus orientation at the time of the encounter and the time-of-flight, or downtrack position of the s/c relative to the nucleus.

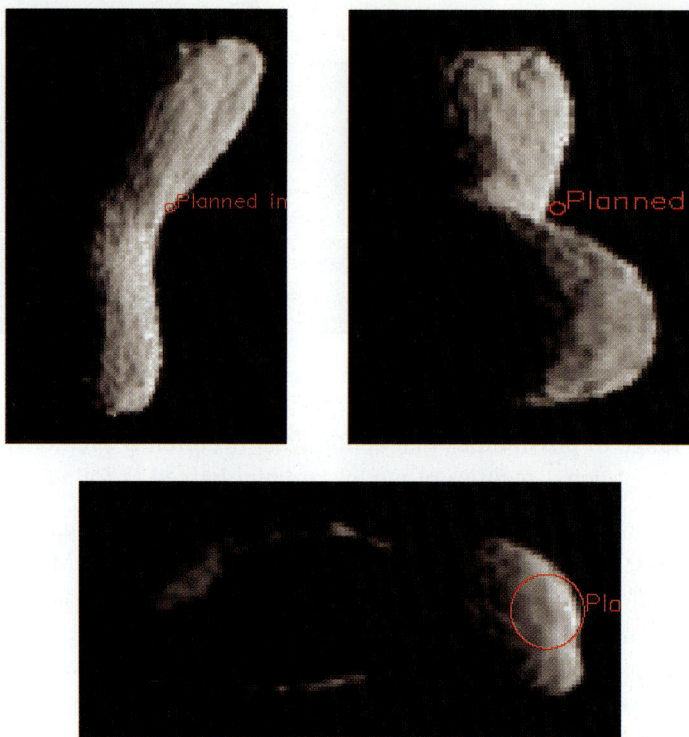

Figure 9. Typical examples of Borrelly-like nucleus models with a 3.5:1 axial ratio. The small circle indicates the center of brightness in the upper left and upper right images and the large circle is the site selected using scene analysis in the bottom image.

The coma brightness is one of the key considerations for a successful encounter. The need to discriminate off-nucleus coma, so as to guide the Impactor towards the nucleus, and on-nucleus foreground coma, so as to select an intrinsically illuminated impact site is paramount. The coma brightness is specified in terms of the brightness ratio between the peak brightness of a jet and the peak brightness of the nucleus when both are fully resolved and is bounded between the most-likely case, which is the current best expectation and the worst-likely case, which represents scenarios whereby all observed dust is created by a single narrow jet. The brightness ratio for the most-likely case coma model is 1/32 and for the worst-likely case coma model it is 1/6. AutoNav computes an autonomous coma cutoff, relative to an "average peak" brightness value per image, which for conservatism is currently set to 35% of the peak or ~2 times larger than the worst-case coma. As the coma cutoff value increases, a larger portion of the nucleus image is discarded as being part of the coma. This has the undesirable side effect of moving the CB closer to the bright limb as well as causing a less-stable CB from image to image, which maps directly into an OD error. Figure 11 illustrates this point for two different values of the coma

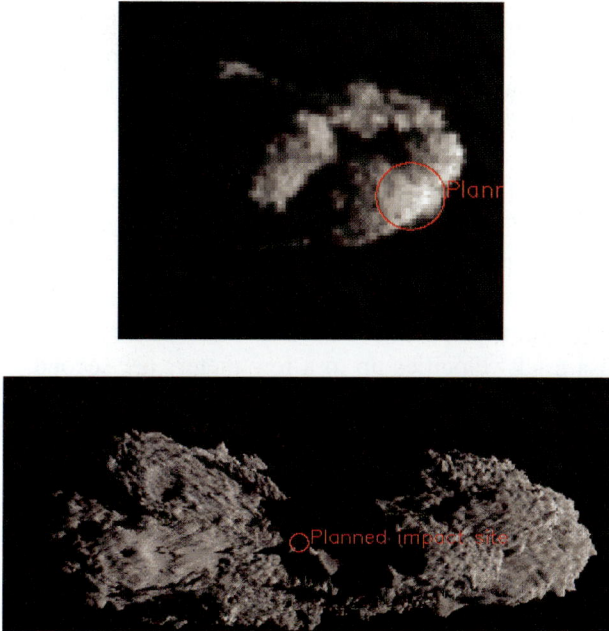

Figure 10. Examples of accretion models at different levels of spatial resolution. The top image is taken at E−11 min and shows the site selected using scene analysis; the bottom image was taken at E−3 min and shows the CB, which is not illuminated.

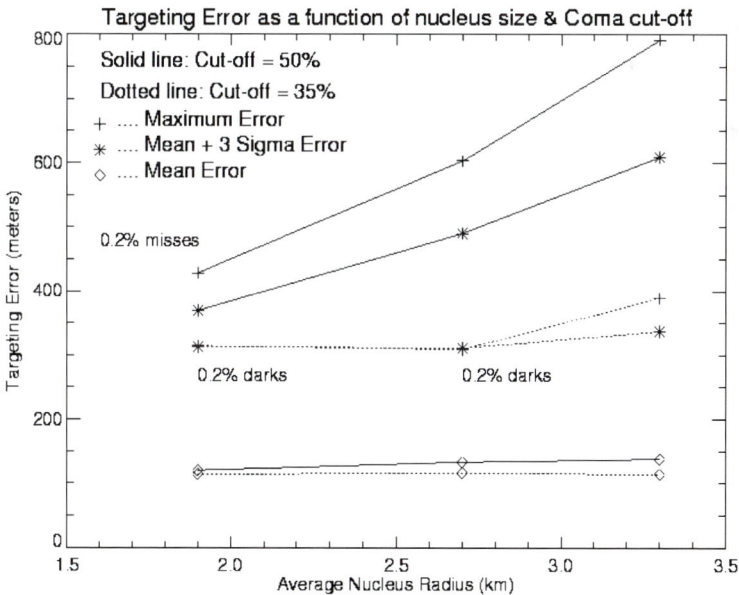

Figure 11. Targeting error as a function of mean radius and coma brightness cutoff for a Halley-like nucleus. While the mean targeting errors of 100–120 m are not sensitive to the size of the body, the maximum and 3σ targeting errors are significantly reduced with a lower coma cutoff.

cutoff. The larger targeting error for a 50% brightness cutoff results in a larger miss distance and a higher probability of a miss, 0.2% in this example, compared with the 35% brightness cutoff, where the probability of a miss is traded with a probability for a dark impact (0.2%). The final trade between completely eliminating all influence of coma without discarding too much of the nucleus, will be made in-flight after the dust environment of Tempel 1 has been analyzed using HRI image data during the approach phase of the mission (Encounter-60 to Encounter-10 days).

The rotation of the nucleus is very important and affects the magnitude of the targeting error by moving the CB about the center of mass. In the presence of significant shadowing, the apparent motion of CB can be quite large, giving rise to apparent cross-track velocities up to a few meters per second, which result in an incorrect orbit solution relative to the true center of mass of the nucleus. This effect becomes very pronounced for very short rotation periods, with an attendant increase in the mean targeting error.

The OD errors described earlier are also proportional to the nucleus size *via* a lever-arm effect: any apparent CB motion in the plane-of-sky is amplified by a larger nucleus, which in-turn proportionally increases the OD error. In this regard, we have the counter-intuitive result that a larger nucleus, although offering a larger cross-sectional area for targeting, causes a targeting error proportional to its size. The end result is that a larger nucleus does not necessarily result in better targeting with the Impactor and degrades pointing for the Flyby s/c. In Figure 12 we show

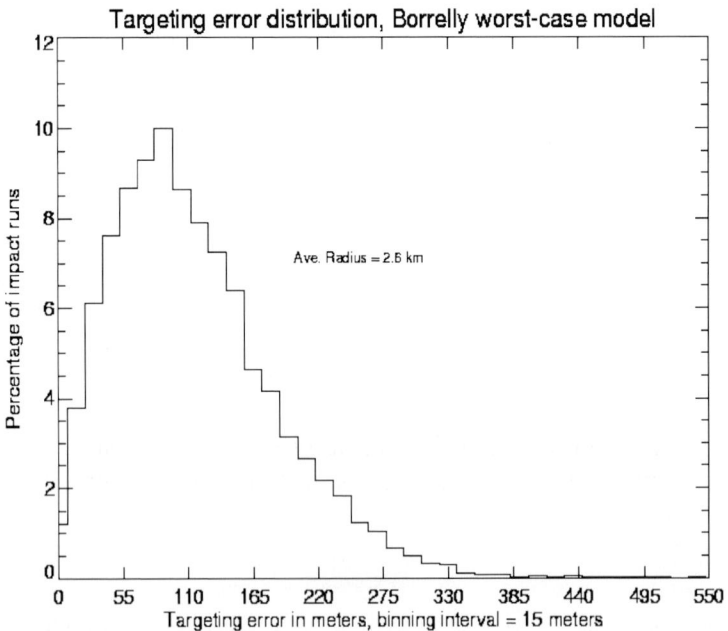

Figure 12. Distribution of targeting errors, which represent a typical example of the skewed distribution seen for all nucleus models examined. Impactor misses occur at the tail-end of the distribution.

a histogram distribution of the absolute value of the targeting errors for a 2.6 km mean nucleus radius. In our parametric studies, we have assumed a range of mean radii between 1.8 and 4 km, which spans the currently predicted mean radius of Tempel 1.

The inertial direction of the rotation pole and the orientation of the nucleus at encounter are treated as unknown in the AutoNav targeting and tracking process. In order to have a realistic picture of how different body orientations affect targeting and tracking, we perform large-scale simulations where the pole spans the celestial sphere at 124 different directions. For each pole direction the nucleus is placed at 72 different orientations, relative to the s/c, each 5° apart, before it begins to rotate for a given simulation run.

Time-of-flight errors are particularly large for cometary encounters compared to other small bodies. Owing to non-gravitational accelerations from outgassing that vary from orbit to orbit, and systematic errors in astrometric observations due to the displaced center of light from the center of mass, long-term predictions of cometary orbits are difficult. As a result, the actual position of the comet along the line-of-sight of the navigation instruments will not be reliably known until the Flyby s/c determines its TOF *via* direct geometric observations during the last 1 h of the encounter. That information will not be available to the Impactor for navigation, since observations from the ITS camera have no strength in establishing TOF due to having zero parallax. TOF errors map into targeting errors, which are to first-order proportional to the magnitude of a given maneuver. The nominal, 1σ, TOF error at E-24 h is expected to be 9 s. However, past experience has shown that TOF errors up to 30 s can be experienced, which for an ITM-3 of $\Delta V = 13.5$ m/s (the 99% figure) will contribute to ~400 m of targeting error.

Finally, direct parallax measurements of the Flyby s/c during nucleus tracking will allow AutoNav to determine its position relative to the nucleus center of mass only to within a few kilometer, because the actual location of the impact site relative to the center of mass of the nucleus is unknown and cannot be established by the observations. This effect introduces pointing errors when viewing the crater at large angles of 30° or more, relative to the approach asymptote. This cannot be compensated for, unless the shape, size and orientation of the body are known prior to the encounter. This type of pointing error increases with the size of the body.

The error sources given earlier, associated with the observations of an unknown nucleus that are used to establish a stable orbit solution, constitute by far the largest contributors to the targeting error for the Impactor and pointing error for the Flyby s/c.

ADCS Attitude Knowledge Errors

The ADCS subsystem, which estimates and controls the spacecraft's attitude, provides AutoNav with the attitude information necessary to determine the inertial pointing of the navigation camera's boresight. Errors in estimation of the spacecraft's attitude map directly into errors in the estimated position and velocity when the s/c state is determined from optical observations from the navigation camera.

The attitude estimation errors are broken down into high-frequency noise, which amounts to 60 μrad (3σ) about any axis, a fixed attitude bias of 150 μrad (3σ) about any axis for the Flyby s/c, a fixed attitude bias of 220 μrad (3σ) about any axis for the Impactor s/c, and bias stability of 50 μrad/h (3σ) on both s/c. These errors scale with range, frequency of observations and span of time over which the observations are used to determine the orbit solution (OD arc length). With the current baseline of a 20 min arc length, one image every 15 s and the last picture prior to ITM-3 at E-11 min, the 3σ ADCS-induced targeting error for the Impactor is ~240 m. The magnitude of this error is quite modest and analysis has shown that AutoNav is tolerant to ADCS attitude noise that is six times larger and a bias drift that is ~3 times larger than given earlier before noticeable performance degradation begins to occur. The estimation of two additional cross-line-of-sight attitude bias drift parameters in the initial state allows AutoNav to tolerate much larger bias drift values up to 1 mrad/h without a performance penalty.

Impactor Targeting Maneuver Execution Errors

Maneuver errors are the largest errors expected to influence targeting on the Impactor s/c. Their behavior is such that the largest component is proportional to the magnitude of the requested ΔV. For conservatism, our assumptions on maneuver execution errors exceed the expected performance figures by 90–35% for maneuvers in the range of 10–25 m/s, respectively. In Figure 13, the targeting error is shown as a function of the maneuver magnitude, ΔV. For a ΔV of 13.5 m/s, (the 99% tile), the 1σ maneuver error contributes ~200 m to the overall targeting error.

IMPACTOR SPACECRAFT TARGETING RESULTS

In Table I we summarize the overall expected Impactor performance based on the assumed ADCS and maneuver error models described in the previous section, and for the range of different nucleus shape models, nucleus sizes, ITS point-spread function (PSF) ranging between 0.8 and 1.7 pixels, OD arc lengths in the 10–40 min range and data weights in the 2–15 pixel range. For comparison, in Table II, we also show one representative set of results for targeting the CB of the nucleus as opposed to targeting the best site based on Scene Analysis.

The range in statistics for the Borrelly–worst-case model is the overall range of results found under the previously described parameter variations. The few misses observed are the result of unfavorable nucleus orientations at encounter that give rise to large OD errors. It is also evident that the simplest approach, CB targeting, is not adequate in meeting the mission's goals, but only for a small range of possible shape and topography models such as those derived from Halley.

TABLE I

Scene analysis targeting results.

Nucleus model	Probability of illuminated impact (%)	Dark impact probability (%)	Probability of a miss (%)	Approximate number of simulations
Halley	99.98	0.02	0	9,000
Borrelly–worst case	99.98–99.88	0.01–0.1	0–0.05	100,000
Accretion	97.28	2.7	0.02	9,000

TABLE II

Impactor CB targeting results.

Nucleus model	Probability of illuminated impact (%)	Dark impact probability (%)	Probability of a miss (%)	Approximate number of simulations
Halley	99.8	0.2	0	1,500
Borrelly–worst case	90.75	6.75	2.75	9,000
Accretion	65	35	0.02	700

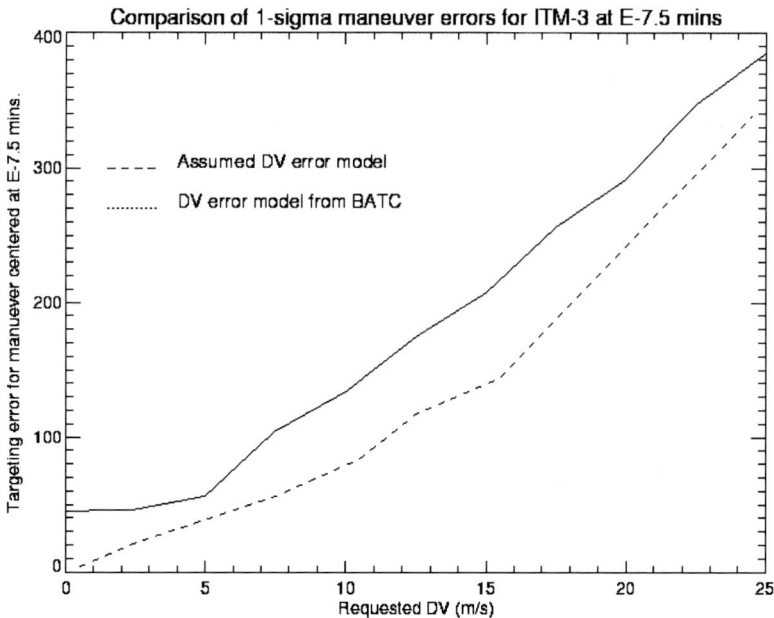

Figure 13. The 1σ Impactor targeting error that results from maneuver execution errors over the range of expected ΔV values.

KEY FLYBY RESULTS

One could break down the Flyby spacecraft's encounter goals into two main categories: 1) establish an accurate trajectory relative to the center of the nucleus; and 2) point the instruments at the impact site and continue tracking the crater during formation and for as long as possible to the end of the post-impact Science observations.

Compared to the Impactor, the Flyby s/c is able to determine its position relative to the nucleus in all three dimensions, including the downtrack direction or TOF, by parallax observations (i.e. measuring the nucleus CB at different view-angles over time). The strength of those observations is proportional to the ratio of the Flyby s/c miss distance to the downtrack distance or range to the point of closest approach. Significant TOF corrections occur during the last 40 min of observations. Determining the TOF is important for the temporal optimization of Science imaging to capture the impact event with the Flyby s/c instrumentation and for optimizing the pre-impact imaging on the Impactor s/c. The TOF is transmitted from the Flyby to the Impactor *via* the S-band cross link at ∼E−5 min and E−3 min. As seen in Figure 14, the initial TOF error converges rapidly towards zero as the s/c closes the range to the comet. Because of the direct observations, the reduction in the TOF error is practically insensitive to the initial conditions as long as the apriori covariance of the comet is sufficiently large to accommodate the magnitude of the necessary adjustments. The TOF requirement allocated to AutoNav is 2.7 s between E-5 min and E-3 min. This requirement is met at the 98% confidence level when the

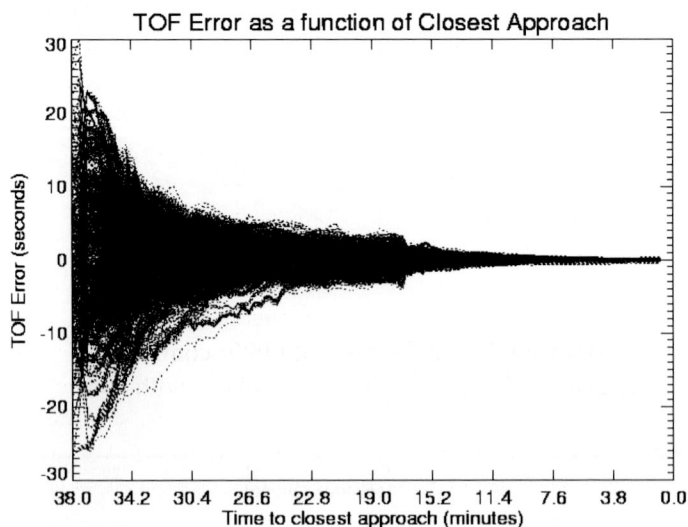

Figure 14. The decrease in the TOF error as a function of time to closest approach of the Flyby s/c from ∼4,500 Monte Carlo simulation runs. The initial conditions simulated a 10 s (1σ) TOF error.

initial TOF error at E-2 h is assumed to be 10 s (1σ) and to the 97% confidence level when the initial TOF error is 30 s (1σ). At the TOI the largest TOF error is 1.6 s. At E+12 min and later the mean TOF error is <0.15 s, or equivalently, the error in the location of the center of mass of the nucleus is <1.5 km in the downtrack direction.

Pointing the instruments at the expected impact site prior to the impact event and to capture the actual crater location following the impact event, requires both knowledge of the nucleus CB and of the impact site relative to the CB. Currently, the impact site chosen *via* Scene Analysis on the Impactor s/c is not communicated to the Flyby. Instead, the Flyby estimates the best impact site independently and by the same process used on the Impactor. This is done on the Flyby s/c using the HRI at a time necessary to match the resolution element of the ITS on the Impactor s/c when the Scene Analysis image is processed, where the figure of merit for the resolution is the full width half maximum (FWHM) of each instrument. Currently, with a 1.5 pixels FWHM for the ITS and 2.25 pixels FWHM for the HRI, the Scene Analysis images on the Flyby will be acquired at ~E-23 min. This independent site selection process is affected by: 1) difference in the scene due to nucleus rotation between the time when the HRI Scene Analysis image is taken and the time when the ITS image is taken; 2) the ~3° nucleus view angle difference due to the Flyby and Impactor being on different trajectories relative to the nucleus; and 3) different noise characteristics between the ITS and HRI detectors.

Consequently, the two s/c do not always select the same impact site. In addition, the actual impact site may differ from the one selected by the Flyby due to targeting errors, such as maneuver execution errors, on the Impactor, which result in an impact site that differs from the one targeted by the Impactor. Figure 15 shows the error distribution in the impact site selection, defined as the difference between the impact site computed on the Flyby and the actual impact site achieved by the Impactor. The figure of merit used to describe the Flyby tracking performance is a site error of 1,750 m, which is equivalent to half the HRI FOV at the time when a 7 m resolution image becomes achievable. The probability of such an error is presently 3.6%, although the mean error of ~400 m is significantly smaller.

The error in the identification of the impact site is the largest contribution to Flyby instrument pointing errors at TOI. The current requirement for a total pointing error at TOI of no more than 100 μrad (3σ), for a 128×128 pixel HRI subframe on the impact site, can be met with a 91% confidence. Using a larger HRI subframe, 512×512 pixels for example, can be met with 99% confidence. Figure 16 shows the pointing error distribution at TOI, in microradians, with a mean value of 61 μ rad.

Following imaging of the impact event and the ejecta plume expansion, the main engineering goal is to track the crater through its expansion and to capture high-resolution images of the crater. The main requirement is to capture a 7 m resolution image of the crater, with the goal of capturing a even higher, 3.4 m resolution image of the fully developed crater.

Parallel Scene Analysis error distribution, Borrelly Worst-Case

Mean Parallel SA error = 400 meters

Max Parallel SA error = 9 km

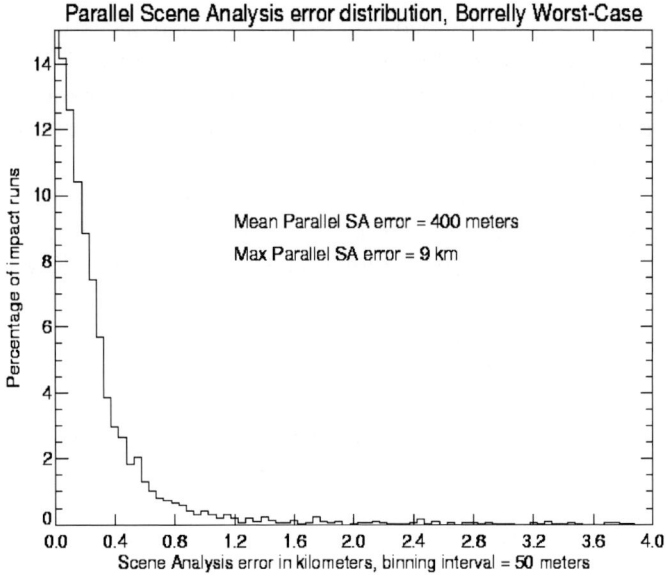

Figure 15. Error distribution in kilometer, between the actual impact site and the expected impact site as computed on the Flyby s/c at E−23 min.

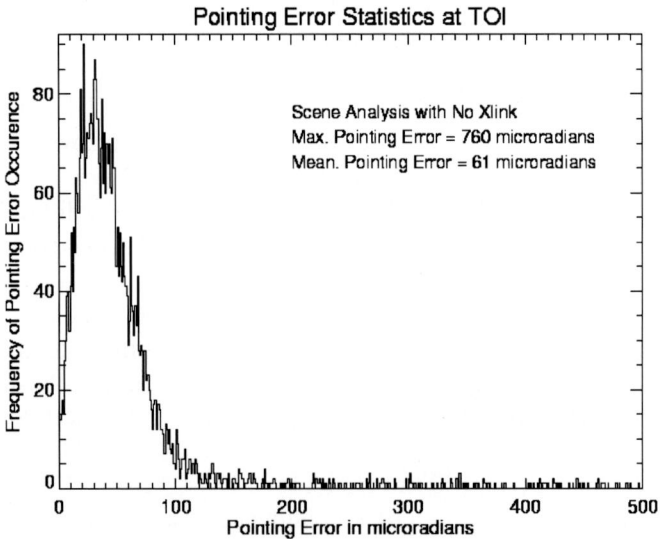

Pointing Error Statistics at TOI

Scene Analysis with No Xlink

Max. Pointing Error = 760 microradians

Mean. Pointing Error = 61 microradians

Figure 16. Pointing error at TOI. The large values at the long "tail" of the distribution are due to misidentification of the actual impact site.

As the Flyby s/c reaches its closest approach point (CA), the pointing errors present at time of impact propagate to later times. In addition, OD errors, as discussed earlier, associated with projection effects of the downtrack direction and by the rapid change of the nucleus brightness profile with the view angle, begin to dominate. As a result, the probability of keeping the crater in the HRI FOV drops sharply after CA-120 sec, i.e. during the last 70 s of imaging. On the other hand, the pixel scale and resolution of the HRI increases linearly with time to closest approach and the competition between these two factors determine the resolution attained at any given time. For the current analysis, the resolution element is the FWHM PSF of the instrument with the addition of the smear induced by the slew of the instrument as it tracks the comet. Smear is very small up to ~CA-120 s and increases rapidly to a mean value of 0.7 HRI pixels for a 50 ms exposure at CA-50 s. The actual crater resolution ultimately depends on crater tracking, the instrument FWHM, and the smear rate and desired exposure, with the exposure duration affecting the amount of smear.

Figure 17 shows a typical example of how the probability for attaining certain crater resolution varies relative to CA. The sharp drop during the last 1 min of imaging is typical of all scenarios examined, although, the actual probability of the crater in the HRI depends on the nucleus model with a more spherical shape giving a higher probability, and the nucleus size with a larger size giving a smaller probability. In this particular example, there is a small difference in the probability curve between the baseline and the worst-case nucleus models. The vertical dashed lines mark the first time that a 3.4 m resolution becomes possible for a certain PSF value of the HRI and for a 20 ms exposure. The two vertical dashed-dotted

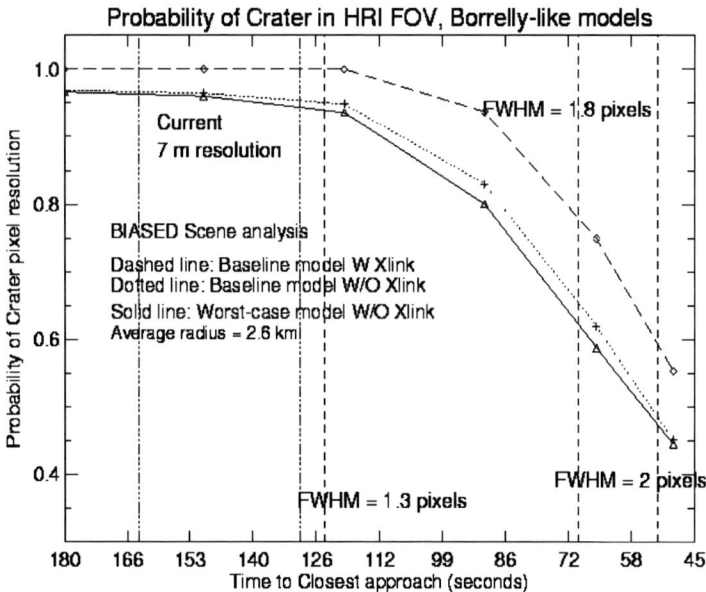

Figure 17. The probability curves for the crater in the HRI FOV as a function of time under a number of different scenarios.

lines mark the first time where a 7 m resolution image becomes possible for PSF values in the range of 2–2.5 pixels FWHM. Currently, the probability of capturing a 7 m resolution image is in the range of 95–97%; the result of propagating errors from performing Scene Analysis independently to the time of ~CA-150 s. The corresponding probability curve, if Impactor were to send its predicted impact site to the Flyby *via* the cross-link (dashed line), shows 100% probability for a 7 m resolution image, because in this case the crater identification is correct to within 120 m (1σ). Currently, the probability for a 3.4 m resolution is quite small: 45% for a 2 pixel FWHM and for larger PSF values a 3.4 m resolution becomes untenable regardless of whether or not AutoNav can track the crater throughout the encounter.

Summary

In this article, we have described an overview of the autonomous navigation (AutoNav) as applied to Deep Impact and we have discussed the expected system performance. Clearly, this is complicated and challenging mission, which relies on the successful interaction of the three instruments (ITS, MRI and HRI), AutoNav and ADCS in a closed-loop fashion. From a navigation standpoint, robustness in the face of unexpected cometary properties, has been one of the focal points of the encounter design, operations of AutoNav and the various trades that are necessary on a cost-constrained mission. Any future work, prior to encounter, on design issues will likely be directed towards alternative scenarios for capturing high-resolution images of the crater. Such scenarios include a closer flyby altitude to increase instrument resolution and a simple one-dimensional mosaic along the slew direction, which is also the direction of the dominant pointing errors. The remainder of the work will be focused on AutoNav testing, both at the test-bench level and in-flight during Earth departure as well as on developing a limited number of encounter contingency scenarios. These are topics, which out of space limitations, are not discussed in this paper.

Acknowledgments

We are grateful to Bob Gaskell of the Optical Navigation Group at JPL for constructing all of the nuclei shape models used for navigation analyses and to George Null of the same group (retired) for establishing many of the essential guidelines and procedures upon which the current AutoNav strategies are based. We would also like to thank the anonymous referee for many helpful comments. The research described in this paper was carried out at the Jet Propulsion Laboratory, California Institute of Technology, under contract with the National Aeronautics and Space Administration for the Deep Impact Project.

References

Bhaskaran, S., Riedel, J. E., and Synnott, S. P.: 1996, Autonomous Optical Navigation for Interplanetary Missions, *Science Spacecraft Control and Tracking in the New Millennium, Proc. SPIE*, pp. 32.

Bhaskaran, S., *et al.*: 1998, Orbit Determination Performance Evaluation of the Deep Space 1 Autonomous Navigation Software, *AAS/AIAA Space Flight Mechanics Meeting*, Monterey, CA.

Kirk, R. L., *et al.*: 2004, *Icarus* **167**, 54.

Riedel, J. E., *et al.*: 2000, Autonomous Optical Navigation Technology Validation Final Report, *Deep Space 1 Technology Validation Symposium*, February 8–9, Pasadena, CA.

Russ, J. C.: 1999, *The Image Processing Handbook*, CRC and IEEE Press.

Stooke, P. and Abergel, A. 2000, Halley Nucleus Shape Model, Personal Communication via M. Belton.

Trochman, W.: 2001, Impactor Spacecraft Attitude Knowledge Performance, BATC System Engineering Report, DI-IMP-ACS-010.

Weidenschilling, S. J.: 1997, *Icarus* **127**, 290.

Zarchan, P.: 1997, *Tactical and Strategic Missile Guidance*, 3rd edn., *Progress in Astronautics and Aeronautics*, Vol. 176, AIAA, Reston, VA.

Zimpfer, D.: 2003, *26th Annual Guidance and Control Conference*, February 6–10, Breckenridge, CO.

References

THE HISTORY AND DYNAMICS OF COMET 9P/TEMPEL 1

DONALD K. YEOMANS*, JON D. GIORGINI and STEVEN R. CHESLEY

301-150, Jet Propulsion Laboratory/Caltech, Pasadena, CA 91109, U.S.A.

*(*Author for correspondence; E-mail: Donald.k.yeomans@jpl.nasa.gov)*

(Received 18 August 2004; Accepted in final form 22 November 2004)

Abstract. Since its discovery in 1867, periodic comet 9P/Tempel 1 has been observed at 10 returns to perihelion, including all its returns since 1967. The observations for the seven apparitions beginning in 1967 have been fit with an orbit that includes only radial and transverse nongravitational accelerations that model the rocket-like thrusting introduced by the outgassing of the cometary nucleus. The successful nongravitational acceleration model did not assume any change in the comet's ability to outgas from one apparition to the next and the outgassing was assumed to reach a maximum at perihelion. The success of this model over the 1967–2003 interval suggests that the comet's spin axis is currently stable. Rough calculations suggest that the collision of the impactor released by the Deep Impact spacecraft will not provide a noticeable perturbation on the comet's orbit nor will any new vent that is opened as a result of the impact provide a noticeable change in the comet's nongravitational acceleration history. The observing geometries prior to, and during, the impact will allow extensive Earth based observations to complement the in situ observations from the impactor and flyby spacecraft.

Keywords: comets, deep impact, space missions, 9P/Tempel 1, cometary dynamics

1. Introduction: Orbital History of Comet 9P/Tempel 1

Comet 9P/Tempel 1 was discovered in the constellation of Libra on April 3, 1867 in Marseille, France by the itinerant German lithographer and part time astronomer, Ernst Wilhelm Leberecht Tempel (see Appendix 1). It was the first discovery of a periodic comet by Tempel and the ninth periodic comet to be recognized as such, as the designation "9P" indicates. Tempel described the comet as having a coma diameter of 4–5' and before the comet's last observation by J. F. J. Schmidt at Athens on August 27, several observers commented upon the distinct or star-like nucleus (Kronk, 2003). On May 4 and again on May 8, 1867, William Huggins observed the comet with a spectroscope and noted that this comet, like comet 55P/Tempel-Tuttle that he had observed one year earlier on January 8, had a continuous spectrum. However for comet Tempel 1, Huggins only suspected the three spectral lines (i.e., C_2 Swan bands) that were observed with the brighter comet 55P/Tempel-Tuttle. Comet 9P/Tempel 1 was the third comet to be observed spectroscopically; the first was yet another Tempel discovery (1864 N1 Tempel), a non-periodic comet spectroscopically observed by G. B. Donati at Florence in August 1864 (Yeomans, 1991).

Space Science Reviews (2005) 117: 123–135
DOI: 10.1007/s11214-005-3392-6

The first elliptic orbit solution for comet 9P/Tempel 1 by K. C. Bruhns demonstrated that the comet was of short period (5.7 years) and a perturbed ephemeris by H. Seeliger allowed the comet to be recovered at Marseille by E. J. M. Stephan on April 4, 1873. It was followed by a number of astronomers to July 1. At its next apparition, Tempel, who was now observing from Arcetri, Italy, recovered the comet on April 25, 1879 (Kronk, 2003). A close Jupiter approach in 1881 (to within 0.56 AU) pushed the comet's perihelion passage distance out to 2.1 AU so that – despite attempts to recover the comet – it was not seen again for nearly a century (see Figure 1). Beginning with orbital elements derived from the nineteenth century observations, Marsden (1963) integrated the perturbed motion of the comet forward to 1972 and noted that due to Jupiter approaches in 1941 and 1953, the perihelion distance evolved back in toward the Earth's orbit with the perihelion distance being about 1.5 AU. Search ephemerides were issued and the comet was again observed during the favorable 1972 return to perihelion. As a result, a single image taken by Elizabeth Roemer on June 8, 1967 was also confirmed as comet Tempel 1. In addition to the nineteenth century returns to perihelion, the comet has been observed at its 1967, 1972, 1978, 1983, 1989, 1994, and 2000 returns.

Figures 1 and 2 show the time evolution of the comet's perihelion and aphelion distances along with the inclination changes with time. Dramatic orbital changes are due to Jupiter close approaches; Table I presents the planetary close approaches over the interval 1600–2400. Plots similar to those in Figures 1–2 were also presented in

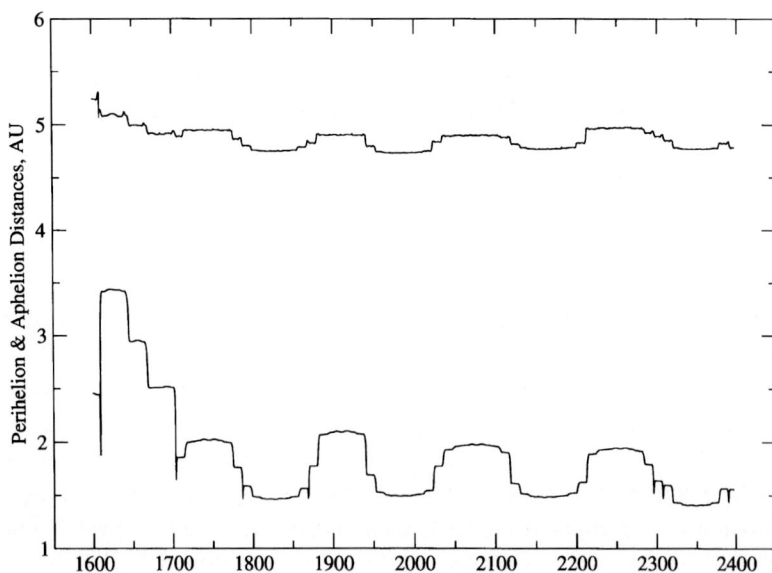

Figure 1. The evolution of the comet's perihelion and aphelion distances are presented over the 1600–2400 interval. The changes result primarily from the Jupiter close approaches noted in Table I. The changes in the perihelion and aphelion distances are due primarily to corresponding changes in the semi-major axis.

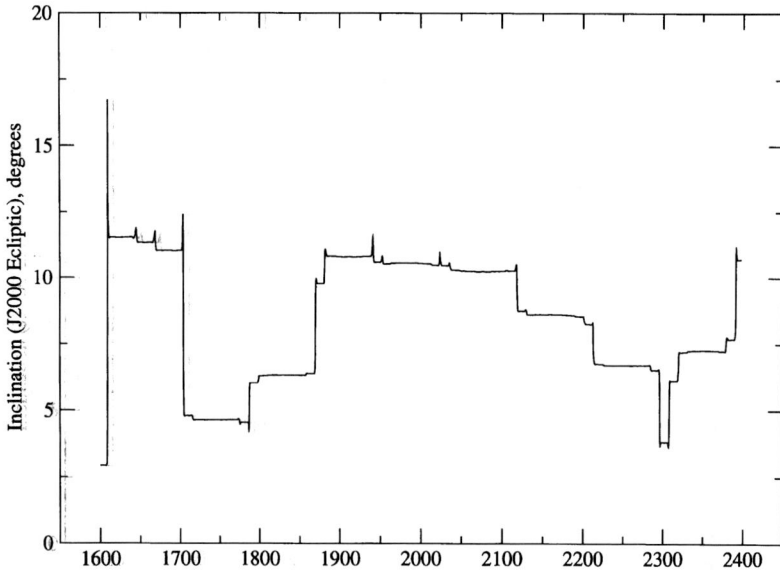

Figure 2. The evolution of the comet's orbital inclination over the 1600–2400 interval.

Carusi *et al.* (1985). It is clear that the evolution of the orbital comet is inextricably linked to its frequent approaches to Jupiter – approaches that move the perihelion distance into 1.5 AU and out to 2 AU with some regularity. Because the orbital period of comet 9P/Tempel 1 is nearly one half that of Jupiter, the comet's current orbital motion is close to a 2:1 resonance with Jupiter and hence its motion is rather stable, without dramatic variations in its orbital evolution. The uncertainties in the comet's nongravitational acceleration model, coupled with rather a close Jupiter approach in 1609, prevented a meaningful extrapolation of the comet's motion prior to the early seventeenth century.

2. Modeling the Nongravitational Acceleration of Comet 9P/Tempel 1

When modeling the motion of active comets, the rocket-like thrusting due to the sublimation of the ices must be taken into account and there have been many attempts to model these so-called nongravitational accelerations (Yeomans *et al.*, in press). The model that is most often employed is based upon the vaporization rate of water ice as a function of heliocentric distance whereby the outgassing is assumed to act symmetrically with respect to perihelion (Marsden *et al.*, 1973). In general, two nongravitational acceleration parameters are included in the orbital solution, where A1 is the acceleration acting in the radial (R), Sun-comet direction at 1 AU from the Sun and A2 is the corresponding transverse (T) acceleration acting in the comet's orbit plane, normal to the radial direction and positive in the

TABLE I

For the interval 1600–2400, planetary close approach distances (CA distance) in AU
are noted, as are the corresponding relative velocities of the encounters in km/s.

Date (CT)	Body	CA distance	V_{rel}
1609 Feb. 20.8	Jupiter	0.106	5.029
1644 Nov. 25.7	Jupiter	0.732	4.906
1668 Jul. 23.8	Jupiter	0.662	4.810
1703 Nov. 02.5	Jupiter	0.199	4.583
1715 Apr. 03.6	Jupiter	0.987	6.679
1775 Jul. 04.4	Jupiter	0.651	5.475
1787 Mar. 10.2	Jupiter	0.348	4.535
1870 Feb. 01.9	Jupiter	0.359	4.471
1881 Oct. 19.8	Jupiter	0.553	4.780
1885 Apr. 23.9	Pallas	0.033	16.888
1941 Oct. 12.6	Jupiter	0.412	5.090
1953 Sep. 12.0	Jupiter	0.759	6.038
2011 Nov. 11.9	Ceres	0.041	10.103
2024 May 26.8	Jupiter	0.551	5.281
2036 Apr. 07.2	Jupiter	0.911	6.169
2119 Nov. 29.0	Jupiter	0.497	4.861
2183 Oct. 17.8	Mars	0.019	6.579
2214 May 10.1	Jupiter	0.469	5.054
2297 Dec. 21.2	Jupiter	0.380	4.262
2309 Dec. 19.0	Jupiter	0.395	4.469
2322 Jan. 08.0	Jupiter	0.880	6.551
2357 Jan. 03.2	Mars	0.034	7.593
2393 Jan. 05.5	Jupiter	0.390	4.490

direction of the comet's orbital motion. A third component (A3), acting normal
to the comet's orbital plane such that $N = R \times T$, is occasionally necessary as
well. A function, $g(r)$, expresses the water ice vaporization rate as a function of
heliocentric distance (r) so that, for example, A1 $g(r)$ gives the radial outgassing
acceleration acting upon the comet's nucleus at a particular heliocentric distance.
In addition, due to seasonal outgassing effects, the outgassing need not reach a
maximum at perihelion. Yeomans and Chodas (1989) introduced an asymmetric
nongravitational acceleration model that allows the outgassing to reach a maximum
a certain number of days (ΔT) before or after perihelion. In the current JPL small
body orbit determination software, any combination of A1, A2, A3, and ΔT can
be solved for in the orbital solutions.

For comet Tempel 1, there are astrometric observations at each of the seven
modern returns to perihelion (1967–2000), and the observations from all of these

apparitions can be successfully included into a single orbital solution using radial and transverse nongravitational accelerations that peak at perihelion. In addition to the nongravitational accelerations, the perturbative actions of the nine planets as well as Ceres, Pallas and Vesta were taken into account at each variable time step in the numerical integration. Successful orbital solutions over the 1967–2003 interval were obtained using $\Delta T = 0$ and constant values for A1 and A2. At first look, this suggests that the comet's ability to outgas is symmetric with respect to perihelion and the ability of the comet to outgas has not changed significantly during the previous seven apparitions. However, Lisse *et al.* (2005) notes that narrow band observations of OH in 1983 and 1994 imply either that the outgassing has a broad peak centered about 2 months prior to perihelion and dropping by a factor of three at perihelion, or the 1994 apparition was systematically a factor of two fainter than the 1983 apparition. In addition, the 1994 gas production rate at perihelion was down by a factor of 1.5–2 compared to the pre-perihelion peak. In light of these results, it seems clear that the nongravitational accelerations affecting the motion of comet 9P/Tempel 1, among the smallest for any active short periodic comet, do not allow meaningful constraints to be placed upon any secular changes in the comet's outgassing. Moreover, there is no appreciable signal in the astrometric data for an out-of-plane (A3) nongravitational acceleration or an asymmetric outgassing with respect to perihelion (ΔT). However, it seems unlikely that the rotation pole has significantly altered its position in space over the 1967–2003 interval since this should have introduced detectable changes in the values of A1 and A2.

Some 706 observations were fit over the interval June 8, 1967 through Dec. 26, 2003 with a resultant weighted RMS residual of $0.85''$. Planetary ephemeris DE405, from JPL, was utilized for the planetary coordinates and masses at each time step (Standish, 1998) (Table II).

TABLE II

The osculating orbital elements for comet 9P/Tempel 1 are presented along with the formal 1-sigma uncertainties in parentheses. JPL orbital solution K058/3 was employed.

EPOCH	2005 July 9.0 E.T.
Eccentricity, e	0.5174906 (0.0000001)
Perihelion distance, q (AU)	1.5061670 (0.0000003) AU
Perihelion passage time, T	2453556.81530 (0.00014) = 2005 July 5.31530 E.T.
Argument of perihelion, ω (°)	178.83893 (0.00007)
Longitude of the ascending node, Ω (°)	68.93732 (0.00006)
Inclination, I (°)	10.53009 (0.00001)
A1 (AU/day^2)	0.0091 (0.0029) $\times$ 10^{-8}
A2 (AU/day^2)	0.00176 (0.00002) $\times$ 10^{-8}
Semi-major axis (AU)	3.1215
Orbital period (years)	5.515

3. Orbital Perturbations as a Result of Deep Impact Itself

The Deep Impact collision with comet 9P/Tempel 1 takes place near the comet's perihelion point and at a relative velocity of 10.2 km/s. Using the formulation outlined by Ahrens and Harris (1994), we can estimate the recoil velocity resulting from an impacting body. We assume the comet's weak structure is in a gravity regime and the collision is partially elastic in the sense that the ejecta causes a momentum transfer above that of an inelastic collision. With a relative velocity of 10.2 km/s and impactor and cometary masses of 360 and 9×10^{13} kg respectively, the impactor will impart a very modest 0.00005 mm/s velocity change in the comet's orbital motion. The comet is traveling at a greater velocity than the impactor, so it will overtake and collide with it. Since the incoming impact direction is $15.2°$ Sunward from the comet's velocity vector and just 1 day prior to perihelion, the impact will introduce a perturbative impulse nearly opposite to the comet's orbital motion. This impulse direction and time is nearly optimal for secularly decreasing the comet's semi-major axis and hence decreasing its orbital energy and period. Figure 3 plots, as a function of time, the position differences between a comet that is perturbed and one that is unperturbed by the impact. By the time of the perihelion passage in 2022, these differences do not even reach 250 m. If the comet were in a strength regime, the resulting differences would be still less. Because the comet is so much larger and more massive than the impactor, the changes imparted in the motion of comet Tempel 1 by Deep Impact are completely negligible, especially

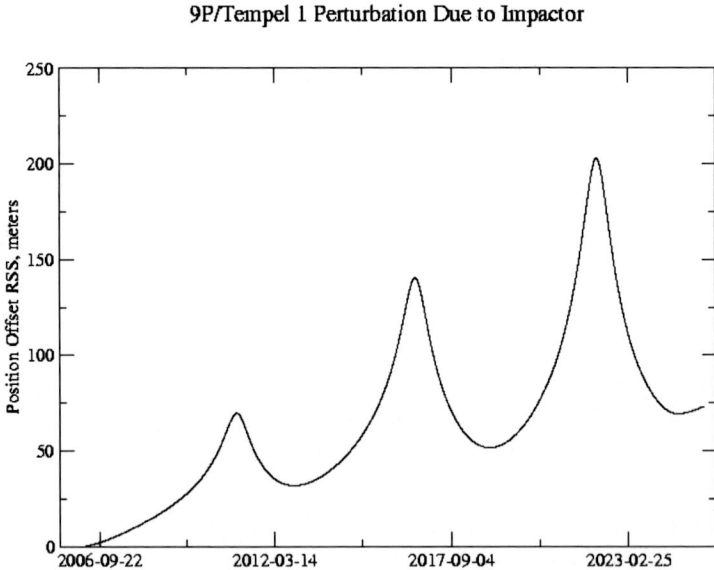

Figure 3. Position differences between a comet perturbed and unperturbed by the spacecraft impact, plotted as a function of time.

when compared to the orbital changes on the comet due to periodic passages near the giant planet Jupiter (e.g., 34 billion meter change due to the passage by Jupiter in 2024).

It has been suggested that one effective technique for deflecting a small comet or asteroid that is on an Earth threatening trajectory would be to run into it with a massive spacecraft at high velocity several years prior to its predicted Earth encounter. The optimal technique for this type of kinetic energy impact would involve a head on crash of a massive spacecraft with the comet near perihelion causing it to lose a bit of its orbital energy and hence change its orbital velocity by a few millimeters per second. Over a period of 10 years' time, a 4 mm/s change in the comet's velocity would modify its orbital position by one Earth radius thus allowing the comet to miss the Earth entirely. Although, the impulse given to the 6 km sized comet 9P/Tempel 1 in 2005 will not materially affect its orbit, this same impact magnitude could substantially affect the trajectory of a much smaller comet. For example, the impulse delivered to comet Tempel 1 in 2005 would be sufficient to move, in 10 years time, a comet of diameter 150 meters by one Earth radius.

4. Possible Orbital Perturbations by a New Vent Being Opened by Deep Impact

While any cometary orbital change due to the impactor's impulse will go unnoticed, the question remains as to whether or not the opening of an active vent on the comet's surface might introduce an observable change in its nongravitational acceleration. The following rough computation suggests that this will be a difficult effect to observe with confidence. For the isotropic outflow of cometary gases at a terminal velocity (V) on the surface of the nucleus with radius R, the gas pressure (P_g) is

$$P_g = \frac{NV}{4\pi R^2}$$

where N is the gas production rate in kg/s. At a heliocentric distance of 1.5 AU, the gas production rate is about 1.5×10^{28} molecules/second (Belton $et\ al.$, 2005). Assuming a mean atomic weight of 20.4 for the (mostly water) gas, a gas terminal velocity of 800 m/s and a nucleus radius of 3.0 km, the gas pressure would then be 3.6×10^{-3} N/m^2. For an impactor crater radius of 50 meters, the crater surface area would be about 7.9×10^3 m^2 with the nongravitational force acting upon the nucleus being 28 N. Dividing this force by the comet's assumed mass then gives the total acceleration acting upon the comet's nucleus as 3.1×10^{-13} m/s^2. This acceleration can then be compared to the computed nongravitational acceleration. This latter acceleration, in the radial direction, is just the value for A1 scaled from 1 to 1.5 AU using the $g(r)$ expression given by Marsden $et\ al.$ (1973). With $g(r) = 0.36$

for $r = 1.5$ AU, the radial nongravitational acceleration computed from the orbit determination process is A1 $g(r) = 6.6 \times 10^{-10}$ m/s². Hence, a new active vent opened as a result of the impactor would only contribute about 1/2000 the value of the pre-existing radial nongravitational acceleration. Even if the computed gas pressure value were a factor of 10 higher because the nucleus had only a 10% active area, the new nongravitational acceleration (due to the impactor) would still be only 1/200 times the value of the pre-existing nongravitational acceleration. Although the pre- and post-encounter solutions for the nongravitational parameters (A1 and A2) will be carefully monitored, a first rough estimate of the likely effects do not suggest that any changes will be noticeable in the orbit determination solutions.

5. Ground Based Observational Circumstances for the Deep Impact Collision

Figure 4 displays the comet Tempel 1 observing conditions for the returns to perihelion in 1983, 1989, 1994, 2000, 2005, and 2011 presented in a rotating coordinate system so that for any particular return to perihelion, the positions of the Earth and Sun are fixed. The comet's positions with respect to the Earth and Sun are plotted as open circles at 30-day intervals before and after the perihelion point, which is denoted as a filled in circle. The 3 o'clock position (vernal equinox) represents the Earth's fixed location for a perihelion passage time of September 21, while the 12, 9 and 6 o'clock positions represents the Earth's locations for comet perihelion passages on December 21, March 21 and June 21 respectively. Using this plot, one

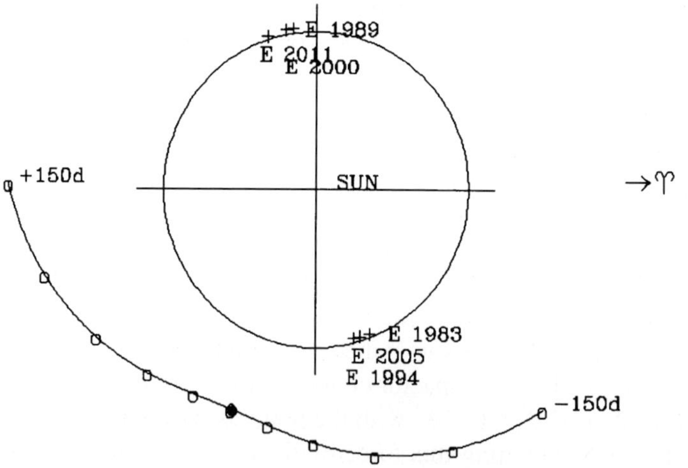

Figure 4. For perihelion passages in 1983, 1989, 1994, 2000, 2005, and 2011, the motion of comet Tempel 1 relative to a fixed Earth is shown in a rotating reference system. The positions of the comet are shown at 30-day intervals from 150 days before perihelion (-150 d) to 150 days after perihelion ($+150$ d).

9P/Tempel 1 Ground Observations

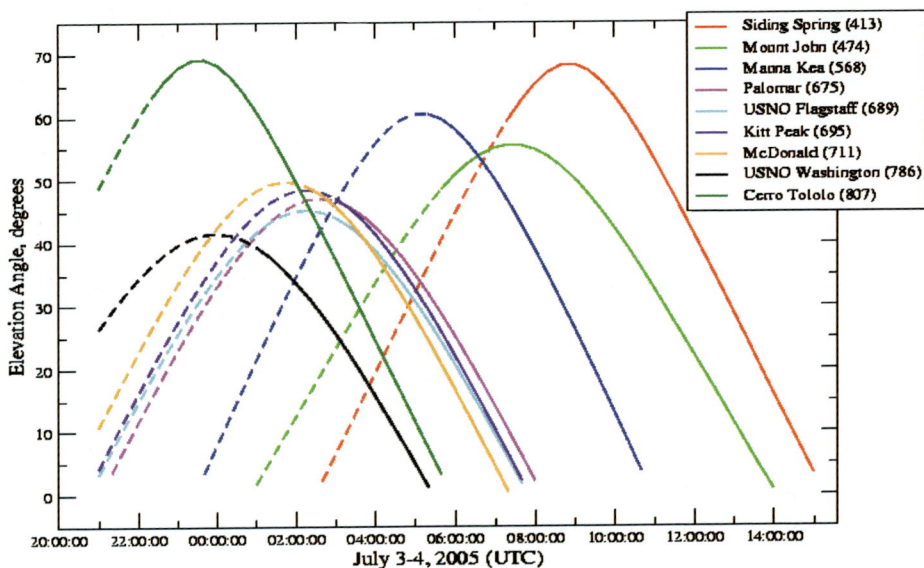

Figure 5. For various observatories, elevation angles for comet Tempel 1 are presented for several hours on either side of the cometary collision, which is scheduled for an Earth receive time of 6:00 UT on July 4, 2005. The actual impact time is 7 min and 26 s earlier. In each case, the dashed curve becomes solid at the end of nautical twilight.

can easily note that while the perihelion passages in 1989, 2000, and 2011 are extremely unfavorable since the comet reaches perihelion near solar conjunction, the 1983, 1994 and 2005 returns to perihelion have nearly identical, and very favorable, viewing conditions since the comet returns to perihelion near opposition. Due to the comet's 5.5-year orbital period, alternate returns to perihelion are favorable for ground-based viewing.

For the interval of time surrounding the impact itself, Figure 5 plots the elevation angle of the comet above the local horizon for nine observatories. In each case, the broken line becomes solid when nautical twilight ends (i.e., the Sun's zenith distance reaches 102°). For ground-based observations of the impact itself, and immediately thereafter, observatories in Hawaii and New Zealand are favored although observatories in the southwestern United States will have low-altitude viewing as well.

Figure 6 shows a sky plot of the comet from early December through the end of July 2005. Both the comet, and nearby Jupiter, will be in retrograde loops during the months of February, March, and April 2005. The comet's pre-impact apparent magnitude will be about 10 with a short tail pointing away from the bright star Spica (0.9 magnitude) less than 4° away in the constellation of Virgo. Depending upon how much dust is ejected as a result of the impactor on July 4, 2005 the comet's

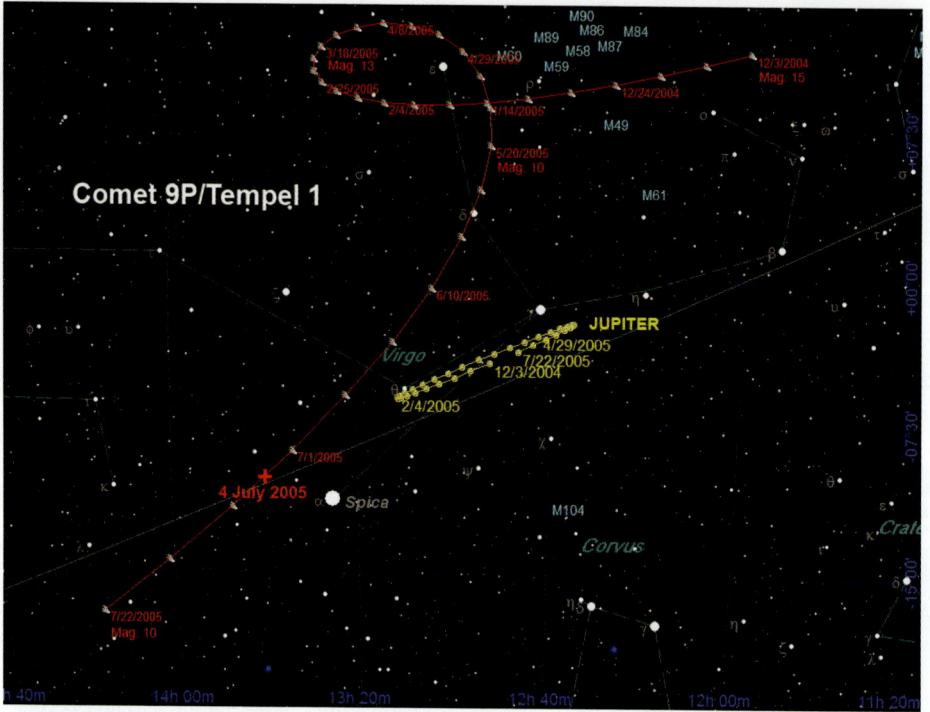

Figure 6. The apparent motion of comet Tempel 1 on the celestial sphere is illustrated from early December 2004 through late July 2005. Both the comet and nearby Jupiter undergo retrograde loops during this interval. Illustration provided by Dale Ireland.

apparent magnitude could increase by a few magnitudes but whether or not the brightening will be sufficient to render the comet a naked eye object remains to be seen.

6. Summary

The orbital period of comet 9P/Tempel 1 is about half that of Jupiter and, as a result, the comet's recent orbital evolution is controlled by its frequent close approaches to that planet; the comet's perihelion distance is periodically lowered to about 1.5 AU, raised to just over 2 AU, and then back again.

The comet was discovered in 1867, observed for two subsequent returns to perihelion in 1873 and 1879 and then, as a result of Jupiter close approaches, the perihelion distance was raised and the comet was not seen again until 1967, a century after its discovery.

As is the case for almost all active periodic comets, a successful orbital solution for comet Tempel 2 required the use of a nongravitational acceleration model to represent the rocket-like thrusting of the cometary outgassing as a function of heliocentric distance. More than 700 astrometric observations over the comet's seven modern apparitions (1967–2003) have been successfully represented with a single orbital solution that included constant values for the nongravitational parameters A1 and A2, where A1 and A2 are the radial and transverse nongravitational accelerations acting upon the comet at 1 AU from the Sun.

The constancy of A1 and A2 over the 1967–2003 interval suggests, but does not prove, that the comet's rotation axis has remained relatively fixed in inertial space. Was this not the case, one would expect the parameters A1 and A2 to have changed their values with time.

Despite the planned collision of the Deep Impact spacecraft with the comet's nucleus on July 4, 2005, there will only be a negligible 200 m, or less, modification to the comet's orbital position after 20 years.

While the spacecraft impact may open up a new active area on the surface of the nucleus, it seems unlikely that this additional activity will be sufficient to cause a measurable effect in the comet's subsequent orbital behavior.

At the time of the impact, and shortly thereafter, the comet will be observable in a dark sky from a number of different locations (e.g., southwestern United States, Hawaii, New Zealand). At the time of the impact, the comet will be 0.89 AU from the Earth with a predicted pre-impact apparent magnitude of about 10. It will be located in the constellation Virgo, <4° from the first magnitude star, Spica.

Subsequent to the impact, and depending upon the amount of dust excavated and thrown into the comet's atmosphere, the comet's apparent magnitude could brighten by several magnitudes.

Acknowledgement

This research was carried out at the Jet Propulsion Laboratory, California Institute of Technology, under contract with the National Aeronautics and Space Administration.

Appendix. The Discoverer of Comet Tempel 1: Ernst Wilhelm Leberecht Tempel (1821–1889)

An artistic free spirit, Ernst Tempel was born on December 4, 1821 in Saxony, one of 12 children. Of poor circumstances, Tempel received only a modest education and became largely self-educated. Upon reaching his twentieth year, Tempel began employment in Copenhagen as a lithographer. After 3 years, he continued his artistic talents in Venice Italy. Having acquired a keen interest in astronomy, Tempel purchased a 4-inch refracting telescope from the Bavarian K.A. von Steinheil and began systematic searches of the heavens from the balcony of a Venetian palace. His first discovery was on April 2, 1859 when he discovered the comet 1859 G1.

In March 1860, Tempel moved to Marseilles, France and remained attached to the Observatory there until the end of 1861 when he resumed work as a lithographer in Marseilles. In January 1871, the German Tempel was expelled from France by the Provisional Government. He traveled to Milan, Italy, where he became an assistant to Giovanni Schiaparelli at the Brera Observatory. Toward the end of 1874, Tempel became the assistant in charge of the Arcetri Observatory in Italy. This observatory had been erected in the years 1869–1872 from the designs of Giovanni Donati. However, after Donati's death in 1873, support for the Observatory declined and Tempel was forced to subsist on a meager salary with no funds to complete or maintain the observatory's two refracting telescopes. These instruments had apertures of 9.4 and 11 in.. Despite his difficulties, Tempel observed and recorded a considerable number of nebulae, often using his artistic skills to produce detailed drawings.

Along with the American Horace Tuttle, Tempel discovered the parent comet of the November Leonid meteors in 1866 (periodic comet 55P/Tempel-Tuttle). Among his cometary discoveries were three more periodic comets, 11P/Tempel-Swift-LINEAR in 1869 and comets 9P/Tempel 1 and 10P/Tempel 2 in 1867 and 1873. Along wyth the discovery of a total of 13 comets, Tempel also discovered five minor planets. For his first two discoveries in March 1861, the names Angelina and Maximiliana were proposed. Angelina was named in remembrance of the astronomical station of Baron F.X. von Zach near Marseilles while Maximiliana was named for Maximilian II, the king of Bavaria. The English astronomers John Herschel and George Airy along with some prominent German astronomers criticized both names because they broke with the tradition of using mythological figures as minor planet names. As a result, the name Maximiliana was changed to Cybele, a nature goddess of the ancient peoples of Asia Minor.

Born 2 years after the death of Tempel, the German surrealist painter, Max Ernst (1891–1976), saw a kindred spirit in Tempel's lust for adventure and the imperturbable joy he took in discovery. Ernst identified with Tempel's difficult life and sympathized with the troubles he encountered in finding suitable work in Germany because he lacked a formal education. One of Max Ernst's last artistic efforts was a collection of 39 lithographs that was dedicated to Tempel and appropriately named "Maximiliana."

References:
http://www.lutz-clausnitzer.de
http://leo.astronomy.cz/tempel/tempel.html

References

Ahrens, T. J. and Harris, A. W.: 1994, in: Gehrels, T. (ed.), Hazards Due to Comets and Asteroids, University of Arizona Press, pp. 897–927.

Belton, M. J. S., Meech, K. J., Groussin, O., McFadden, L., Lisse, Ca., Fernandez, Y. *et al.*: 2005, *Space Sci. Rev.* **117**, 137–160.

Carusi, A., Kresák, L., Perozzi, E., and Vealsecchi, G. B.: 1985, Long-Term Evolution of Short-Period Comets. Adam Hilger Ltd., Bristol England.

Kronk, G. W.: 2003, Cometography: A Catalog of Comets, Vol. 2: 1800–1899. Cambridge University press, Cambridge.

Lisse, C., A'Hearn, M. F., Farnham, T. L., Groussin, O., Meech, K. J., Fink, U. *et al.*: 2005, *Space Sci. Rev.* **117**, 161–192.

Marsden, B. G.: 1963, *Astronom. J.* **68**, 795.

Marsden, B. G., Sekanina, Z., and Yeomans, D. K.: 1973, *Astronom. J.* **78**, 211.

Standish, E. M.: 1998, JPL Planetary and Lunar Ephemerides, DE405/LE405. Jet Propulsion Laboratory Interoffice Memorandum 312.F-98-048 dated August 26, 1998.

Yeomans, D. K.: 1991, *A Chronological History of Comets*, John Wiley, New York.

Yeomans, D. K. and Chodas, P. W.: 1989, *Astronom. J.* **98**, 1083.

Yeomans, D. K., Chodas, P. W., Sitarski, G., Szutowicz, S., and Królikowska, M.: in press, Cometary Orbit Determination and Nongravitational Forces. in: Festou, M.(ed.), *Comets II*, University of Arizona Press.

DEEP IMPACT: WORKING PROPERTIES FOR THE TARGET NUCLEUS – COMET 9P/TEMPEL 1

MICHAEL J. S. BELTON[1,*], KAREN J. MEECH[2], MICHAEL F. A'HEARN[3],
OLIVIER GROUSSIN[3], LUCY MCFADDEN[3], CAREY LISSE[3],
YANGA R. FERNÁNDEZ[2], JANA PITTICHOVÁ[2], HENRY HSIEH[2],
JOCHEN KISSEL[4], KENNETH KLAASEN[5], PHILIPPE LAMY[6], DINA PRIALNIK[7],
JESSICA SUNSHINE[8], PETER THOMAS[9] and IMRE TOTH[10]

[1] *Belton Space Exploration Initiatives, LLC, Tucson, AZ, U.S.A.*
[2] *Institute for Astronomy, University of Hawaii, Honolulu, HI, U.S.A.*
[3] *University of Maryland, College Park, MD, U.S.A.*
[4] *Max-Planck-Institut für Sonnensystemforschung, Katlenburg-Lindau, Germany*
[5] *Jet Propulsion Laboratory, Pasadena, CA, U.S.A.*
[6] *Laboratoire d'Astronomie Spatiale CNRS, Marseille, France*
[7] *Department of Geophysics and Planetary Sciences, Tel Aviv University, Israel*
[8] *Science Applications International Corporation, Chantilly, VA, U.S.A.*
[9] *Center for Radiophysics and Space Research, Cornell University, Ithaca, NY, U.S.A.*
[10] *Konkoly Observatory, Budapest, Hungary*
(*Author for correspondence; E-mail: michaelbelton@beltonspace.com)

(Received 21 August 2004; Accepted in final form 14 December 2004)

Abstract. In 1998, Comet 9P/Tempel 1 was chosen as the target of the *Deep Impact* mission (A'Hearn, M. F., Belton, M. J. S., and Delamere, A., *Space Sci. Rev.*, 2005) even though very little was known about its physical properties. Efforts were immediately begun to improve this situation by the *Deep Impact* Science Team leading to the founding of a worldwide observing campaign (Meech *et al.*, *Space Sci. Rev.*, 2005a). This campaign has already produced a great deal of information on the global properties of the comet's nucleus (summarized in Table I) that is vital to the planning and the assessment of the chances of success at the impact and encounter. Since the mission was begun the successful encounters of the *Deep Space 1* spacecraft at Comet 19P/Borrelly and the *Stardust* spacecraft at Comet 81P/Wild 2 have occurred yielding new information on the state of the nuclei of these two comets. This information, together with earlier results on the nucleus of comet 1P/Halley from the European Space Agency's *Giotto*, the Soviet *Vega* mission, and various ground-based observational and theoretical studies, is used as a basis for conjectures on the morphological, geological, mechanical, and compositional properties of the surface and subsurface that *Deep Impact* may find at 9P/Tempel 1. We adopt the following *working* values (*circa* December 2004) for the nucleus parameters of prime importance to *Deep Impact* as follows: mean effective radius = 3.25 ± 0.2 km, shape – irregular triaxial ellipsoid with $a/b = 3.2 \pm 0.4$ and overall dimensions of $\sim 14.4 \times 4.4 \times 4.4$ km, principal axis rotation with period = 41.85 ± 0.1 hr, pole directions (RA, Dec, J2000) = $46 \pm 10, 73 \pm 10$ deg (Pole 1) or $287 \pm 14, 16.5 \pm 10$ deg (Pole 2) (the two poles are photometrically, but not geometrically, equivalent), Kron-Cousins (V-R) color = 0.56 ± 0.02, V-band geometric albedo = 0.04 ± 0.01, R-band geometric albedo = 0.05 ± 0.01, R-band $H(1, 1, 0) = 14.441 \pm 0.067$, and mass $\sim 7 \times 10^{13}$ kg assuming a bulk density of $500 \, \text{kg m}^{-3}$. As these are *working* values, i.e., based on preliminary analyses, it is expected that adjustments to their values may be made before encounter as improved estimates become available through further analysis of the large database being made available by the *Deep Impact* observing campaign. Given the parameters listed above the impact will occur in an environment where the local gravity is estimated at 0.027–$0.04 \, \text{cm s}^{-2}$ and the escape velocity

Space Science Reviews (2005) 117: 137–160
DOI: 10.1007/s11214-005-3389-1

between 1.4 and $2\,\mathrm{m\,s^{-1}}$. For both of the rotation poles found here, the *Deep Impact* spacecraft on approach to encounter will find the rotation axis close to the plane of the sky (aspect angles 82.2 and 69.7 deg. for pole 1 and 2, respectively). However, until the rotation period estimate is substantially improved, it will remain uncertain whether the impactor will collide with the broadside or the ends of the nucleus.

Keywords: comets, space missions, nucleus, 9P/Tempel 1

1. Introduction

Comet 9P/Tempel 1 was chosen from a short list of possible targets in 1998, not because a great deal was known about the properties of its nucleus, but because of its orbital properties (Yeomans *et al.*, 2005). As described by A'Hearn *et al.* (2005), the comet provided an acceptable launch date, an acceptable solar phase angle on approach, and an impact at a time when the comet was observable from the vicinity of the Earth. All that was known about the nucleus properties at the time it was chosen was that it appeared to be a typical Jupiter family comet with perhaps only a small fraction of its surface active (A'Hearn *et al.*, 1995). Later, as a consequence of *Hubble Space Telescope* (HST) data taken in December 1997 (Lamy *et al.*, 2001), information on the effective radius of the nucleus (~ 3 km) became available. There was also an indication that the rotational period was about a day or longer. As we shall develop in this paper, preliminary results from the *Deep Impact* observing campaign (Meech *et al.*, 2005a) now indicate an R-band geometric albedo, phase law, and light curve modulations for 9P/Tempel 1 that are substantial improvements over what had to be assumed at the beginning of the project.

Although information available in 1998 was sufficient to make 9P/Tempel 1 an acceptable target, it was clear that a far more precise body of information would be needed to both aid in the technical development of the mission (particularly in the areas of navigation of the impactor and the assessment of the hazard due to dust in the coma) and to help interpret the data acquired during the encounter in the immediate post-impact period. *Deep Impact* is a fast (10.2 km/s) dual spacecraft flyby/impactor mission and during the short time interval when the nucleus is spatially resolved its instruments will only view the nucleus at a limited number of viewing angles and rotational phases (Klaasen *et al.*, 2005). *Optimum* mission planning greatly benefits from prior knowledge of the nucleus shape, spin state, and global surface properties. To provide estimates of these, the *Deep Impact* science team organized an observing campaign that included investigations with some of the most powerful telescopes, both on the ground and in space. Since the initiation of the project, this campaign has, as described in Meech *et al.* (2005a), produced high quality ground-based data on the nucleus and its coma at all apparitions of the comet from 1999 through 2004. It has also utilized powerful facilities such as the Keck telescope (Fernández *et al.*, 2003), the European Southern Observatory Very Large Telescope, the *Hubble Space Telescope*, the *Spitzer Space Telescope* infra-red facility (Lisse *et al.*, 2005a), several 4 m telescopes, and the University of Hawaii 2.2 m telescope (Fernández *et al.*, 2003).

2. Global Properties of the Nucleus

Table I contains values of parameters that we have adopted as 'working' estimates of the global properties of the nucleus of 9P/Tempel 1 for the purposes of contemporary mission planning. Because the analysis of the large data set obtained during the *Deep Impact* observing campaign is still incomplete, some of the values in this table can be expected to undergo revision as the analysis continues. However, at this stage most of the parameters are sufficiently well established to provide a useful and quantitative picture of the nucleus.

2.1. MEAN DENSITY

Prominent in Table I is the uncertainty in the estimate of the mean bulk density of the nucleus, which we take to be $500 \pm 400 \, \mathrm{kg \, m^{-3}}$. Even after three space encounters with cometary nuclei this parameter remains beyond our ability to measure directly. Indirect attempts involving the analysis of orbital motion in response to reaction forces applied as the result of cometary activity (Rickman, 1986, 1989; Sagdeev *et al.*, 1988, 1989) or, e.g., in the specific case of comet 1P/Halley with a correct rotational model by Samarasinha and Belton (1995), tidal forces (e.g., the case of Comet D/Shoemaker-Levy 9 by Asphaug and Benz, 1996), and rotational stability (Jewitt and Meech, 1988; Lowry and Weissman, 2003) provide confidence that a typical mean bulk density lies somewhere between 100 and $1000 \, \mathrm{kg \, m^{-3}}$. In addition, an estimate of $300 \, \mathrm{kg \, m^{-3}}$ for the mean bulk density is provided by the interstellar aggregate dust model (Greenberg *et al.*, 1995). In a recent review, apparently following work by Skorov and Rickman (1999), Weissman *et al.* (2004) recommend a bulk density in the range $500{-}1200 \, \mathrm{kg \, m^{-3}}$; our assessment, which is more strongly weighted by the numerical experiments of Samarasinha and Belton (1995), overlaps this range. During encounter *Deep Impact* will attempt to measure the mass and, by implication, the mean density of 9P/Tempel 1 by following the trajectories of any pieces of large ejecta or clumps of smaller ejecta that may be released into ballistic orbits by the impact. The dimensions and shape of the nucleus will also be determined.

2.2. DIMENSIONS OF THE NUCLEUS

Investigations of the effective radius and shape of the nucleus were the focus of work by Lowry *et al.* (1999), Weissman *et al.* (1999), Lamy *et al.* (2001), and later by Fernández *et al.* (2003), the latter employing the Keck and University of Hawaii

TABLE I

Working properties for the nucleus of Comet 9P/Tempel 1

Property	Nominal values	Reference	Notes
Mean density (kg/m^3)	500 ± 400	See text	No direct measurements available
Mean mass (kg)	$\sim 7 \times 10^{13}$	See text	Range: 0.05–1.3 $\times 10^{14}$ kg
Mean radius (km)	3.25 ± 0.2	Lisse *et al.* (2005a)	Spitzer data
Shape	Elongated, irregular	This work	Non-sinusoidal light curve
Axial ratio (*a/b*)	3.2 ± 0.4	This work	From pole solution
Dimensions (km)	$a = 7.2 \pm 0.9$; $b = c = 2.2 \pm 0.3$	This work	Uses $a/b = 3.2 \pm 0.4$ and the tabulated mean radius
Estimated volume (km^3)	151 ± 60	This work	Derived from shape
Estimated Surface area (km^2)	143 ± 18	This work	Derived from shape
Spin state	Principal axis rotation	This work	Preliminary result
Rotation period	41.85 ± 0.1 hr	This work	Preliminary result
Northern Pole direction (RA [deg], Dec [deg]. Epoch J2000)	Pole 1: $(46 \pm 10, 73 \pm 10)$ Pole 2: $(287 \pm 14, 16.5 \pm 10)$	This work	Photometrically, but not geometrically, equivalent poles
Surface gravity (cm s^{-2})	Pole: ~ 0.04 Long end ~ 0.027	This work	Density $= 500\,\mathrm{kg\,m^{-3}}$; Rotation has little effect
Escape velocity (m s^{-1})	Pole: ~ 2.2 Long end: ~ 1.4	This work	Density $= 500\,\mathrm{kg\,m^{-3}}$; Rotation has little effect
Mean R-band Absolute Magnitude	14.441 ± 0.067	Hsieh and Meech (2004)	Preliminary result
Estimated mean phase law: $\Delta H\,(\alpha)$ mag. (α = phase angle in degrees)	$-0.0180955 - 0.2502604\alpha + 0.03062\alpha^2$ $- 0.0021805\alpha^3 + 0.0000798\alpha^4 - 0.0000015\alpha^5$		Preliminary result see text. Good for $\alpha < 15$ deg
Estimated V-band Hapke parameters	$\varpi_0 = 0.018, h = 0.0135, S(0) = 0.2,$ $g = -0.525, \langle\theta\rangle = 20$ deg.		Preliminary result. This work.

(*Continued on next page*)

TABLE I

(*Continued*)

Property	Nominal Values	Reference	Notes
Geometric Albedo (V, R-bands)	0.04 ± 0.01, 0.05 ± 0.01	(V-band) Lisse *et al.* (2005a) and (R-band) this work	Based on tabulated Abs. Mag. And mean radius
Color (V-R) [Kron – Cousins system]	0.56 ± 0.02	Hsieh and Meech (2004)	Preliminary result
Spectral Reflectance	~P-type asteroid	Abell *et al.* (2003)	Inferred from spectrum of other comets
Composition	Silicate dust (~33%); Organics (33%); Ices (33%)	See text	Huebner (2003)
Gas Production Rate (Q_{OH})	1.7–2.15×10^{28} mol s^{-1}	A'Hearn *et al.* (1995)	At perihelion
Dust Production Rate ($Af\rho$)	293 cm	A'Hearn *et al.* (1995) (at perihelion)	Marked asymmetry about perihelion
Active fraction of surface (%)	9 ± 2	A'Hearn *et al.* (1995) Lisse *et al.* (2005a)	From OH production rate and estimated surface area
Thermal inertia	$0 - 100$ J/K/m^2/s$^{1/2}$	Lisse *et al.* (2005a)	Preliminary result
Surface Morphology	Rugged at the scale of 100 m but physically weak. Possibility of impact craters		By analogy to 19P/Borrelly and 81P/Wild 2
Subsurface	Volatile differentiation to several tens of meters		Working hypothesis based on theoretical models
Interior mass distribution	Effectively homogeneous	Belton *et al.* (1991)	By analogy to 1P/Halley
Interior structure	Weak accumulation of cometesimals	Asphaug and Benz (1996)	Analogy to D/Shoemaker-Levy 9
Large scale tensile strength	500 ± 400 dynes cm^{-2}	See text	See text
Crustal (small scale) strength	Up to 10^7 dynes cm^{-2}	Smoluchowski (1989)	Crustal development (Kührt and Keller, 1994)
Insolation history	Orbit regularly varying between perihelia near 1.5 and 2.0 AU	Yeomans *et al.* (2005)	Steady non-gravitational forces

2.2 m telescope on Mauna Kea. These early results are summarized in Lamy *et al.* (2004) who suggest an effective radius of 3.1 km, a minimum axial ratio of 1.40, and rotation period of 41 h. More recently, observations in the thermal infra-red have been acquired in support of *Deep Impact* from the newly launched *Spitzer Space Telescope* (Lisse *et al.*, 2005a), and results, based on thermal measurements, are now available. In addition, the *Deep Impact* ground-based campaign (Meech *et al.*, 2005a) has produced a large database of photometric R-band and visual observations that characterize the comet in all phases of its orbit. This has allowed accurate values for the absolute magnitude, phase law coefficients, color, pole direction, and axial ratio to be determined.

The observations of Weissman *et al.* (1999) and the observations of Lamy *et al.* (2001) using the HST were taken when the comet was near aphelion (4.74 AU). The authors, in their interpretation of these observations were required to make the assumption that it was inactive at such heliocentric distances in order to determine an effective radius for the nucleus. Today this assumption is suspect and should be justified in each application. At least one other Jupiter family comet, 2P/Encke, has been shown to be active at its aphelion (4.09 AU) even in the absence of a detectable coma (Meech *et al.*, 2001; Belton *et al.*, 2005). In this paper, we justify the use of this assumption by appealing to the behavior of 9P/Tempel 1's R-band light curve beyond 4 AU. In Figure 1 we show a light curve based on R-band data

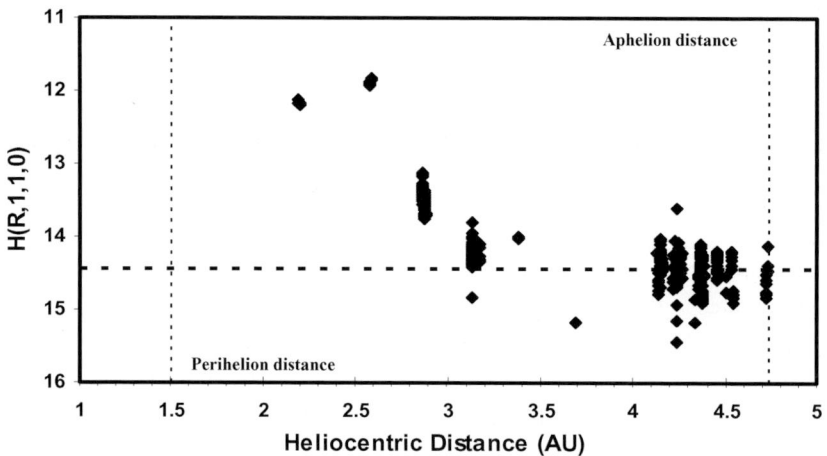

Figure 1. R-band absolute magnitudes reduced to zero solar phase angle covering active and, possibly, inactive phases of comet 9P/Tempel 1. The phase law used is the polynomial shown in Table I (cf. Figure 4). These data, taken between 1997 and 2002 are part of the database generated during a worldwide observing campaign (Meech *et al.*, 2004a) to support the *Deep Impact* mission. Beyond 4 AU the light curve is dominated by periodic variations associated with nucleus rotation and the mean brightness is essentially independent of heliocentric distance (cf. Figures 2 and 3). The horizontal dashed line is at $H(R, 1, 1, 0) = 14.441$, the mean absolute magnitude at zero phase deduced by Hsieh and Meech (2004, and Table I). At distances closer to the sun than 3.5 AU the lightcurve begins to increase in brightness as a result of the volatization of surface materials and the development of a coma.

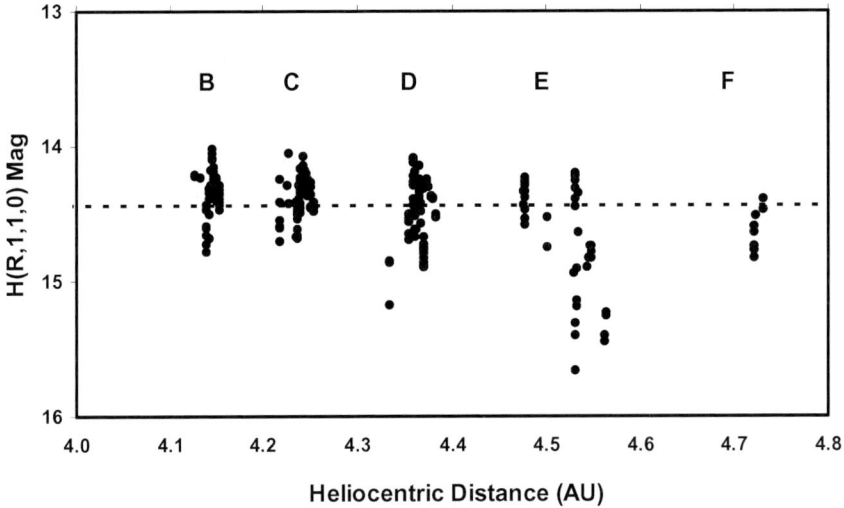

Figure 2. An expanded view of the data in Figure 1 beyond 4 AU. It shows how the data are clustered in groups. So far, three of these, groups B, C, and D, have been used to estimate the rotational period. The horizontal dashed line is at $H(R, 1, 1, 0) = 14.441$, the mean absolute magnitude at zero phase (Table I).

obtained during the Deep Impact observing campaign that covers both its active phase near perihelion and its, possibly, inactive phase near aphelion. The steep brightening at heliocentric distances less than 3.5 AU contrasts with the behavior beyond 4 AU where large rotational variations are seen (cf. Figures 2 and 3 for more detail) and that are superposed on a mean value that is essentially independent of heliocentric distance. While some active contribution cannot be discounted, the observed behavior is consistent with the assumption that the light reflected off the nucleus is the dominant contributor to the light curve. This should be compared with the quite different behavior seen in 2P/Encke at 4 AU where activity contributions range over several magnitudes and the nucleus component can be shown to be usually only a few 10s of percent of the total light (Meech *et al.*, 2001; Belton *et al.*, 2005). With the assumption that 9P/Tempel 1 was inactive at 4.48 AU, Lamy *et al.*, (2001) found a mean radius of 3.06 ± 0.25 km with the additional assumptions that the geometric albedo was 0.04 and a linear phase coefficient of 0.04 mag/deg applied. In these observations, only a fraction of the rotational variation of the light curve was observed, which made it difficult to ensure how well this result applies to an estimate of a *mean* effective radius (i.e., averaged over a complete rotational cycle). An earlier study by Lowry *et al.* (1999) suggested an effective radius of 2.3 ± 0.5 km under similar albedo and phase coefficient assumptions, but in this case coma was definitely present and the result is much less certain.

The *Deep Impact* related investigations of Fernández *et al.* (2003) focused on thermal infrared and R-band observations from co-located telescopes in order to

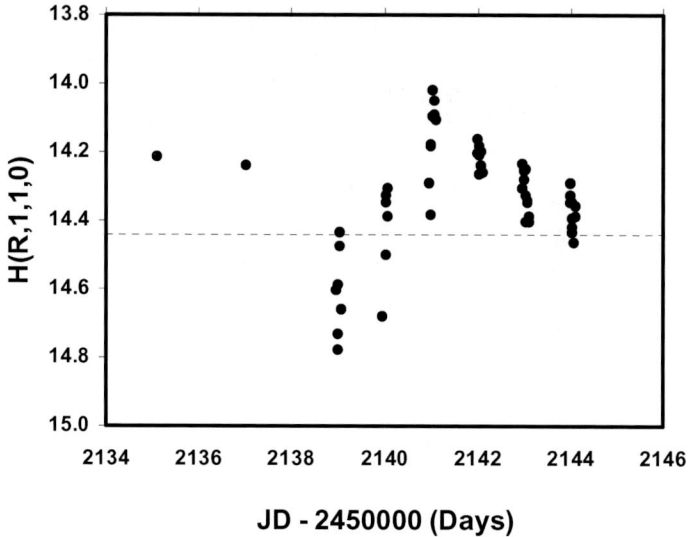

Figure 3. An expanded view of Group B (cf. Figure 2). The figure shows clearly how the underlying periodicity in the data is nearly commensurate with the diurnal sampling period. This makes the period of the comet and the magnitude range of the light curve difficult to pin down. The probable error of each point is ±0.03 magnitudes. The photometric uncertainty in the individual data points is not shown to maintain the clarity of the figure. The horizontal dashed line is at $H(R, 1, 1, 0) = 14.441$, the mean absolute magnitude at zero phase (Table I).

remove the necessity of making assumptions about the geometric albedo and phase coefficient when deducing the effective radius of the nucleus. That work also attempted to bring together several other datasets to obtain a more global picture of the nucleus' properties. While less than perfect weather and the presence of coma presented real challenges, these investigators found that the effective radius *at the time* of the observation was 3.0 ± 0.2 km. To derive the *mean* effective radius (averaged over a rotation period), they had to make two important assumptions: first, that the observations referred to the maximum light in the rotational cycle (for which there was some observational evidence), and second, that the peak-to-valley range of 0.6 ± 0.2 magnitudes that is typical of most comets applied to 9P/Tempel 1's light curve as well. Given these assumptions, they found that the mean effective radius would be 2.6 ± 0.2 km. The geometric albedo was derived by first incorporating the Lamy *et al.* (2001) and Lowry *et al.* (1999) results with photometry from the database of Meech *et al.* (2004a). This let them derive a photometric characterization that was consistent with as many datasets as possible. They deduced an R-band absolute magnitude, H, of about 14.5 mag. and a phase darkening parameter G of roughly -0.2 to obtain an R-band geometric albedo of 0.072 ± 0.016.

The surprisingly high value for the R-band geometric albedo from this study and the uncertainty in rotational phase led, in part, to a further investigation in the thermal infra-red (at several wavelengths between 8 and 22 microns) that utilized the newly launched *Spitzer Space Telescope* (Lisse *et al.*, 2005a). While final results are not yet available, the full rotational light-curve variation was successfully covered and most of the uncertainty associated with rotational phase was eliminated. Preliminary results give a mean effective radius of 3.25 ± 0.2 km and a V-band geometric albedo of 0.04 ± 0.01 (Lisse *et al.*, 2005a). Working with the large database of ground-based R-band observations Hsieh and Meech (2004) have determined preliminary values for the R-band absolute magnitude (14.441 ± 0.067 mag) and together with the above result on mean effective radius we find a mean R-band geometric albedo of 0.05 ± 0.01. The run of absolute magnitude with solar phase angle, α, is shown in Figure 4. From this information we have been able to make a preliminary estimate for the phase law polynomial, $\Delta H(\alpha)$, at small phase angles where $H(1, 1, 0) = H(1, 1, \alpha) + \Delta H(\alpha)$ and values for globally averaged Hapke parameters ϖ_0, h,

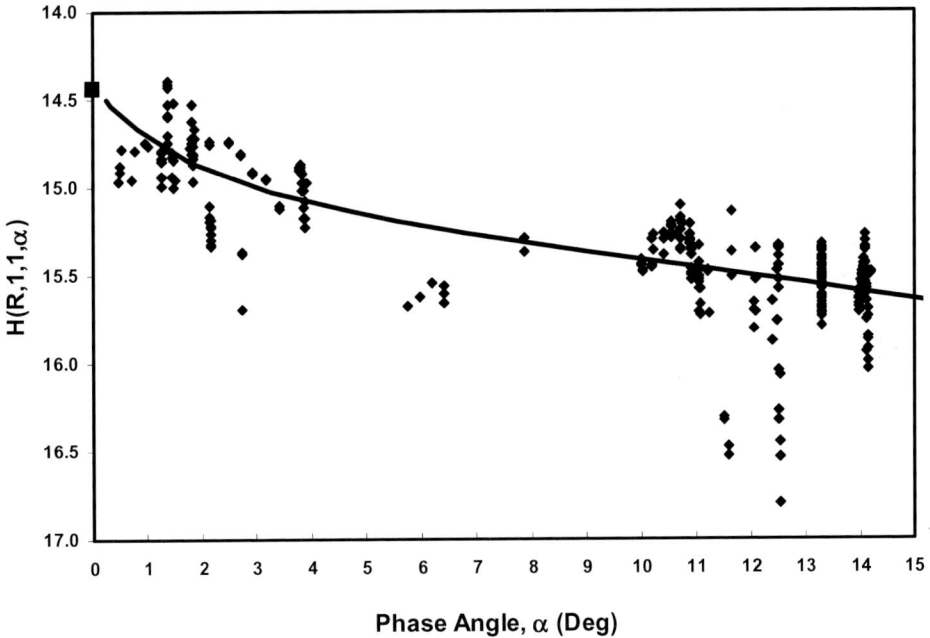

Figure 4. Dependence of absolute magnitude on solar phase angle, α. Most of the apparent scatter of the points is due to rotational variations (typically ~ 0.6 mag peak-to-peak). The faint points near ~ 12 deg phase angle are an enigma. Possibly they are the result of irregularities in the shape of the nucleus casting large shadows near minimum light. The solid curve is for a sphere and was used to estimate the Hapke parameters ($\varpi_0 = 0.018$, $h = 0.0135$, $S(0) = 0.2$, $g = -0.525$, and $\langle \theta \rangle = 20$ deg. (V-band)). The effective radius of the sphere needed for this fit was slightly larger (3.43 km) than for the model nucleus with its axial ratio of 3.2 and mean effective radius of 3.25 km (such dependencies of photometric parameters on shape have been discussed by Helfenstein and Veverka (1989)). The filled square is at $H(R, 1, 1, 0) = 14.441$ the value found by Meech and Hsieh (2004) and the values of ϖ_0, $S(0)$, and g are forced to be consistent with the V-band geometric albedo of 0.04 using a relationship developed by Hapke (1984). The color V-R $= 0.56$ is assumed independent of phase angle in this analysis.

$S(0)$, g, and $\langle\theta\rangle$ (Helfenstein and Veverka, 1989). We find $\varpi_0 = 0.018$, $h = 0.0135$, $S(0) = 0.2$, $g = -0.525$, and $\langle\theta\rangle = 20$ deg. (V-band), values that are similar to those found by Buratti *et al.* (2004) for 19P/Borrelly. The phase law polynomial (valid for $\alpha < 15$ deg) is: $\Delta H(\alpha) = -0.0180955 - 0.2502604\alpha + 0.0306201\alpha^2 - 0.0021805\alpha^3 + 0.0000798\alpha^4 - 0.0000015\alpha^5$ mag.

2.3. THE SHAPE OF THE NUCLEUS

Appraisal of the nucleus' shape is intimately involved with the assessment of the spin state of the nucleus through its periodic brightness variations. If it can be assured that the lightcurve is free of the influence of cometary activity, as justified above, and that the variations are a reflection of the changing areal cross-section of the rotating nucleus, the peak-to-valley magnitude range gives an approximate value for the lower limit to the axial ratio (there is an implicit assumption that deviations from axial symmetry are not grotesquely large). If the direction of the rotational angular momentum vector (assumed fixed in space or at most slowly varying) is also known, then observations of the light curve range at different epochs can be translated into a rough estimate of the dimension of the nucleus. The magnitude range in Lamy *et al.* (2001) observations placed the earliest constraint, a lower limit to the axial ratio, $a/b = 1.4$. An elongated nucleus was thereby established. Ground-based R-band data (Figure 3), taken for the *Deep Impact* campaign during October, 2001, when the comet was at 4.15 AU and phase angle of 14.1 deg, yielded an even greater magnitude range, 0.72 magnitudes, raising the lower limit of the nucleus axial ratio to 1.9. In Figure 5 these data have been phased for the rotational

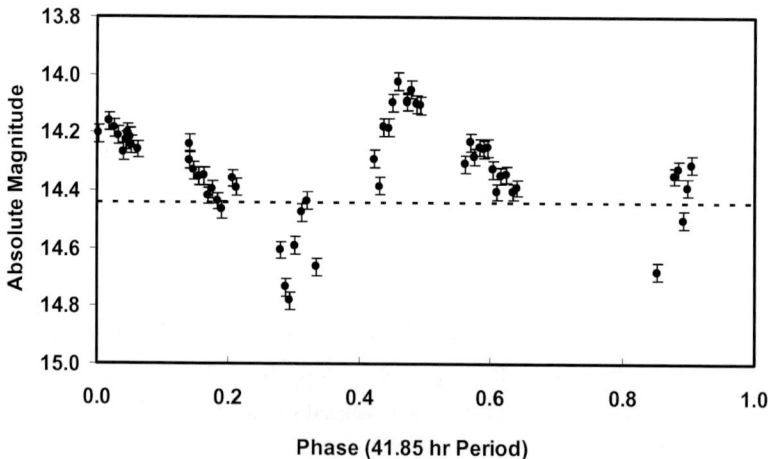

Figure 5. The data for group B has been phased to the rotational period of 41.85 hrs. The "saw-tooth" shape, double-peaked nature of the light curve and the possibility of unequal maxima and minima is clearly seen. Experience with this kind of light curve suggests an elongated nucleus with considerable irregularity. The error bars are the current estimate for the probable error of each point.

period of 41.85 hr (see below) and the slightly asymmetric, double-peaked, "saw-tooth" shape of the light curve is clearly seen. Our experience with modeling such curves suggests that the nucleus of 9P/Tempel 1 is likely to be elongated and has considerable irregularities in its shape.

2.4. THE ROTATION STATE

The set of ground-based photometric data that was primarily used in this work was taken beyond 4 AU and consists of several groupings as shown in Figure 2. One of these groups, B, is shown in detail in Figure 3. Taken all together, these data show strong evidence of periodic variations with their mean (reduced to absolute magnitude) roughly independent of heliocentric distance. The analysis of these data is ongoing using the WindowClean algorithm and phase dispersion minimization based on simple harmonic wave model fits (for an up-to-date review of these techniques see Fernández *et al.*, 2005), but by themselves they seem to be ambiguous about the actual spin period. We show a WindowClean periodogram for Group B in Figure 6. While a rotational period of 41.85 ± 0.1 hr provides a good accounting of the data for groups B, C, and D, we also found that there is also a case for a period of, roughly, half that value ($\sim$22 h) in other parts of the ground-based data. Published investigations provide little guidance with Weissman *et al.* (1999) suggesting a period in the range of 14–39 h and Lamy *et al.* (2001) a period in a range from 25 to 33 h. To settle this issue we have obtained a new time series of nucleus

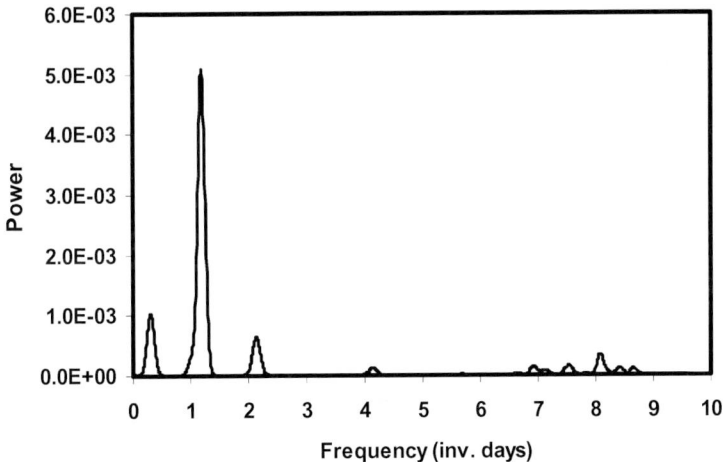

Figure 6. A WindowClean power spectrum (analogous to a periodogram) of the data in group B (cf. Figure 3). The main peak is at 1.20 ± 0.08 inverse days, i.e., a periodicity of 20 ± 1.4 hrs. For a double peak lightcurve that is expected for an elongated object, this corresponds to a rotation period of 40 ± 2.8 hrs. Preliminary analysis of groups B, C, and D together yields a rotation period of 41.85 ± 0.1 hrs (cf. Table I and Figure 5).

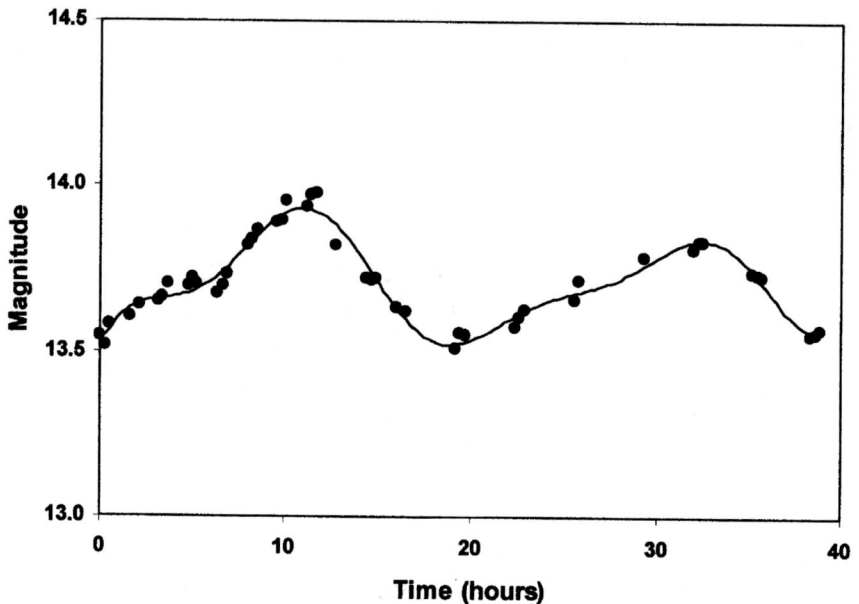

Figure 7. A preliminary reduction of Hubble Space Telescope V-band data showing a double peaked lightcurve with considerable asymmetry. The magnitudes shown here are uncalibrated and represent the total light in the images. A much improved reduction that involves the extraction of the nucleus signal from any coma using knowledge of the point spread function is ongoing. The fit to the light curve is based on the provisionally adopted rotational period of 41.85 hr and is the sum of the fundamental and the first two harmonics.

magnitudes with the *HST* when the comet was at 3.53 AU. Although the data in Figure 1 might suggest that there is still some possibility for a residual level of activity at this heliocentric distance, the HST data show no evidence of coma. The closely sampled data covers a sufficient length of time (~40 h) to ensure a clear discrimination between the candidate rotational periods near 22 and 42 hr. Preliminary reductions (Figure 7) yield a smooth, but asymmetric, double-peaked light curve with a period of near 40 h settling the issue of the rotation period and again suggesting an elongated and irregular shape. This result is reinforced by the newly acquired *Spitzer Space Telescope* data which also shows a double peaked light curve with a period ~40 h. The *Spitzer* and *HST* data, separated by ~43 days or roughly 25 cycles, can be combined to obtain a more accurate estimate of the period but, because of the large gap between the data sets the estimate is susceptible to aliasing. A preliminary study yields a preferred periodicity at 39.6 ± 0.15 hr, however, as seen in the phase plots of Figure 8, the period of 41.85 ± 0.1 hr obtained from ground-based data appears to give a better account of the light curve and we adopt this value as the rotational period of the nucleus. Eventually we expect that by combining all of the ground and space-based data sets, an effort that is ongoing, it should be possible to ultimately estimate the period with an accuracy approaching

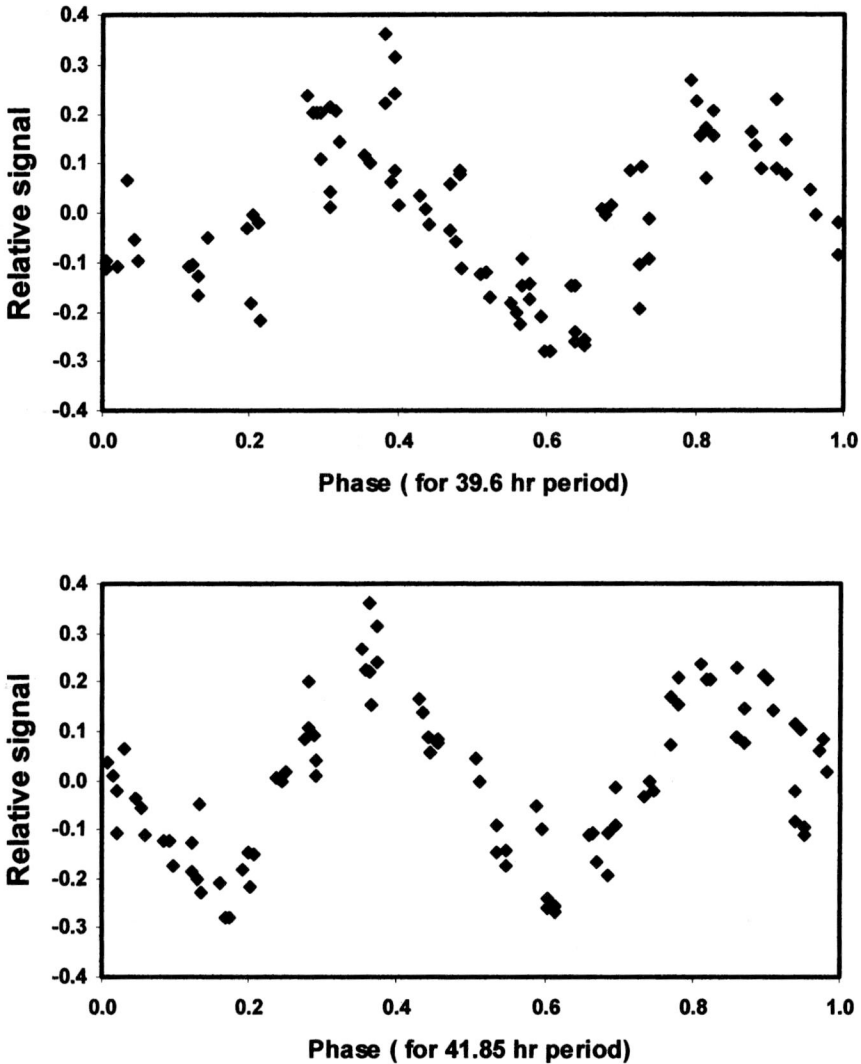

Figure 8. Normalized versions of the signal in the 2004 *Spitzer* and *HST* data combined in phase plots for periods at 39.6 and 41.85 hr to show the sensitivity of the shape of the phased lightcurve with assumed rotation period. The 41.85 hr period provides a smoother accounting of the combined lightcurve and is our currently adopted rotation period. A detailed assessment of these data, which includes an error analysis, is ongoing.

± 10 s, *i.e.*, good enough to be able to predict the general orientation of the nucleus at encounter before launch of the spacecraft on January 8, 2004.

In our studies so far we have found no evidence (e.g., a second periodicity non-commensurate with 41.85 hr) for an excited spin state in this comet and we presume that the nucleus must spin in a state close or equal to that of principal axis rotation.

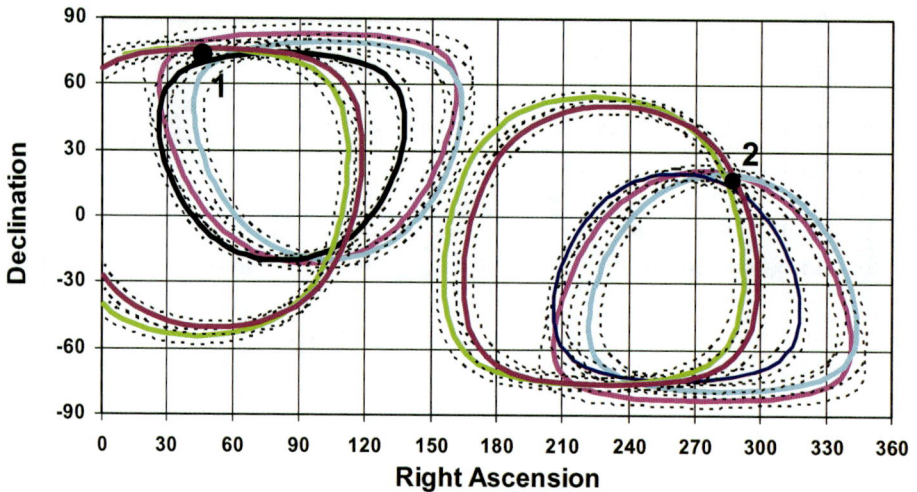

Figure 9. Determination of the rotation pole. The colored small circles represent the locus of the possible pole positions for an assumed value for the axial ratio of the model nucleus. Each color represents a separate group of observations taken under different geometrical conditions. The case for the axial ratio ($a/b = 3.2$), which gives the smallest spread in the locations of the intersections of the sky, is shown. Where the intersections of the small circles cluster together is a candidate position for the direction of the rotation axis on the sky. There are four such places that represent opposite ends of two photometrically equivalent poles. In Table I we quote only the directions that are in the Northern hemisphere and that are marked 1 and 2. The dashed lines that follow each small circle reflect the uncertainty in the observed amplitude of the lightcurve and help to better define the uncertainty in the pole solutions.

2.5. THE DIRECTION OF THE ROTATION AXIS

To achieve a better estimate of the overall dimensions of the nucleus, knowledge of the spin pole direction is required. To achieve this, observations of the magnitude range of the nucleus lightcurve at three or more different viewing geometries are needed. In a preliminary study, we have used six independent light curves including the recently acquired *HST* and *Spitzer Space Telescope* observations. We used the "amplitude" method, which yields two distinct, but essentially photometrically equivalent, solutions for the pole (Magnusson *et al.*, 1989). In Figure 9 we show a series of intersecting small circles, each belonging to a single group of observations that traces out the locus of possible pole positions for an assumed axial ratio. The position on the sky where all of these small circles intersect at a common value is a solution for the direction of the rotation axis. Four such positions are identified which represent opposite ends of two photometrically equivalent pole positions. A good accounting of the various observations plus the best definition of the common point of intersection on the sky is obtained for an axial ratio $a/b = 3.2 \pm 0.4$. This quantity is not, however well defined. The directions of the poles in the northern hemisphere are (RA, Dec, J2000) = 46 ± 10, 73 ± 10 deg (Pole 1) and 287 ± 14,

16.5 ± 10 deg (Pole 2), and, while it is not possible to be sure which is the true rotation axis, based on the scatter of intersection points in Figure 9, pole 1 seems to give a better account of the data. The errors for the pole positions are based on the spread of the various intersection points on the sky and the underlying uncertainty of the location of the small circles as indicated in the figure. They are purposely given conservative values. The sense of spin remains unknown. By combining the axial ratio with the mean effective radius we find the overall dimensions of the nucleus to be approximately 14.4 × 4.4 × 4.4. Using the pole directions found here we can calculate the aspect angle of the rotation axis as seen by the spacecraft on approach to encounter. As seen from the comet, the spacecraft is at RA, Dec = 20.4, −25.6 on approach to encounter. The aspect angles for the two poles are therefore 82.2 deg (Pole 1) and 69.7 deg (Pole 2) – so the pole direction will be close to the plane of the sky. The predicted magnitude range for the light curve on approach is ∼1.2 mag, *i.e.*, large. However, until the rotational period can be substantially improved, it is not yet possible to predict whether the impactor will have a broadside or end-on collision with the nucleus.

2.6. ESTIMATED MASS, LOCAL GRAVITY, AND ESCAPE VELOCITY

Estimates of the value of these parameters may be important in understanding the details of the impact and the behavior of the ejecta. Using the nominal bulk density of $500 \, kg \, m^{-3}$ and approximating the shape with a prolate ellipsoid, we find a volume of $151 \pm 60 \, km^3$ and a mass of $\sim 7 \times 10^{13}$ kg. The acceleration due to gravity at the long end of the object is $\sim 0.027 \, cm \, s^{-2}$ and at the waist $\sim 0.044 \, cm \, s^{-2}$. The corresponding escape velocities are 1.4 and $2.2 \, m \, s^{-1}$, respectively. Because of the long rotation period, the centrifugal and coriolis contributions are negligible. Because of the large uncertainty in the bulk density, the values quoted in this section are similarly uncertain.

2.7. COLOR AND SPECTRAL PROPERTIES

The V-R color of the bare nucleus of 9P/Tempel 1 in the Kron-Cousins photometric system is 0.56 ± 0.02 and was obtained when the comet was beyond 4AU from the sun from the data described in Meech *et al.* (2005a). Observations at smaller heliocentric distances when there was coma present, give colors (V-R) between 0.40 < V-R < 0.51. The color of the active comet reflects a combination of composition and particle size, and is therefore more difficult to interpret. The solar color is V-R = 0.36 (Fernie, 1973; Livingstone, 2000), and the average nucleus color is about V-R = 0.42 ± 0.02 (Meech *et al.*, 2004b) for a sample of 17 comets. Thus, cometary nuclei are in general red with respect to the sun (although there is a large range colors, with some being bluer) and among comet nuclei, 9P/Tempel 1 is quite red. We know there is a trend of reddening as a function of increasing heliocentric distance in the asteroid belt, and that Transneptunian Objects (TNOs) and Centaurs can be extremely red, and have a large diversity of surface colors, possibly due to the presence of organic material and ion irradiation and other weathering processes. Also, cometary activity is likely to alter surface composition by redistributing dust or creating lag deposits on localized areas of the surface, or by removing weathered (possibly reddened) materials. As discussed in Meech *et al.* (2004b), comet colors on average are shifted toward bluer colors with respect to Kuiper Belt Objects (KBOs) and Centaur colors, which have average V-R between 0.54 and 0.63 (Hainaut and Delsanti, 2002), however, the range or distribution of colors is as broad as those of the more distant objects.

As no visible or near infrared spectrum of the comet's nucleus is available, the spectral properties of the nucleus surface are unknown. Our expectations to explore the near-IR spectrum (1–5 μ) of 9P/Tempel 1 are discussed in a companion paper (Sunshine *et al.*, 2005). Abell *et al.* (2003) obtained a high quality spectrum of the surface of comet C/2001 OG108 (LONEOS) and, aside from *Deep Space 1*'s partially saturated spectrum of 19P/Borrelly (Soderblom *et al.*, 2004b), is perhaps the best representation of the near-infrared reflectance spectrum of a comet nucleus available. The spectrum is featureless and mimics that of a P-type asteroid (slightly red, largely featureless, and is distinguished by a low geometric albedo in the optical range) quite closely. We anticipate that the near-infrared spectrum of 9P/Tempel 1 will be similar. A reasonable explanation for the spectrum is thought to be the dominant role of low albedo, highly absorbing, mix of organic material in the reflectivity of the surface material. Therefore most spectral signatures of silicates and ices that might be expected to constitute the primary materials in the crust can expected to be only weakly evident. The *Deep Space* 1 (1.3–2.6 μ) spectrum of 19P/Borrelly shows a spectral reflectance that increases with increasing wavelength after removal of an appreciable thermal component beyond 1.8 micron. There is also an unidentified absorption band near 2.4 micron which may possibly be present in the spectrum of 9P/Tempel 1.

A thermal spectrum has been obtained with the *Spitzer Space Telescope* but its calibration is at present uncertain.

3. Anticipated Properties of the Surface and Subsurface

3.1. ACTIVE REGIONS

An important issue for the *Deep Impact* mission is the identification of possible correlations between localized areas (active areas) and jet activity in the inner coma. As an example, the results of such a study for comet 81P/Wild 2 have recently been reported by Sekanina *et al.* (2004). They find that cometary activity can flourish perhaps even in unilluminated regions – already an important result that needs confirmation. In the case of 9P/Tempel 1 there are many pieces of evidence suggesting that only a small fraction of the total surface area can actively release dust and gas. Early estimates of the active area required to produce the observed OH release yielded an active area of roughly 4 km^2 (A'Hearn *et al.*, 1995), indicating that only 4% of the total surface area as deduced by Lamy *et al.* (2001), was active (according to Groussin [personal communication], this result contains a numerical error of a factor of two and the value should be increased to 8%). Lisse *et al.* (2005a), using a mean effective radius of 3.25 km, derived from a thermal model, an active fraction of $9 \pm 2\%$ at the time of perihelion *i.e.*, an area of roughly 17 km^2. Both of these estimates indicate a relatively low level of activity compared to comet nuclei like 46P/Wirtanen ($\sim$100% at perihelion, Groussin and Lamy, 2003), 22P/Kopff ($>$35%, Lamy *et al.*, 2002), or 103P/Hartley 2 ($\sim$100% at perihelion, Groussin *et al.*, 2004). The shape of the comet's heliocentric light curve may also imply discrete active areas because it peaks about three months prior to perihelion and the activity has dropped by a factor of two by the time of perihelion Light curves by C. H. Morris can be found on the web at http://encke.jpl.nasa.gov/comets_short/9P.html). Such behavior may require a heterogeneous outer layer for the nucleus and could be interpreted as a seasonal effect due to a small number of discrete active areas. In Figure 10 we show direct evidence of discrete active areas: drawings of the inner

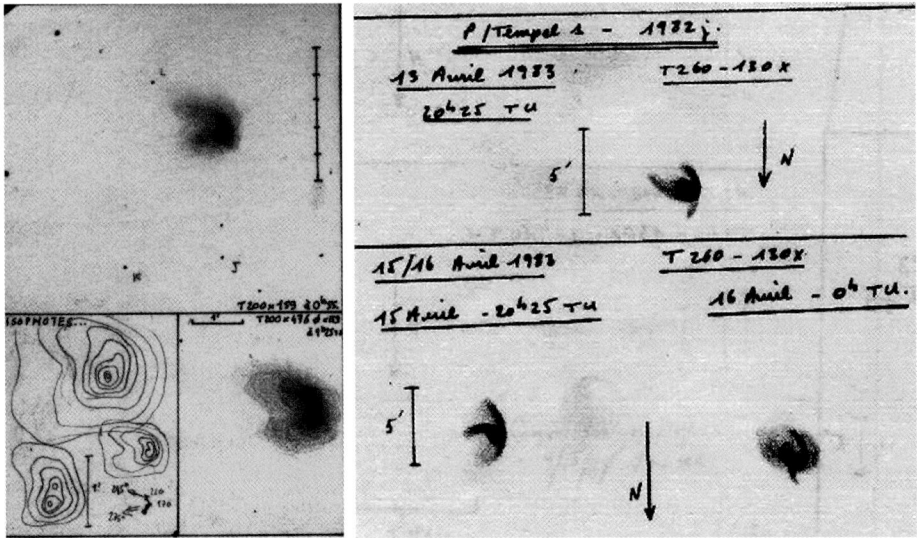

Figure 10. Visual observations of the telescopic appearance comet 9P/Tempel 1 made in April 1994 (left) and April 1983 (right). The geometric circumstances of the comet seen from Earth repeat almost exactly every 11 years. The similarity of the structures in the coma suggests that the rotational pole is reasonably stable and that we will likely see similar structures from Earth in April 2005. The existence of the structures suggests that only a small fraction of the nuclear surface is active. The drawing of the comet in 1983 was first published in the *International Comet Quarterly* and reprinted here by courtesy of J.-C. Merlin. The drawing of the comet in 1994 is by courtesy of N. Biver. North is down and East is to the right. Such drawings should eventually be helpful in locating active areas on the nucleus.

regions of the comet's coma in April, 1983, and April, 1994 show jet structures that have generally a similar morphology at the two different times. Since the orbital period of the comet is almost exactly 5.5 years and assuming the active areas are constant from one apparition to the next, the geometry of Earth-based observations will repeat every 11 years. The jet structures may therefore be expected to repeat, for a given rotational phase. We expect that, at encounter in July 2005, we will observe several discrete active areas on the nuclear surface, much as did the *Stardust* mission during its recent encounter with comet 81P/Wild 2 (Brownlee *et al.*, 2004; Sekanina *et al.*, 2004).

3.2. SURFACE MORPHOLOGY

Images obtained during the recent encounters of *Stardust* with 81P/Wild 2 and *Deep Space 1* with 19P/Borrelly provide the most detailed information that we have on the surface morphology on cometary nuclei. These images with resolutions as high as 14 m/pixel (Brownlee *et al.*, 2004; Soderblom *et al.*, 2004a) show two apparently distinct types of surface topography that are possibly related to the sublimation history of the two objects (Brownlee *et al.*, 2004). Nevertheless, both exhibit rugged topography at the scale of ∼100m. In the case of 81P/Wild 2 we find large, quasi-circular, features, some of which may have an impact origin, steep cliffs, spires, and evidence for overhanging walls; in the case of 19P/Borrelly, the scene is more subdued (possibly partly due to the lower resolution at 19P/Borrelly) but stereoscopic evidence is clearly there for mesa-like structures and large scale fracture systems separating distinct elements of the comet's nucleus. At smaller scales pits and ridges are seen. There is also stereo-photometric evidence for appreciable variations in surface roughness from one place to another (Kirk *et al.*, 2004). In a companion paper (Thomas *et al.*, 2005) examine the question of what geological features might be expected in the case of 9P/Tempel 1 and we refer detailed discussion to that paper. Suffice it to say that the surface geology is expected to be dominated by processes associated with the ablation of volatile materials and the formation of lag deposits. The characteristics of these features are expected to be shaped by inhomogeneities in the local mix of materials, variations in slope, shadowing *etc.* Craters, particularly small craters, are not expected to dominate the landscape since erosion associated with ablation is expected to be rapid process. The increase in surface resolution by a factor of four or five, to ∼3 m/pxl and possibly even higher, plus the acquisition of very high spatial resolution (∼10 m/pxl) color maps should yield new discoveries about the surface morphology. More significantly, the combination of resolution, stereography, and color that is expected to be achieved by *Deep Impact* imaging should advance the exploration of cometary surfaces well beyond simple photo-geological reconnaissance and possibly reach a level where the nature of ongoing geological processes can be understood.

3.3. COMPOSITION

The surface composition and the bulk composition of the comet are expected to be different as a result of volatile depletion in the surface layers (see below). The bulk composition for any comet has yet to be measured, but theoretical ideas abound. As a working hypothesis, we chose to base our estimate on the analysis of Huebner (2003) that yields a roughly equal split (based on mass fraction) between volatile ices, carbonaceous material, and silicates. While accurate measures of various species may be available for observables in the coma (Lisse *et al.*, 2005b), model studies show that the extrapolation of these numbers into the source regions in the interior of the nucleus is fraught with uncertainty (Huebner and Benkhoff, 1999).

So far direct mass spectrometric data sets are available for two comets: The first is for 1P/Halley, obtained in 1986, from the PUMA 1&2 and PIA instruments on the *VEGA* 1&2 and *Giotto* missions (Kissel *et al.*, 1986a,b). More recently, Jan 4, 2004, the CIDA instrument on the *Stardust* mission (cf. Kissel *et al.*, 2003) provided similar data for 81P/Wild 2 (Kissel *et al.*, 2004a). While the data for 1P/Halley were obtained at a flyby speed of over 67 km /s, the relative speed at 81P/Wild 2 was only 6.1 km/s. At the higher speed mainly atomic positive ions were measured, which lessened the chance to be able to characterize the organic material (Kissel and Krueger, 1987) using molecular ions. At the lower speed the molecular ions dominate, while mineral ions are almost undetectable (Kissel *et al.*, 2004). The two sets of measurements were complemented by measurements of the interstellar dust during cruise (Krueger *et al.*, 2004). While the 1P/Halley data suggested the presence of a chemical class of PAHs with O and N as hetero-atoms, the interstellar sample contained a specific class of PQQ (Pyrroloquinoline quinine: 4,5-dihydro-4,5-dioxo-1H-pyrrolo-[2,3-*f*]quinoline-2,7,9-tricarboxylic acid) like molecules (Krueger *et al.*, 2004). In addition, a relative enhancement of N bearing molecules was found for 81P/Wild 2. These could be derived from the interstellar molecules by the elimination of H_2O and CO. Not only are the PQQ-like molecules likely to release H_2O and CO upon initial heating, but they can also store substantial amounts of latent energy from bombardment with cosmic rays. Oberc (1993, 1996) argues, on the basis of measurements with VEGA 2 and the global properties of the organic material reported by Krueger *et al.* (1991) that the latent energy could be as high as 16 kJ/mol. Due to the latent energy potential, possibly buried in the cometary organic material, we should perhaps not be surprised to find the impact to liberate a substantial amount of material by this energy source and not just from the impact energy alone.

3.4. MATERIAL STRENGTH

The strength of the material in the nucleus is very uncertain, no matter whether we consider tensile strength, shear strength, or compressive strength. The tidal disruption of comet D/Shoemaker-Levy 9 (S-L 9) by Jupiter sets a rough upper limit to the tensile strength of about $1000\,\mathrm{dyn\,cm^{-2}}$ for spatial scales near 1 km and allows for strengths approaching zero (e.g., Asphaug and Benz, 1996). The apparent success of a rubble-pile interior model in predicting the distribution of the fragments of S-L 9, suggests that the tensile strength in 9P/Tempel 1 may be similarly low at spatial scales much less than 1 km. Nevertheless, the existence of overhangs of 100–150 m in the *Stardust* images of 81P/Wild 2 (Brownlee *et al.* 2004) requires that the strength on these shorter scales not be zero. The gravity is probably so weak on these objects that a tensile strength of ~100 dyn $\mathrm{cm^{-2}}$ on scales of 10–100 m could be sufficient to support such structures. Because gravity acting on a coherent overhang will produce a torque – stretching the upper surface of the overhang from the wall while compressing the lower surface against the wall – this estimate will represent some combination of compressive and tensile strength. Davidsson (2001) estimated the lower limits of comet tensile strengths to escape rotational breakup for a selection of comets with a range of possible densities at between 10 and $530\,\mathrm{dyne\,cm^{-2}}$. A further and comparative study has been performed for comets and asteroids to estimate the tensile strengths and bulk densities by Toth (2001) and it was found that the vast majority of observed comets are stable against rotational breakup because they are in the stable domain in the rotational period – effective radius diagram. Finally, the interstellar aggregate dust model (Greenberg *et al.*, 1995), predicts a bulk material strength at $\sim0.88\rho$ (SI units), where ρ is the mean bulk density of the nucleus. This leads to an estimate of $440\,\mathrm{dyne\,cm^{-2}}$. For *Deep Impact* we adopt a value of $500\pm400\,\mathrm{dyne\,cm^{-2}}$ for the bulk strength of the material on the nucleus of 9P/Tempel 1.

Near the surface of the nucleus, particularly at small linear scales, the material strength may be very much higher as a result of processes such as radiation sintering (Smoluchowski, 1989) as the material structure has evolved into a crust under the influence of solar insolation (Kührt and Keller, 1994). Their estimates of the local strength (*i.e* at small scales) are as high as 10^7 dynes $\mathrm{cm^{-2}}$. As suggested by Kührt and Keller (1994), the build-up of a strong crust, perhaps a few meters thick, may be essential to understanding the observed existence of widespread regions of surface inactivity. Since the *Deep Impact* impactor spacecraft is like to strike the surface in such an inactive area, these concepts may become essential elements in understanding the nature of the crater forming process that is observed.

3.5. SUBSURFACE AND INTERIOR STRUCTURE

Comet 9P/Tempel 1 appears to be a typical Jupiter family comet. Its orbital history suggests that it has been in a relatively stable configuration with non-gravitational parameters (and by implication spin state and outgassing rate) that are "unusually constant from one apparition to the next" (Yeomans *et al.*, 2000). It follows an orbit that, at least for the last 400 years (the time for which orbital integrations are considered reliable), is close to a 2:1 resonance with Jupiter. The orbit oscillates with extended periods where the perihelion distances is either near 1.5 or 2 AU (Yeomans *et al.*, 2005). This situation should have provided a reasonably stable environment for aging of the surface and for subsurface differentiation of volatiles. At formation, a comet nucleus is expected to include many different volatile species. If cometary water ice is crystalline, these volatiles will be frozen out as separate phases; otherwise, volatiles maybe trapped in the amorphous ice. Clathrate hydrates are now considered a less likely possibility as their formation in impure ice at low temperatures is unlikely (Jenniskins and Blake, 1996). The nature of cometary ice is still uncertain; the *Deep Impact* mission is bound to shed light on this puzzle. As the heat absorbed at the surface penetrates the porous nucleus, ices will evaporate from the pore walls and the gas will flow, in part, to the surface and out of the nucleus, and in part, to the colder interior where it will refreeze. Since evaporation rates are strongly temperature-dependent and vary widely among gas species, several distinct evaporation fronts are expected to form and also, several separate layers of refrozen gases, although the ice mixture may have been homogeneous initially. If the water ice is amorphous, the trapped gas will escape when the ice crystallizes. In this case, all species will escape together and, generally, at higher temperatures than those typical of evaporation. However, once they are released from the ice, these gases will behave similarly to gases that evaporate from the pore walls. Again, a chemically differentiated layered structure will emerge (Prialnik *et al.*, 2004).

Thus models show that a layered structure will occur regardless of the nature of the water ice, starting at a depth of 10 cm to a few tens of meters (Prialnik and Mekler, 1991; Strazzula and Johnson, 1991; Benkhoff and Huebner, 1995; Klinger, 1996). A more accurate picture is difficult to predict, since it depends on physical properties, as well as thermal history. In Jupiter family comets the water is expected to be crystalline above this differentiated layer, even if it was amorphous at formation. The outer layer of crystalline ice – essentially, the orbital skin depth of the nucleus – will be depleted of more volatile ices. Models also show that at an exposed (active) area, the porosity is bound to be high near the surface; in contrast, a dust-covered area may have a crust of dense (low-porosity) ice just beneath it. The dust layer covering inactive regions may be as thin as a few cm (Rickman *et al.* 1990, Prialnik and Mekler, 1991; Coradini *et al.*, 1997), which is sufficient for insulating the ice-rich material beneath it and reducing sublimation by a large factor.

The interior structure on a global scale is yet another mystery. Models range from a fluffy mix of largely interstellar material to accumulations of cometesimals rather like the asteroidal "rubble pile' model (for an up-to-date review of our understanding of cometary interiors see the reviews of Donn, 1991; Weissman *et al.*, 2004). The only quantitative information that might bear on its interior are the number of pieces that were observed as a result of the tidal break-up of comet D/Shoemaker-Levy-9 near Jupiter (Asphaug and Benz, 1996) and the expectation for near homogeneity of its internal mass distribution based on results for 1P/Halley. In this case the internal mass distribution of the nucleus is constrained by a combination of observations of its observed shape and the two ratios of principal moment of inertia that follow from the observed excited spin state (Belton *et al.*, 1991). At the surface, as noted above, the steep walls, spires, and overhangs seen on 81P/Wild 2 (Brownlee *et al.*, 2004) provide qualitative evidence for an increase in strength (and perhaps sealing in of volatiles) in the upper layers of the nucleus that may have occurred as a result of a protracted period of compositional and structural evolution under the influence of sunlight (Kührt and Keller, 1994).

4. Conclusions

Many diverse lines of evidence, including the results of missions to 1P/Halley, 19P/Borrelly, and 81P/Wild 2, allow us to formulate a reasonably consistent picture of conditions on 9P/Tempel 1. Also, major recent advances in our observational characterization of the nucleus of 9P/Tempel 1 have occurred as a result of a massive and highly successful Earth-based observational campaign organized by the *Deep Impact* science team (Lisse *et al.*, 2005a; Meech *et al.*, 2005a). Preliminary estimates, partially based on this database, allow us to specify a set of working parameters (Table I), which will be used by the team in planning for the encounter and impact. They will also provide a guide in the initial data analysis phase. The value of many of these quantities can be expected to be modified, or updated, as a result of this ongoing campaign, the ongoing data analysis, and the results of further observations taken from the *Deep Impact* spacecraft and from the vicinity of Earth around the time of impact. For now we must be satisfied with the set of working parameters in this paper.

References

Abell, P. A., Fernandez, Y. R., Pravec, P., French, L. M., Farnham, T. L., Gaffey, M. J., *et al.*: 2003, 34th *Lunar & Planet. Sci. Conf.* (Abstract).

A'Hearn, M. F., Millis, R. L., Schleicher, D. G., Osip, D. J., and Birch, P. V.: 1995, *Icarus* **118**, 223.

A'Hearn, M. F., Belton, M. J. S., and Delamere, A.: 2005, *Space Sci. Rev.*, this issue.

Asphaug, E. and Benz, W.: 1996, *Icarus* **121**, 225.

Belton, M. J. S., Julian, W. H., Anderson, A. J., and Mueller, B. E. A.: 1991, *Icarus* **93**, 183.

Belton, M. J. S., Samarasinha, N. H., Fernandez, Y. R., and Meech, K. J.: 2005, Accepted for publication in *Icarus*, Nov. 2004.

Benkhoff, J., and Huebner, W.: 1995, *Icarus* **114**, 348.

Brownlee, D. E., Horz, F., Newburn, R. L., Zolensky, M., Duxbury, T. C., Sandford, S., *et al.*: 2004, *Science* **304**, 1764.

Buratti, B. J., Hicks, M. D., Soderblom, L. A., Britt, D., Oberst, J., and Hillier, J. K.: 2004, *Icarus* **167**, 16.

Coradini, A., Cappacioni, F., Capria, M. T., De Sanctis, M. C., Espinasse, S., Orosei, R., *et al.*: 1997, *Icarus* **129**, 337.

Davidsson, B. J.: 2001, *Icarus* **149**, 375.

Donn, B.: 1991, in: Newburn, R. L., Neugebauer, M., and Rahe, J., (Eds.), *Comets in the Post-Halley Era*, Kluwer Academic, Dordrecht, The Netherlands.

Fernández, Y. R., Meech, K. J., Lisse, C. M., A'Hearn, M. F., Pittichová, J., and Belton, M. J. S.: 2003, *Icarus* **164**, 481.

Fernandez, Y. R., Lowry, S. C., Weissman, P. R., Mueller, B. E. A., Belton, M. J. S., and Meech, K. J.: 2005, Accepted for publication in *Icarus*, Nov. 2004.

Fernie, J. D.: 1973. *Pub. Astron. Soc. Pac.* **95**, 782.

Greenberg, J. M., Mizutani, H., and Yamamoto, T.: 1995, *Astron. Astrophys.* **295**, L35.

Groussin, O. and Lamy, P.: 2003, *Astron. Astrophys.* **412**, 879.

Groussin, O., Lamy, P., Jorda, L., and Toth, I.: 2004, *Astron. Astrophys.* **419**, 375.

Hainaut, O. R., and Delsanti, A. C.: 2002, *Astron. Astrophys.* 1.

Hapke, B.: 1984, *Icarus* **59**, 41.

Helfenstein, P., and Veverka, J.: 1989, in: Binzel, R. P., Gehrels, T., and Mathews, M. S. (eds.), *Asteroids II*, University of Arizona Press, Tucson.

Huebner, W. F., and Benkhoff, J.: 1999, *Space Sci. Rev.* **90**, 117.

Huebner, W. F.: 2003, in: Boehnardt, H., *et al.* (eds.), Kluwer Academic, Dordrecht, The Netherlands.

Jenniskins, P. and Blake, D. F.: 1996, *Astrophys. J* **473**, 1104.

Jewitt, D. C. and Meech, K. J.: 1988, *Astrophys. J.* **328**, 974.

Kirk, R. L., Howington-Krauss, E., Soderblom, L. A., Giese, B., and Oberst, J.: 2004, *Icarus* **167**, 54.

Kissel, J. and Krueger, F. R.: 1987, *Nature* **326**, 755.

Kissel, J., Brownlee, D. E., Büchler, K., Clark, B. C., Fechtig, H., Grün, E., *et al.*: 1986a, *Nature* **321**, 336.

Kissel, J., Sagdeev, R. Z., Bertaux, J. L., Angarov, V. N., Audouze, J., Blamont, J. E., *et al.*: 1986b, *Nature* **321**, 280.

Kissel, J., Glasmachers, A., Grün, E., Henkel, H., Höfner, H., Haerendel, G., *et al.*: 2003, *J.Geophys. Res.* **108**, 401.

Kissel, J., Krueger, F. R., Silén, J., and Clark, B. C.: 2004, *Science* **304**, 1774.

Klaasen, K. P., Carcich, B., Carcich, G., and Grayzeck, E.: 2005, *Space Sci. Rev.*, this issue.

Klinger, J.: 1996, in: H. Rickman *et al.* (eds.), *Proceedings of the First Comet Nucleus Surface Properties Workshop*. Unpublished.

Krueger, F. R., Korth, A., and Kissel, J.: 1991, *Space Sci. Rev.* **56**, 167.

Krueger, F. R., Werther, W., Kissel, J., and Schmid, E. J.: 2004, *Rapid Commun. Mass Spectrom.* **18**, 103.

Kührt, E. and Keller, H. U.: 1994, *Icarus* **109**, 121.

Lamy, P. L., Toth, I., A'Hearn, M. F., Weaver, H. A., and Weissman, P.: 2001, *Icarus* **154**, 337.

Lamy, P., Toth, I., Jorda, L., Groussin, O., A'Hearn, M. F., and Weaver, H. A.: 2002, *Icarus* **156**, 442.

Lamy, P., Toth, I., Fernández, Y. R., and Weaver, H. A.: 2004, in: M. C. Festou *et al.* (eds.), *Comets II*, University of Arizona Press, Tucson.

Lisse, C., A'Hearn, M. F., Groussin, O., Fernández, Y. R., Belton, M. J. S., Van Cleve, J. E., *et al.*: 2005a, Submitted to *Astrophys. J. Lett.*

Lisse, C., A'Hearn, M. F., Farnham, T., Groussin, O., Meech, K. J., and Fink, U., *et al.*: 2005b, *Space Sci. Rev.*, this issue.

Livingstone, W. C.: 2000. in: Cox, A.N. (ed.), *Allen's Astrophysical Quanitities*, Springer, New York.

Lowry, S. C., Fitzsimmons, A., Cartwright, I. M., and Williams, I. P.:1999, *Astron. Astrophys* **349**, 649.

Lowry, S. C. and Weissman, P. R.: 2003, *Icarus* **164**, 492.

Magnusson, P., Barucci, M. A., Drummond, J. D., Lumme, K., Ostro, S. J., Surdej, J., *et al.*: 1989, in: Binzel, R. P., Gehrels, T., and Mathews, M. S. (eds), *Asteriods II*, University of Arizona Press, Tucson.

Meech, K. J., *et al.*: 2005a, *Space Sci. Rev.*, this issue.

Meech, K. J., Hainaut, O. R., and Marsden, B. G.: 2004b, *Icarus* **170**, 463.

Meech, K. J., Fernandez, Y., and Pittichova, J.: 2001. *Bull. Amer. Astron. Soc.* **33**, 1075.

Prialnik, D. and Mekler, Y.: 1991, *Astrophys. J.* **366**, 318.

Prialnik, D., Benkhoff, J., and Podolak, M.: 2004, in Festou, M. (ed.), *Comets II*, University of Arizona Press, Tucson.

Rickman, H., Fernandez, J. A., and Gustafson, B. A. S.: 1990, *Astron. Astrophys.* **237**, 524.

Rickman, H.: 1986, *ESA SP-249*, 195.

Rickman, H.: 1989, *Adv. Space Res.* **9**, 59.

Sagdeev, R. Z., Elyasberg, P. E., and Moroz, V. I.: 1988, *Nature* **331**, 240.

Samarasinha, N. H. and Belton, M. J. S.: 1995, *Icarus* **116**, 340.

Sekanina, Z., Brownlee, D. E., Economou, T., Tuzzolino, A. J., and Green, S. F.: 2004, *Science* **304**, 1769.

Skorov, Y. V., and Rickman, H.: 1999, *Planet. Space Sci.* **47**, 935.

Smoluchowski, R.: 1989, *Astron. J.* **97**, 241.

Soderblom, L. A., Boice, D. C., Britt, D. T., Brown, R. H., Buratti, B. J., Kirk, R. L., *et al.*: 2004a, *Icarus* **167**, 4.

Soderblom, L. A., Britt, D. T., Brown, R. H., Buratti, B. J., Kirk, R. L., Owen, T. C., *et al.*: 2004b, *Icarus* **167**, 100.

Strazzula, G., and Johnson, R. E.: 1991, in: Newburn, R.L., Neugebauer, M., and Rahe, J. (eds.), *Comets in the Post-Halley Era*, Kluwer Academic, Dordrecht, The Netherlands.

Sunshine, J., A'Hearn, M. F., Groussin, O., McFadden, L.-A., Klaasen, K. P., Schultz, P. H., *et al.*: 2005, *Space Sci. Rev.*, this issue.

Thomas, P. C., Veverka, J., A'Hearn, M. F., McFadden, L., and Belton, M. J. S.: 2005, *Space Sci. Rev.*, this issue.

Toth, I.: 2001, in: *Asteroids 2001: From Piazzi to the 3rd Millennium*, Abstr. No. VIII. 11, p. 313. Palermo, Italy.

Weissman, P. R., Doressoundiram, A., Hicks, M., Chamberlin, A., and Hergenrother, C.: 1999, *Bull. Am. Astron. Soc.* **31**, 1121.

Weissman, P. R., Asphaug, E., and Lowry, S. C.: 2004, in: Festou, M. (ed.), *Comets II*, University of Arizona Press, Tucson.

Yeomans, D. K., Chamberlin, A. B., Chodas, P. W., Keesey, M. S., and Wimberly, R. N.: 2000, *Jet Propulsion Laboratory Interoffice Memorandum*.

Yeomans, D. K., Giorgini, J., and Chesley, S.: 2005, *Space Sci. Rev.*, this issue.

THE COMA OF COMET 9P/TEMPEL 1

C. M. LISSE[1,2,*], M. F. A'HEARN[2], T. L. FARNHAM[2], O. GROUSSIN[2],
K. J. MEECH[3], U. FINK[4] and D. G. SCHLEICHER[5]

[1] *Planetary Exploration Group, Space Department, Johns Hopkins University Applied Physics Laboratory, 11100 Johns Hopkins Road, Laurel, MD 20723, U.S.A.*
[2] *Department of Astronomy, University of Maryland, College Park, MD 20742, U.S.A.*
[3] *Institute for Astronomy, University of Hawaii, 2680 Woodlawn Drive, Honolulu, HI 96822, U.S.A.*
[4] *Department of Planetary Sciences, University of Arizona, LPL, Tucson, AZ 85721, U.S.A.*
[5] *Lowell Observatory, 1400 West Mars Hill Road, Flagstaff, AZ 86001, U.S.A.*
(*Author for correspondence; E-mails: carey.lisse@jhuapl.edu, lisse@astro.umd.edu*)

(Received 13 September 2004; Accepted in final form 28 December 2004)

Abstract. As comet 9P/Tempel 1 approaches the Sun in 2004–2005, a temporary atmosphere, or "coma," will form, composed of molecules and dust expelled from the nucleus as its component icy volatiles sublimate. Driven mainly by water ice sublimation at surface temperatures $T > 200$ K, this coma is a gravitationally unbound atmosphere in free adiabatic expansion. Near the nucleus ($\leq 10^2$ km), it is in collisional equilibrium, at larger distances ($\geq 10^4$ km) it is in free molecular flow. Ultimately the coma components are swept into the comet's plasma and dust tails or simply dissipate into interplanetary space. Clues to the nature of the cometary nucleus are contained in the chemistry and physics of the coma, as well as with its variability with time, orbital position, and heliocentric distance.

The DI instrument payload includes CCD cameras with broadband filters covering the optical spectrum, allowing for sensitive measurement of dust in the comet's coma, and a number of narrowband filters for studying the spatial distribution of several gas species. DI also carries the first near-infrared spectrometer to a comet flyby since the VEGA mission to Halley in 1986. This spectrograph will allow detection of gas emission lines from the coma in unprecedented detail. Here we discuss the current state of understanding of the 9P/Tempel 1 coma, our expectations for the measurements DI will obtain, and the predicted hazards that the coma presents for the spacecraft.

Keywords: comets, coma, dust, gas, composition, jets, 9P/Tempel 1, Deep Impact

Introduction

The coma of comet 9P/Tempel 1 is of interest to the Deep Impact (DI) mission as a phenomenon itself, as a component of the measured flux in the DI images, and as an obstacle that the DI spacecraft must traverse. More importantly, however, the coma represents the end product, and thus the measurable entity, of the material excavated during the impact. The coma is also the primary characteristic seen in ground-based observations, and is thus important for revealing the long-term changes that result from the impact, as well as for providing comparisons to other comets. We have attempted to collect and interpret all currently available information concerning the dust and gas emitted by 9P/Tempel 1. Because of its nearly exact 11:2 orbital commensurability with Earth, Tempel 1 alternates between favorable and unfavorable apparitions for Earth-based observing, and the state of our understanding of the coma is moderate at best. Nearly all near-perihelion data of which we are aware are from the favorable apparitions in 1983 and 1994. There are, in addition, numerous images of the dust coma, obtained as part of the Deep Impact project, from the 2000 apparition, but because this was an unfavorable apparition, all the data were obtained when the comet was far from perihelion.

It will be convenient to separate our discussion into the gaseous and dusty components of the coma. In this article, we will focus on what we believe the state of the coma will be near the time of the encounter, as this is particularly relevant to the measurements that will be made by Deep Impact. However, because the impact may produce changes in the long-term behavior of the comet, we will also discuss the properties of the coma as observed over the full range of heliocentric distances.

It is important to note that the ensuing discussion of the current knowledge of Tempel 1's coma is inherently biased. The outer regions of cometary comae, due to observability considerations, are relatively well studied (cf. in *Comets II* by Schleicher and Farnham, 2005; Bockelée-Morvan *et al.*, 2005; Feldman *et al.*, 2005; Crifo *et al.*, 2005; Boice and Huebner, 2005; Combi *et al.*, 2005). Here the density of the outflow is low enough and the mean free path between collisions is large enough that the gas is in the molecular flow regime; photochemistry of volatile species, reactions via H-atom addition, and collisional excitation and charge exchange with the solar wind dominate. Dust particles have reached a terminal velocity from the coma expansion and enter a regime where the dominant forces are solar radiation pressure and gravity. By contrast, in the innermost regions of the coma, from the nuclear surface to $\sim 10^3$ km, gas densities range upwards of 10^6 mol cm^{-3}, collisions are frequent, and the outflow is turbulent and complicated. Multi-body chemistry driven by collisions occurs in the near nucleus zone, and transient chemical species are common. Narrow jets of material are formed, though it is not clear, even after the Deep Space-1 (DS-1) and STARDUST flybys of comets Borrelly and Wild 2, whether the focusing of outflow material is due to surface and subsurface structures in the cometary nucleus or to hydrodynamic outflow behavior in the inner coma, or both. It is within this region that dust particles released by volatile sublimation are swept up and accelerated by the outflowing gas to typical terminal velocities of 1–500 m s^{-1}. Plasma species, created by photoionization of outflowing volatiles and by the in-streaming solar wind, are unable to penetrate into this region due to mass loading and magnetic field concentration. Because it is difficult to resolve from the ground, the inner coma is poorly understood and its characteristics have mostly been inferred from studies of the coma as a whole.

As discussed in the companion article by Hampton *et al.* (this volume), the cameras on Deep Impact include seven intermediate-band filters covering the optical spectrum (all seven are included in the HRI filter wheel, while a subset of four are in the MRI filter wheel). These filters will allow sensitive measurements of dust in the comet's coma. The MRI also carries three narrowband filters for isolating the emission bands of CN, OH, and C_2 gaseous species. One of the most exciting potential returns of the DI mission is the possibility of chemical maps of the little-studied inner coma region at $\sim$10 m resolution. Investigating the chemistry of any jet structures that are present and tracing them back to the nucleus will greatly increase our understanding of the nature of the active regions that produce jets. The HRI near-infrared spectrometer will allow detection of emission lines from neutral gas species in the coma in unprecedented spatial and spectral detail and will also provide novel results on the size, composition and spatial variations of the coma dust. Although the spectroscopic measurements are an important part of coma studies in the DI mission, we will only summarize them here. They are discussed in detail by Sunshine *et al.* (2005) in a companion paper in this volume.

Current Knowledge of Tempel 1's Coma

In the next two sections, we discuss what is known from previous observations about the composition and structure of Tempel 1's coma. Unfortunately, the total amount of archival data is small, and much of this is unpublished, because the low-activity comet 9P/Tempel 1 is rarely bright as seen from the Earth and thus is not a favorite target of observers. Fortunately, the 2005 apparition is the next good one, so extensive ground-based observations can be obtained in support of the DI encounter.

The Gas Coma

Abundances of gaseous species are available from four independent programs – ultraviolet spectroscopy with IUE in 1983 and 1994, optical spectrophotometry at McDonald Observatory using the Image Dissector Scanner (IDS) and the Long-Slit Spectrograph (LSS) at both the 1983 (IDS) and 1994 (LSS) apparitions, optical spectrophometry at Mt. Lemmon during the 1994 apparition, and narrowband photometry in 1983 and 1994 from the program centered at Lowell Observatory. Not all of the data have been previously published. We include here newly available results from the 1994 Mt. Lemmon program (an example is shown in Figure 1), for which most of the earlier results were published by Fink and Hicks (1996). We also include newly available results from the observations during the 1994 apparition from McDonald Observatory courtesy of A. Cochran (private communication) and a new reduction of the narrowband photometry from 1983 (Osip *et al.*, 1992; Farnham *et al.*, 2000; Farnham and Schleicher, 2005), for which

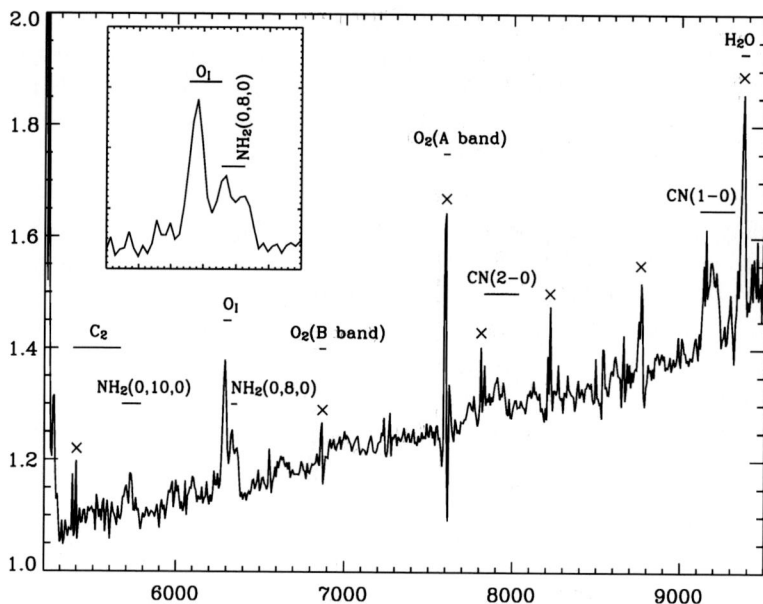

Figure 1. Archival optical spectrophotomteric observation of 9P/Tempel 1. Optical spectrum of Tempel 1 on 12 April 1994, 89 days before perihelion (Fink *et al.* unpublished). The integration time was 6960 s and the vertical axis is in relative units. The prominent cometary emission lines used for analysis have been labeled. Removal of the continuum leads to increased noise and incomplete cancellation of telluric absorption, particularly in the O2 A and B bands. The locations of these spectral artifacts are marked by an 'X'. Note that the zero level for the abscissa has been set much lower than the origin on the vertical scale, in order to emphasize the gaseous coma line emissions. Compared to other comets, Tempel 1 demonstrated relatively strong dust continuum emission relative to weak coma gas line emission.

improved techniques for separating continuum from emission and improved models of atmospheric attenuation in the ultraviolet have now been used, and we include previously unpublished narrowband photometry from 1994. The individual datasets are internally consistent, and variations with time within a given dataset and comparisons with other comets observed by the same groups are appropriate and provide insight into the comet's behavior. Due to the fact that the various groups used different parameters for analyzing the data, the results from different datasets should not be directly compared with each other. We have adjusted the results of Cochran *et al.* (1992) to change from their physically based assumption of expansion velocity to the simpler assumption of $1 \, \text{km s}^{-1}$ that was used by the other investigators. This removes the largest source of discrepancies between the datasets, but other smaller discrepancies remain. A complete reanalysis of all the data using consistent parameters is beyond the scope of this paper, and we expect that future publications will describe the datasets in much more detail.

COMPOSITION

Reanalysis of the photometric data from the 1983 apparition (Osip *et al.*, 1992) and inclusion of the more recent (unpublished) photometric data from the 1994 apparition indicates that, relative to CN, the abundances of the carbon-chain radicals C_2 and C_3 are near the lower edge of the band that defines "typical" cometary chemistry (A'Hearn *et al.*, 1995). Cochran *et al.* (1992) found that the carbon-chain radicals, C_2 and C_3, as well as CH have abundances relative to CN that are very near the mean value for all comets in their survey, though we note that their survey included several comets that A'Hearn *et al.* found to be in the "depleted" class. The photometric data also show that, at least in 1994, the abundances of all radicals were low relative to H_2O.

Two unpublished spectra obtained by Fink (as a continuation of the survey program described by Fink and Hicks, 1996) show the abundance of C_2 relative to CN to be low (Figure 1). Using the analysis methods Fink *et al.* applied to their spectroscopic dataset, the average ratios of C_2 and CN with respect to water (parts per thousand) are respectively 0.66 and 1.60. These numbers put comet Tempel 1 in the middle of an intermediate group which they term the "Borrelly type," with CN ratios like P/Halley but with low C_2 ratios like P/Giacobini-Zinner, the prototypical "depleted" comet.

On the other hand, Fink's unpublished 1994 spectra yield NH_2 abundances relative to CN that put Tempel 1 roughly in the middle of the group of 39 comets studied by Fink and Hicks. While this conclusion does not seem to agree with the data in Table VII of Fink and Hicks (1996), newer g-factors require the abundances of NH_2 in that table to be doubled. Cochran *et al.* find in their single measurement of NH_2 during the 1983 apparition that the abundance is very low relative to CN. A'Hearn *et al.* found that the closely related species NH is normal to relatively rich with respect to CN, consistent with the result from Fink's spectra.

Another key aspect of the gaseous abundances is the dust-to-gas ratio. The dust release is measured through optical measurements of the parameter Albedo × Fill factor × Aperture radius ($Af\rho$; A'Hearn *et al.*, 1984), which is sensitive to the number density, scattering properties and the size distribution of the dust. At optical wavelengths it provides an estimate only of the amount of small dust particles, with diameters within an order of magnitude of the wavelength of observation. It is thus not a good measure of the total mass of dust in the coma, since the total mass is dominated by the largest particles for plausible size distributions, but it is a good measure of the small dust. Within these limitations, the dust-to-gas ratio in Tempel 1 appears to be moderate. The ratio of $Af\rho$ to the production of OH is 2.3×10^{-26} cm s molecule^{-1}. For comparison, in dust-poor comets like 2P/Encke and 6P/d'Arrest the ratio is $<10^{-26}$ cm s molecule^{-1}, and in dust-rich comets (mostly comets with perihelia beyond 2 AU) it is $>10^{-25}$ cm s molecule^{-1}. Examination of the lightcurves in Figure 2 shows also that, at least over the period for

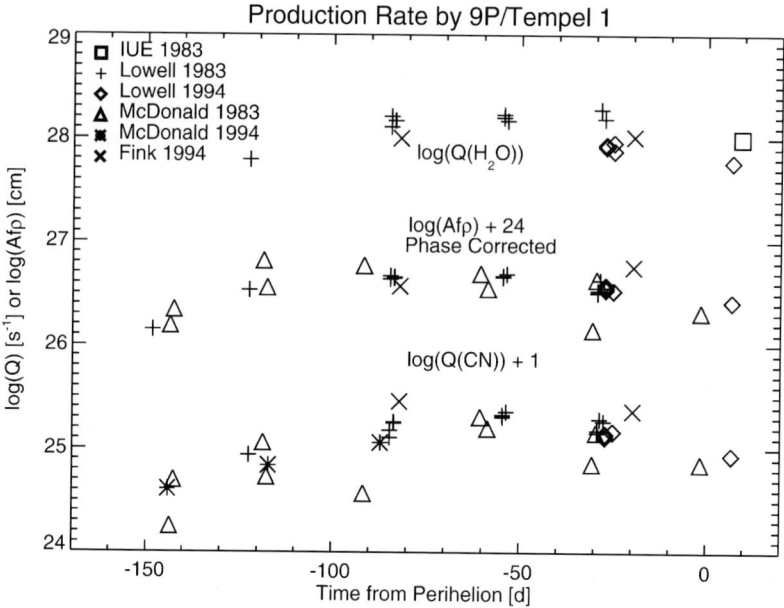

Figure 2. Production rates for OH, CN and dust, from Lowell Observatory narrowband photometry, IUE measurements and IDS spectra from McDonald Observatory. The middle set of data points shows the phase-corrected values of *Afρ* from the narrowband photometry as a reference. The lower points show the variation of production of CN from both narrowband photometry and spectrophotometry. The CN and the dust curves both show excellent agreement between the 1983 and 1994 apparitions. The upper points show the production rates of H_2O deduced from narrowband photometry of OH at both apparitions, from IUE spectra of OH (reduced with the same parameters) in 1983 and from ground-based spectrophotometry of $O(_1D)$ in 1994. Unlike the other species (including those not shown here), there may be a factor of two reduction in water production from 1983 to 1994.

which measurements of the gas are available, the gas-to-dust ratio is approximately constant in any apparition.

A comparison with the other comets visited by spacecraft – Wild 2, Borrelly, and Halley – as well as with typical cometary parameters, is given in Table I. Values are nominally for the conditions at perihelion and are based on the work of A'Hearn *et al.* (1995; i.e. not using the reanalysis by Schleicher which does not yet include the other comets), except for the two entries for NH_2, which are from Cochran *et al.* (1992) and from Fink's unpublished spectra. The column headed "typical" is the average of all typical comets in all cases except that of the NH_2 value from Cochran *et al.*, for which it is the average of all comets.

The only other species for which we can compare across a database of many comets is CS, which has been observed routinely in spectra of comets taken with the International Ultraviolet Explorer (IUE) and subsequently with the Hubble Space Telescope (HST). Feldman and Festou (1991) reported on a spectrum taken at the 1983 apparition, and the abundance of CS relative to water is very similar to that

TABLE I

Water production rates and abundance ratios.

	Tempel 1	Wild 2	Borrelly	"Typical"	Halley
$Q(H_2O)$ mol s^{-1}	$\sim 1 \times 10^{28}$	1.3×10^{28}	2.1×10^{28}		1.3×10^{30}
Taxonomy	Typical	Depleted	Depleted	Typical	Typical
Q_{CN}/Q_{OH}	.0015	.0032	.0023	.0032	.0040
Q_{C2}/Q_{CN}	0.81	0.62	0.44	1.1	1.3
$Af\rho/Q_{OH}$	2.3×10^{-26}	4.8×10^{-26}	3.5×10^{-26}	1.5×10^{-26}	4.9×10^{-26}
Q_{NH}/Q_{CN}	1.9	1.5	1.6	1.3	1.6
Q_{NH2}/Q_{CN} C	0.41			0.83	
Q_{NH2}/Q_{CN} F	0.62			0.55	0.85

in comets Kopff and Tempel 2, also presented in that paper, as well as to the values observed in most other comets.

In summary, what we currently know about the chemical composition of the Tempel 1 coma is the following: carbon-chain molecules are at the low end of the "typical" class relative to CN, most radicals are below average relative to water in 1983, and NH$_2$ (and thus NH) is likely high relative to CN. The dust-to-gas ratio is also well within the range of the "typical" comets, at least as dust can be measured at optical wavelengths. Tempel 1 is thus not clearly anomalous in any way and it is reasonable to assume that, to first order, the composition of its coma in other species not yet observed is also close to A'Hearn et al.'s "typical" comet. Though we are not able to measure all of these species with the instruments on Deep Impact, the spatial distributions of the species that we are able to measure should be invaluable in unraveling the appropriate parameters for analyzing Earth-based observations.

VARIABILITY

Figure 2 shows the variation of certain gaseous production rates from comet Tempel 1. We have included the parameter $Af\rho$, plotted with a large offset from zero, as a point of comparison for the shape of the curves of gases. The data for the dust show good agreement among investigators and good agreement from one apparition to another.

The data on gas from Fink have been adjusted to the scale lengths used by A'Hearn et al. (see Fink and Combi, 2004, a factor 0.6) and the data from McDonald have been corrected to the assumed expansion velocity of A'Hearn et al. (variable but generally between 2 and 3x). In general, the different sources agree well with each other and the shape of the curve repeats from one apparition to the next. For CN, it appears that the scatter is much larger in the data from McDonald than in the data from Lowell, and it is difficult to say whether this is indicative of either temporal or spatial variability, to which the small apertures of the IDS are more sensitive than the large apertures of the narrowband photometry, or due to some other factor. We note that the dust measurements of Storrs *et al.* (1992; from the same spectra as the results for gas by Cochran *et al.*), for which there are no significant dependences on uncertain parameters, also show a large scatter compared to the results from narrowband photometry (see dust coma section, below). It is clear that the 1983 and 1994 apparitions agree well with each other for the CN, suggesting again that there is little secular change in the comet's activity.

The behavior of water, however, may be quite different. We have converted the narrowband photometric observations of OH to production rates of H_2O (assuming the usual vectorial model) and plotted them in Figure 2, together with the value deduced from the IUE observation in 1983 (Feldman and Festou, 1991) and the two results from Fink (based on $O(^1D)$ and adjusted for differences in scale length). These data suggest a possible factor of 2 decrease in the production rate of H_2O from 1983 to 1994 and thus a corresponding increase in the ratio of all species to water between the 1983 and 1994 apparitions. If this effect is real (the limitations in both ground-based and IUE datasets prevent us from asserting confidently that the variation is real), there are no data that would allow us to distinguish between a secular decrease and random fluctuations from one apparition to the next.

Another observation of Tempel 1 with IUE, not shown in Figure 2, was obtained on July 14, 1994, 11 days post-perihelion, less than 2 days later relative to perihelion than the observation in 1983 (Haken, A'Hearn, and Feldman, unpublished). At that time, approaching the end of IUE's lifetime, there was a large amount of scattered light in all IUE observations taken at moderate solar elongation and this had significant deleterious effects on the tracking, which at that point was already being done in a 2-gyro mode. The resultant fluxes, therefore, are rather uncertain. A best effort to analyze these data yielded a water production rate roughly a factor of 2 lower than that deduced from the observation in 1983, for which both the complete set of geometrical circumstances and even the solar activity were nearly the same. Although there are always large uncertainties in correcting ground-based observations of OH for atmospheric extinction, and although the second IUE measurement was uncertain for instrumental reasons, we have some confidence that the factor of 2 change in water release between the two apparitions is real. The change shows up independently in comparing ground-based observations in June 1983 to those in June 1994, in comparing space-based observations in July 1983 with those in 1994, and in comparing the ground-based measurements in July 1994 with the space-based measurements in July 1983. All three comparisons show the water production to be higher in 1983 than the 1994, making it unlikely that the difference is due to calibration uncertainties or instrumental problems. On the other hand, all indications are that the nucleus is in a state of simple rotation (the repeatability of seasonal variations, similarity in features at a given point in the orbit, etc.), so the progression of solar illumination on the nucleus follows the same pattern on each perihelion passage, reducing the possibility that different sources are illuminated on different apparitions. We can present no compelling scenario to explain why the water production drops by a factor of 2 from one apparition to the next while production of other species remains essentially constant. Clearly, an understanding of the spatial distribution of the water in the inner coma relative to the distribution of other species will be an important goal of Deep Impact's encounter.

The Dust Coma

Useful observations of dust in cometary comae have been obtained from UV through sub-mm wavelengths and include spatial imaging, photometry and spectroscopic measurements. For low-activity comets like Tempel 1, where flux limitations are a concern, only broadband optical photometric observations are commonly obtained. Fortunately for studies of the dust, the observed flux in these broadband filters is dominated by the continuum, and these observations work well to characterize the long-term behavior of the comet's activity. To obtain information about the dust composition and particle size distribution (PSD), which are critical for quantifying the mission risk during the Deep Impact encounter, longer wavelengths – IR and beyond – are used. Specifically, the 10–$1000\,\mu$m wavelength regime is most sensitive to the grains larger than $100\,\mu$m that pose the largest impact and navigation hazards to the spacecraft.

The dust coma data fall into three broad categories – temporal (lightcurves), morphological, and spectrophotometric. We discuss each in turn below.

TEMPORAL BEHAVIOR (LIGHTCURVES)

The long-term gross behavior of the Tempel 1 coma can be elucidated from photometric measurements obtained around the comet's orbit (Figure 3). We use the parameter $Af\rho$ (A'Hearn et al., 1984) for interpreting the long-term optical photometry, because it removes the variations introduced by changing geocentric and heliocentric distance, aperture size differences, etc. In this work we also refer to the parameter $\varepsilon f\rho$ as an analogous quantity for interpreting infrared observations of the dust. For $\varepsilon f\rho$, where the measured flux is assumed to be the thermal emission from a blackbody at the local equilibrium temperature instead of the incident light scattered at the observed phase angle.

A number of observations of Tempel 1 exist as unpublished observations and as individual astronomical circulars. The sum of these scattered observations has led to phenomenological claims of variable dust and gas production rates pre- and post-perihelion. To further investigate this issue, we adopted instead the assumption that the comet's temporal behavior is the same from one apparition to another and assembled a long-term lightcurve covering a two-year period around perihelion for the best observed species, dust. For this lightcurve, we combined several data sets to provide good temporal coverage of the perihelion portion of the comet's orbit. These data include the 1983 InfraRed Astronomy Satellite (IRAS) measurements (re-analyzed as discussed below), the 1983 optical photometry of Cochran et al.

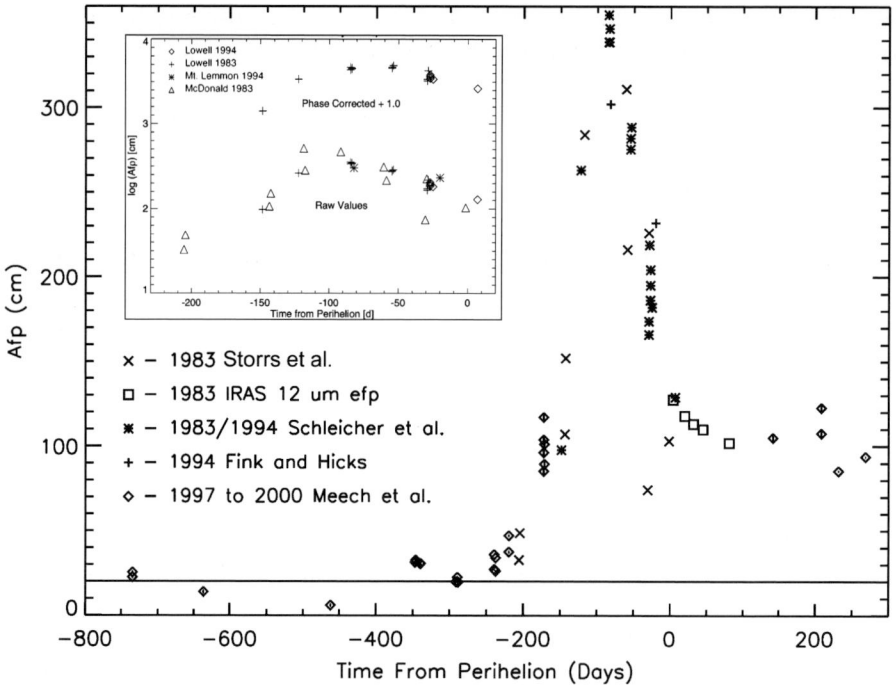

Figure 3. As-observed long term lightcurves for 9P/Tempel 1 derived by overlaying the $Af\rho$ and $\varepsilon f\rho$ measures from the 1983, 1994, and 1999 apparitions, plotted with respect to time of perihelion. The optical $Af\rho$ measurements of Osip *et al.* (triangles) and Meech *et al.* (diamonds) have estimated errors of 15%. The IRAS $\varepsilon f\rho$ points (squares) have estimated errors of 25%, including systematics. The temporal behavior is generally consistent between the three apparitions, suggesting the comet's behavior is repeatable from orbit to orbit and that the $\varepsilon f\rho$ behavior reflects the trends seen in $Af\rho$. The emission rate of dust is quite different before and after perihelion, suggesting seasonal illumination of the actively emitting surface regions. The solid line denotes the $Af\rho$ value expected for a bare nucleus with no dust emission. Inset – $Af\rho$ values corrected for the phase-angle variation of scattering. The narrowband photometry was obtained at λ 4845 Åfor most of the observations, though a few were obtained at λ 5240 Å. McDonald Observatory measurements (Storrs *et al.*, 1992) were obtained at λ 4800 Å while those from Mt. Lemmon are at λ 6250 Å. The narrowband photometry shows the best internal consistency, possibly because it averages out spatial and temporal variations on small scales to which the spectroscopic techniques are sensitive. The top curve shows the $Af\rho$ values that are derived from the narrowband photometry when corrected for the solar phase angle (using the scattering function of Ney and Merrill (1976)). The correction makes the curve much broader with the peak production near −60 days rather than near −85 days, bringing it into agreement with the maxima in the H_2O and CN gas production rates (Figure 1).

taken from the McDonald Observatory, the extensive 1999–2000 optical photometry from Meech *et al.* (2005), and the narrowband optical photometry of Schleicher *et al.* for the 1983 and 1994 apparitions. (We note that the scatter in the data from the IDS at McDonald Observatory is much larger than the scatter in the data from the narrowband photometry. It is not clear why this is the case, but we suggest that it is due at least in part to the relatively small effective aperture in IDS, and to the fact that some of the observations were offset from the nucleus, which makes those measurements much more sensitive to both temporal and spatial variability in the outgassing).

The resulting composite lightcurve indicates that the observations of dust in 1994 and 1997–2000 are, for the most part, consistent with the observations from 1983, suggesting there is little secular variation in the comet. The lightcurve rises rapidly, starting about 8 months before perihelion, apparently peaking around −90 days, and falls off more slowly through and after perihelion. There is still evidence of coma nearly a year after perihelion, at which time the comet is beyond 3 AU from the sun. This continued level of activity is likely due to very slow dissipation of slow-moving dust as the gas production decreases near the ice line, with a concomitant decrease of the terminal velocity of the dust and increase in the dust lifetime in the coma. Another possible cause of the post-perihelion persistence of emission is the time lag for the turnoff of volatile emission created by the propagation of the thermal wave driven by the comet's perihelion passage. The comet interior will be on the whole warmer 6 months after perihelion passage than 6 months before. Without high precision imaging of the gas coma with the comet at the ice line, or detailed nucleus thermal measurements, we cannot easily distinguish between these two possibilities.

The derived lightcurve also explains the phenomenological reports of a variable pre-/post-perihelion asymmetry. By assuming a peak in activity around perihelion, which our lightcurve clearly shows is not the case, discordant values of the activity are found. As can be seen from Figure 3, sampling the Tempel 1 lightcurve at times spaced equally before and after perihelion would lead to an estimate of the ratio of post-/pre-perihelion activity ranging from a value of 0.5 up to 4.0.

The apparent peak in production 90 days before perihelion is misleading because these data have not been corrected for phase angle variations. In order to account for this, we have taken the dataset from Lowell Observatory, since it shows so much less scatter than the data from McDonald, and applied a phase correction following the work of Ney and Merrill (1976). The result is plotted in the inset of Figure 3, displaced upward by 1.0 in $\log(Af\rho)$. The Ney and Merrill phase function produces the largest change to the figure of the several phase functions we have tabulated in planning the Deep Impact mission, and the true phase correction is probably not this large. However, it is clear that the curve is now much flatter than without correction, and if one tries to pick the peak, it is probably near 60 days pre-perihelion rather than the earlier time deduced from the uncorrected lightcurve. In either case, the early peak indicates that there is some kind of seasonal variation affecting the production rates.

Even with this offset in the peak activity, however, there seems to be little influence from non-gravitational forces on the orbit, since the force parameters are relatively small (Marsden and Williams, 2003; Yeomans et al., 2005). This implies that either the gas and dust emission are not consistently directed in a preferred direction, or else the force from the directed emission is very small relative to the mass of the nucleus.

DUST MORPHOLOGY

The dust coma is created when hydrodynamic drag by the outflowing gas lifts dust grains off the surface of the nucleus. At a distance of 10–100 nuclear radii, the gas and dust decouple, leaving the dust particles in their own independent orbits around the sun, where the opposing forces of solar gravitation and radiation pressure govern the particle motions (Combi *et al.*, 2005). In the most basic scenario, a nucleus undergoing constant, isotropic dust emission will produce a coma that starts out spherical, but elongates with time as the grains are swept back into the tail by radiation pressure. The efficiency of radiation pressure on the grains is related to the ratio of cross-sectional area to mass, so small grains are accelerated down the tail much more rapidly than large grains, effectively acting to "sort" the different particle sizes (Finson and Probstein, 1968). Because the basic forces on the dust are purely radial with respect to the Sun, the coma and tail spread out along the comet's orbital plane, though there will be some finite thickness above and below the plane due to the dust's out-of-plane velocity at the time of the gas/dust decoupling.

Many comets, on the other hand, are not dominated by isotropic emission but also have isolated active regions that produce very narrow jets (cf. comets Borrelly and Wild 2). These jets are superposed over the symmetric emission morphology and appear as structure or features in the coma. (We also point out that the isotropic emission in a comet may actually be produced by lateral diffusion of jet material, by the contributions of a large number of distributed weak jets, or by some other mechanism, but the result can be indistinguishable from isotropic emission when observed at large distances.) Depending on the strength of the jet or jets, it may be necessary to process images of the coma to clearly reveal these features, but strong jets can be clearly visible in raw images. Here we apply systematic techniques described by Schleicher and Farnham (2005) to search for features produced by directed emission from the nucleus.

Previous studies of Tempel 1 in the literature (Campins *et al.*, 1990; Lynch *et al.*, 1995; Fernandez *et al.*, 2003) reported what appears to be a featureless, smooth coma, suggesting the emission is isotropic or comes from multiple sources that mimic isotropic outflow. However, even if jets are present, they may not be resolved on the alternate unfavorable apparitions. For example, the most recent apparition, in 2000, was a poor one, and even with the wealth of CCD data that was obtained by the Deep Impact project for the newly designated mission target, no features have yet been detected (Figure 4a). On the other hand, out of only a few CCD images that are available from the 1994 favorable apparition, a broad fan jet is clearly distinguishable (Figure 4b). This feature is visible in all the available images, from mid-March, mid-April, and mid-June (-108, -82 and -21 days from perihelion). It is clear from both the March and June images that the fan cannot be modeled exclusively using radiation pressure forces. Thus it is not a dust tail but instead must be produced, at least in part, by directed emission from an active source on the nucleus. The surface brightness in the jet is about a factor of 2 higher than the average coma surface brightness (at any given distance), so, given the margin of error in the hazard analysis (see the discussion below), the increased dust density within the jet is not a significant concern for the DI spacecraft. No calibration

(a)

Figure 4. Archival coma images of Tempel 1. a) UH 88″ R-band image of Tempel 1 on 21 August 2000, at 202 days after perihelion, $r = 2.54$ AU, $\Delta = 1.67$ AU. North is up and East is to the left. The coma is relatively active in this image, but appears featureless, though a weak tail is apparent. b) Clear filter (3500–10000 Å) image of Tempel 1 on 17 March 1994 (from Fink *et al.*), enhanced by removing a ρ^{-1} profile to show the fan jet. North is at the top, East is to the left and the field of view is 78000 km × 78000 km. The jet is visible pointing to the southwest. The most likely cause is directed emission from the nucleus. Syndynes (dashed) and synchrones (solid) have been overlaid to show the motion of dust under the influence of solar radiation pressure. The fact that the fan is not aligned along the syndyne/synchrone direction indicates that the feature is not produced by the interactions of solar gravity and radiation pressure (e.g., it is not the dust tail). Although the jet appears to lie in the anti-solar direction, this is likely a projection effect. The solar phase angle on this date was only 14.9°, so nearly the entire Earth-facing side of the nucleus is illuminated. c) IRAS survey imaging of 9P/Tempel 1 on 28 July 1983, $r = 1.56$ AU, $\Delta = 1.27$ AU. This 12-μm reconstructed survey image with 24″ per pixel resolution has the highest SNR and largest extent of the four IRAS passbands. South is to the top and West is also to the left. The comet is moving towards the lower right, and the projected direction to the Sun is to the lower right. There is little structure evidenced by the coma, other than some extension in the anti-solar direction, as expected from dynamical modeling of a low-inclination comet. This image, taken near perigee in 1983, also shows a hint of the comet's linear trail structure, composed of large, heavy dust particles. Inset – large scale 12 μm image of the comet, 1.5° × 1.5°, clearly showing the large extent of the comet's trail. The comet nucleus is the bright point in the center, and the comet trail extends mainly north and west in the anti-velocity direction, i.e. along the orbital path the comet has recently traveled.

(*Continued on next page*)

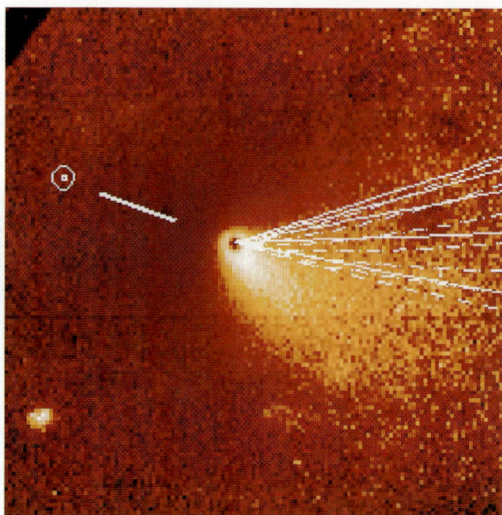

(b)

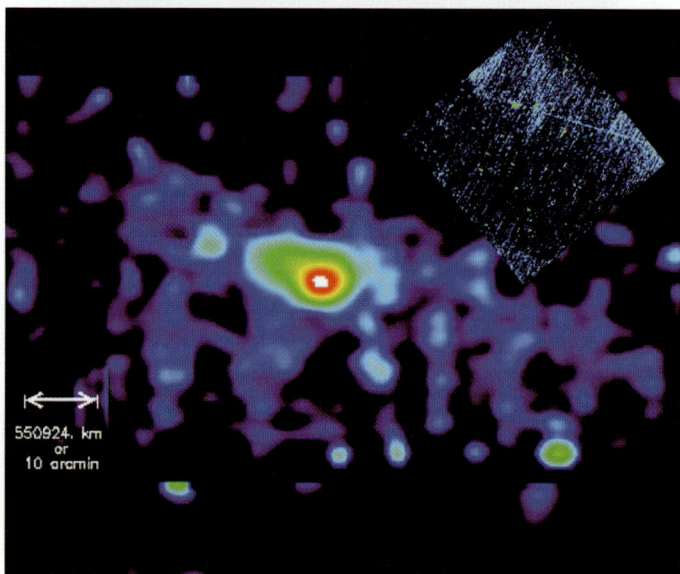

(c)

Figure 4. (*Continued*)

information is available for these images, so it is not known exactly how the absolute surface brightness changes from month to month, but the jet is much more well-defined in March and April, when the coma activity was increasing towards its peak, than in June, when the activity was dropping. In addition to the CCD images, at least two experienced visual observers have detected features in the coma. Both of these detections, in April 1983 (J.C. Merlin, private communication) and May 1994

(N. Biver, private communication; see Belton *et al.*, 2005, Figure 8), were obtained when the comet was relatively close to Earth, providing good opportunities for resolving features.

These three sets of observations are currently the only ones known to show coma morphology. The existence of the jet also indicates that the nucleus has isolated active areas, which is consistent with the result from the models of Groussin *et al.* (2004) that conclude only about 8% of the surface area is active. Interestingly, the features have been observed in the coma only at times when the production rates were near their pre-perihelion peak (Figure 3) and this suggests that the jet and the time of peak emission may be related (e.g., at the time of peak activity, the sun is at its highest point in the sky as seen from the source that produces the jet). We qualify this conclusion with the caveat that the dearth of high-quality optical imaging data in the post-perihelion time frame may introduce significant selection effects.

The shape and appearance of a jet in the coma is dependent on a number of factors: the size and location of the source, the rotation state of the nucleus, the motion of the comet in its orbit, and changes in the viewing geometry of the comet. With sufficient observations, it is possible to use coma models to extract many of these properties, helping to constrain the nature of the nucleus. Unfortunately, the number of currently available images of Tempel 1 that show features is too small to constrain the models. The 2005 apparition, however, will be a favorable one, and the conditions will be such that high-resolution images of the coma can be obtained for several months before perihelion. With the observing campaign in support of the Deep Impact mission, a massive amount of data will be obtained before and after perihelion (=impact), and these data will help to characterize the features that are seen.

Finally, there is another morphological feature that should be mentioned. In July and August 1983, IRAS observed Tempel 1 at 12, 25, 60, and 100 μm. A sample 12-μm image, chosen because the 12 μm passband tends to have the highest signal-to-noise, is shown in Figure 4c. In these images, the coma and tail are only slightly more extended than the instrumental point spread function, with little of the extension along the Sun-comet vector that would be expected for small dust particles highly accelerated by solar radiation pressure (Lisse *et al.*, 1998). It may be that the dust production had dropped significantly so that the small grains were barely detectable. However, the IRAS images do show evidence of a faint dust trail consisting of large dust particles, which can be seen more clearly in large-scale (3° × 3°) IRAS Sky Survey images (Sykes, 1992: Inset, Figure 4c). For the biggest grains emitted from the nucleus (>1000 μm), the emission velocity is small, and radiation pressure is essentially negligible, so they have orbital paths very close to that of the nucleus. Over time, these large grains spread out along the comet's orbit to form a trail that can be observed when the Earth is near the comet's orbit plane. The existence of this trail indicates that large grains could be a danger for the flyby spacecraft, though the actual size and number density of these particles is difficult to predict precisely – only the total particle surface area can be determind for the IRAS measurements. To minimize the hazard from these particles, the DI encounter was designed to pass through the comet's orbital plane (where the large grains are concentrated) after the point of closest approach, and thus after the main sequence of observations have been obtained.

DUST SPECTROPHOTOMETRY

Spectrophotometry of comet 9P/Tempel 1 was obtained by IRAS in four broad bandpasses at 12, 25, 60, and 100 μm and by the IRAS 8–21 μm Low Resolution Spectrometer (LRS) in July–August 1983. These observations, along with 5240 Å narrowband continuum photometry taken contemporaneously (Osip et al., 1992), can be combined to produce a 0.5–100 μm spectral energy distribution (SED; Figure 5), after correcting for the different aperture sizes of the instruments assuming a ρ^{-1} coma brightness profile.

While relatively sparsely sampled in wavelength compared to other comets we have studied, the combination of thermal infrared and optical observations is quite sensitive to the 0.1–100 μm range of dust particle sizes emitted by comets (Lisse et al., 1998). The lack of a strong 8–13 μm silicate emission feature is an indicator of the abundance of relatively large dust grains ($\geq 10 \mu$m), as is a spectral color temperature of $\sim$245 K, within 10% of the local equilibrium temperature, and close agreement to greybody behavior at 30–100 μm (Harker et al., 1999; Mason et al., 1998). The total albedo for scattering, defined as the ratio of the scattered luminosity to the total integrated luminosity over all wavelengths, is $8 \pm 2\%$ (2σ) for Tempel 1, similar to the $\sim$6% found for the large cometary dust particles emitted by comets C/Austin 1990 V (Lisse et al., 1994, 1998) and 2P/Encke in 1997 (Lisse et al., 2004).

Modeling of the scattering and thermal emission spectrum (Lisse et al., 1998, 2004), using the energy balance between absorbed visible photons from the 5750 K solar radiation field and the thermal infrared radiation emitted by the dust was used to further constrain the silicate: refractory carbonaceous material composition ratio and particle size distribution of the dust. Following the results of the Halley encounters (Jessberger et al., 1988; Jessberger and Kissel, 1991), the particles contributing appreciably to the SED are assumed to be spherical with a radius of 0.1μm $< \mathbf{a} < 10^4 \mu$m. We also assume a porosity and density $\rho(\mathbf{a})$ for the dust as a function of particle size, so that the mass of a particle as a function of particle radius is given by $m(\mathbf{a}) = 4/3 \times \rho(\mathbf{a})\mathbf{a}^3$. The dust composition is modeled as a material with index of refraction given by a Bruggeman mixture (Lien, 1990) of the indices for silicates, water ice, amorphous carbonaceous material, and vacuum with density $\rho = \rho_o(a/0.1 \mu m)^{(D-3)/D}$ (where D is the fractal dimension of the dust and $\rho_o = 2.5 \mathrm{g\,cm}^{-3}$). The velocity of emission is derived from dynamical modeling of the coma and tail morphologies of other short-period comets (Finson and Probstein, 1968; Sykes et al., 1990; Lisse et al., 1998) and is on the order of 0.3 km s^{-1} X $\sqrt{(10^{-4}/a \text{ (cm)})}$. Following Fernandez et al. (2003) and Lisse et al. (2005), the nucleus was assumed to have thermal emission following a standard thermal model (Lebofsky and Spencer, 1989), $r_{eff} \leq 5$ km, emissivity $= 0.90$, beaming parameter $= 0.90$, geometric albedo $= 0.05 \pm 0.02$, a visual phase coefficient of 0.065 magnitudes, and an infrared phase coefficient of 0.01 magnitudes deg^{-1}. In all

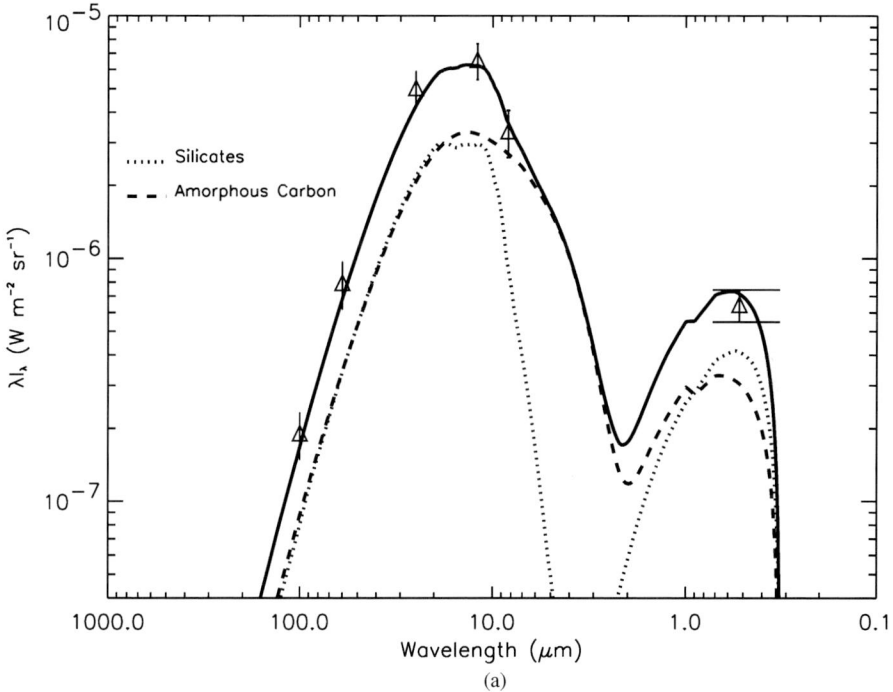

Figure 5. Dust coma spectral energy distribution and spectral model fits for 9P/Tempel 1. a) The coma spectral energy distribution is derived from recalibrated broadband IRAS flux measurements taken near perigee on 12 July 1983. All error bars in this and subsequent plots are $\pm 2\sigma$. The recalibrated fluxes are consistent with a greybody out to 100 μm. Solid curve – Best fit modified Mie model to the data, with 50% astronomical silicate, 50% amorphous carbon content, $\rho = \rho_o$ (m/10^{-14} g)$^{-0.10}$ density law, and a m$^{-0.75}$ particle mass distribution. Of the tested plausible size distributions, only power law particle size distributions produce spectral energy distributions that fit the data adequately. Dotted line – flux emission due to silicate grains, which dominates at wavelengths $>8\,\mu$m and is responsible for the emission features at 8–13 μm and the long wavelength falloff. Dashed line – flux emission due to amorphous carbon grains, which dominates at 3–8 μm and is responsible for the "superthermal emission" at 3–5 μm. The derived Q_{dust} for this model is 220 kg s^{-1} with an albedo of 9%. The 95% confidence level ranges for these parameters are 190–400 kg s^{-1}, and 7–10% albedo. b) χ^2 surface for allowed models with density $= \rho_o$ (m/10^{-14} g)$^{-0.05}$. The abscissa coordinate is the percentage of silicaceous material in the silicate:refractory carbonaceous material composition used to model the spectral energy distribution (SED). The ordinate is the power law index in the particle mass distribution $dn/d\log m \sim m^{-\alpha}$ used to model the SED. c) χ^2 surface for allowed models with density $= \rho_o$ (m/10^{-14} g)$^{-0.10}$. While the allowed power law index is limited to the range $0.64 < \alpha < 0.75$, and the dust density index to the range $0.05 < \eta < 0.10$, the material composition of the dust is poorly constrained by the data.

(*Continued on next page*)

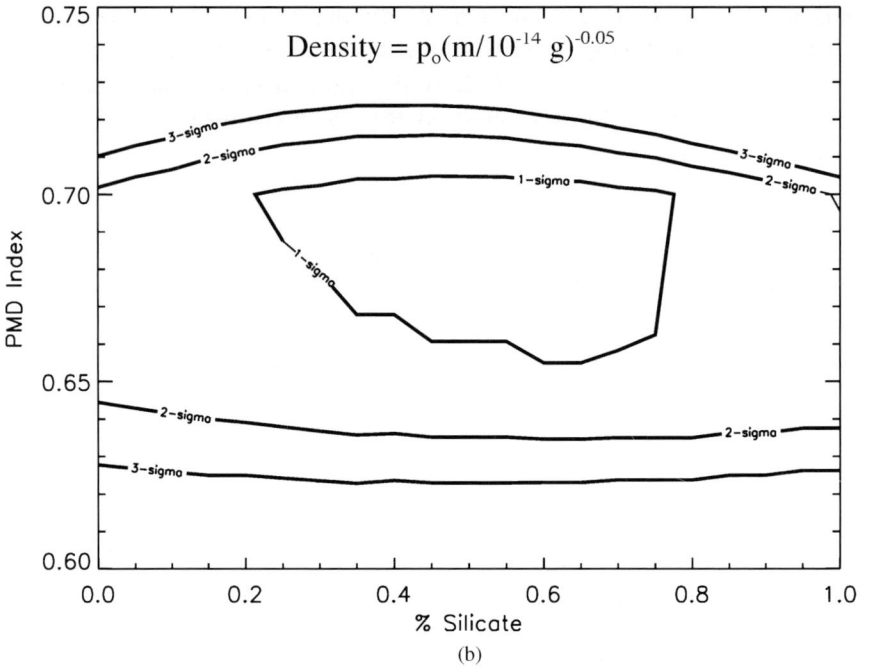

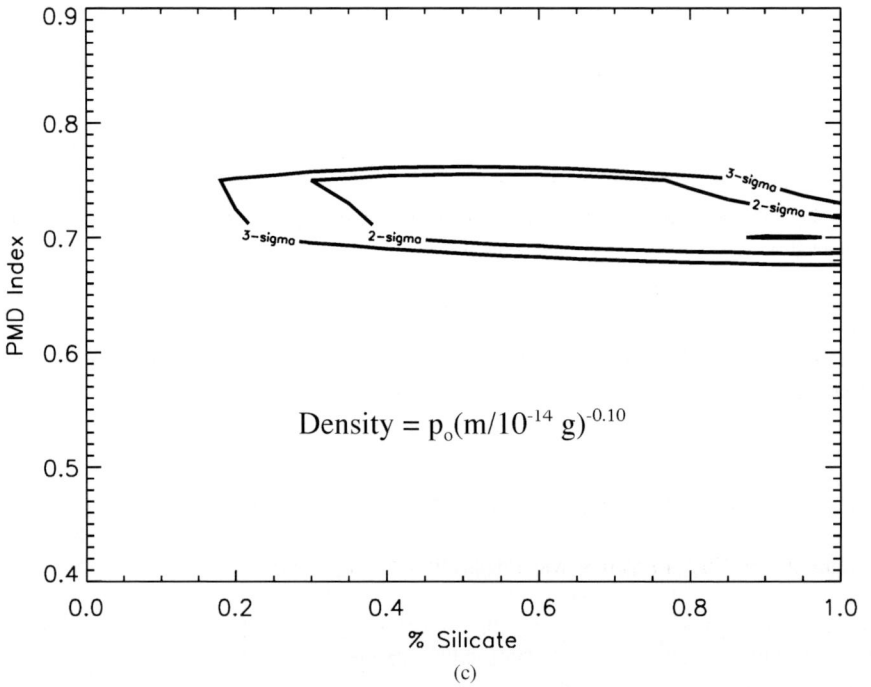

Figure 5. (*Continued*)

models, it is found that the nucleus flux contribution is trivial compared to the dust flux. This is consistent with the longterm lightcurve shown in Figure 3 – at the time of the IRAS observations (0 to +60 days) the signal from the dust is still a factor of 5–7 times higher than the asymptotic flux found when the comet is at aphelion (depicted by the horizontal solid line) that represents the contribution from the nucleus.

The allowed range of dust models (Figures 5b and 5c) is strongly constrained by the combination of mid-IR and optical photometry – for size distributions with an underabundance of small particles, little scattering occurs and the optical flux is too low; for size distributions with an overabundance of small particles, the predicted optical flux is too high, the dust is too warm, and a strong silicate feature (for non-zero silicate compositions) is predicted. At the 95% confidence level only models with size distribution $dn/d\log m \sim m^{-\alpha}$, $0.65 < \alpha < 0.75$, density $\rho = \rho_o(a/0.1\,\mu m)^{-0.25\,\text{to}\,-0.40}$ (fractal dimension $D = 2.76 \pm 0.06$), and nucleus of radius r_{eff} between 2 and 3.5 km, were consistent with the observed thermal emission spectrum. (This implies coma dust particles with effective density ranging from $2.5\,\text{g cm}^{-3}$ at radius $0.1\,\mu m$ to $0.44\,\text{g cm}^{-3}$ at radius 1 cm). The silicate:carbonaceous composition ratio was poorly determined by the essentially featureless SED. No water ice was required for the best-fit models, as expected for warm outer coma dust inside the solar system ice line at ~ 3 AU (Lien, 1990). The allowed particle size distributions (PSDs) are surface-area and mass-dominated by the largest particles ($a > 100\,\mu m$). The emission rate of dust released by the comet was found to be in the range $100–390\,\text{kg s}^{-1}$ ($\pm 2\sigma$), depending on the model adopted.

Comparison to 11 other comets we have photometrically surveyed for dust properties (Lisse *et al.*, 2002, 2004; Figure 6) shows Tempel 1 to be a typical short-period comet of average dust activity, composition, and size, emitting mainly large (10–1000 μm), dark (average albedo $= 8 \pm 2\%$) particles. The higher albedo found for the coma dust (8%) versus the nucleus surface (5%) suggests a relative enrichment in dark absorbing material on the surface of the nucleus as compared to the composition of the coma dust, an effect which could be caused by surface mantling.

Practical Implications for the Deep Impact Mission

SPACECRAFT TRAJECTORY AND TARGET ACQUISITION

The dust optical depth for the coma (and its associated trail) are low from the best-fit greybody models for the observed SED's, less than 3×10^{-7}, so no obscuration of the nucleus by the dust is expected in the Deep Impact flyby images during encounter. We can, however, expect that the pixels detecting emission from the nucleus will also be detecting coma emission. On the spacecraft's

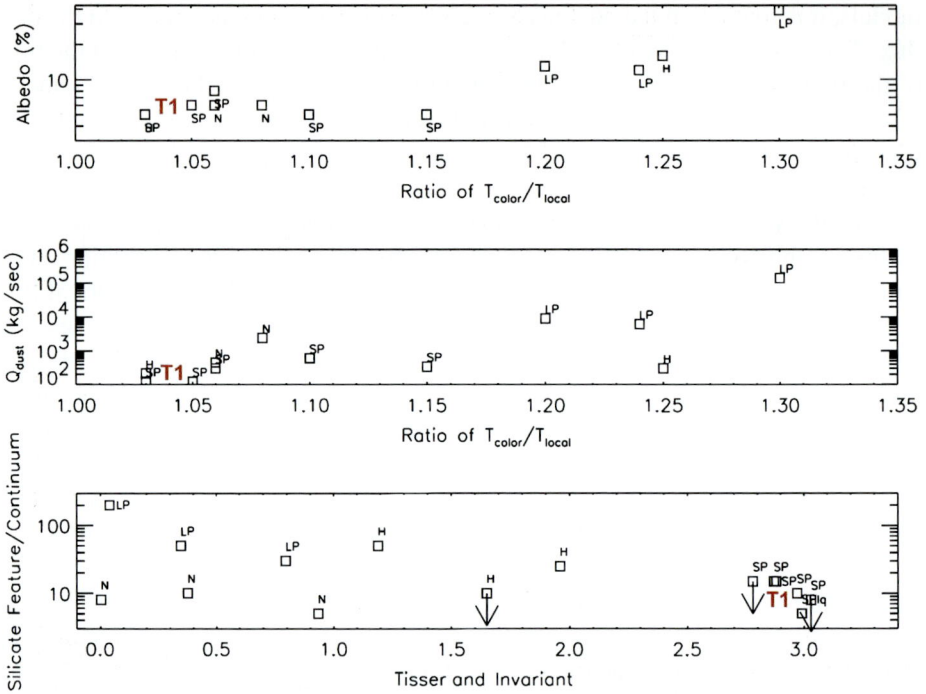

Figure 6. Comparison of Tempel 1 dust properties to those of other comets. Correlations in the observations of emitted cometary dust by albedo, silicate feature amplitude/continuum ratio, dust color temperature, Q_{dust}, and Tisserand invariant of the parent comet from the Lisse *et al.* (2002) survey. Long-period comets produce the most dust with the hottest color temperature, highest albedo, and largest silicate feature. The new and short-period comets produce little dust, and what they do produce is cold and low albedo, with little or no silicate feature. The Halley comets lie between the two extremes. Tempel 1 fits in this scheme as a typical short-period comet of average dust activity albedo, and particle size. (T1's location in the dust survey is denoted by the red T1 label.)

approach, the coma emission will dominate over the nuclear emission until the spacecraft is close enough to the comet that the effective dust surface area in a projected pixel area drops below that of the nucleus surface area contained in the pixel.

An estimate of the measured signal per pixel is given in Figure 7. Here we have assumed an 8% dust scattering albedo and a Halley-like phase law for the dust, a 5% geometric albedo for nuclear scattering, a phase-angle dependence for nuclear scattering found for 2P/Encke 1997 (Fernandez *et al.*, 2000), a dust production rate of $230 \, \text{kg s}^{-1}$, and DI High-Resolution Imager pixel dimensions of $4 \, \mu\text{rad}^2$. The nucleus signal flattens out when the nucleus becomes resolvable, i.e., when the signal per pixel becomes a nucleus surface brightness issue, and not the distance to a point source. Coincidentally, but independently, the signal for the coma begins to flatten out when we start to traverse inside the coma, at about

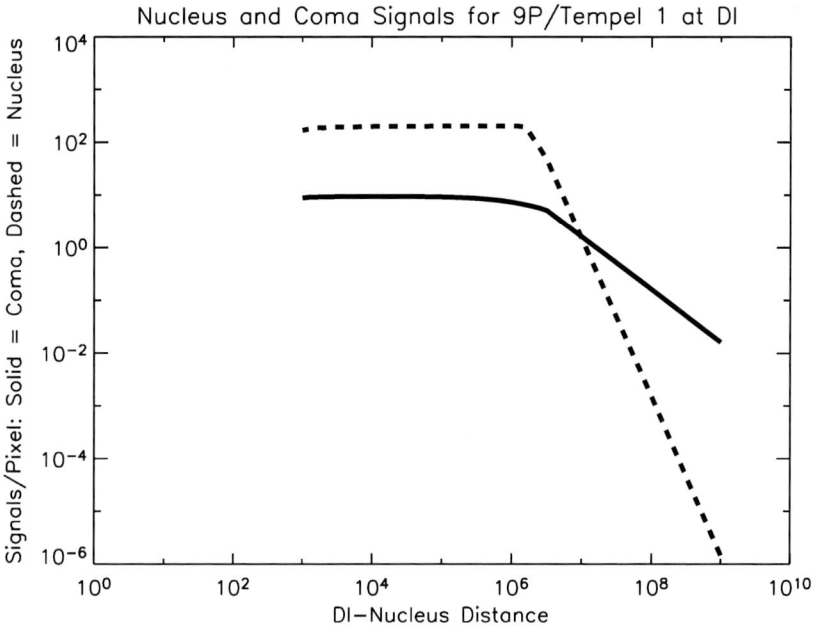

Figure 7. Nucleus/Coma Contrast. Estimated brightest pixel V-band fluxes for 9P/Tempel 1 in the HRI imager during the Deep Impact encounter, as a function of the spacecraft–nucleus distance. The large changes in slope of the two curves occur when the nucleus and coma become resolvable. Each curve has been corrected for the estimated scattering phase function. The horizontal axis has units of km, and the vertical axis has units of number of photons $cm^{-2} s^{-1} \mu m^{-1}$.

1×10^6 km distance to the nucleus. The nuclear signal becomes comparable to the coma signal at a distance of $\sim 8 \times 10^6$ km or $\sim 8 \times 10^5$ s (9.1 days) before impact.

As a check of this calculation, we scaled our Tempel 1 model to the circumstances for the Deep Space 1 flyby of P/Borrelly on September 22, 2001 [$v_{encounter} = 16.5$ km s^{-1}, $Af\rho = 646$ cm at perihelion/encounter, flyby distance = 2000 km, DS-1 pixels ~ 5 times larger than the DI HRI pixels, $r_{nucleus} = (4$ km $\times 2$ km)], and found that the nucleus signal should have been comparable to the coma signal at about 1×10^5 s or 1.6×10^6 km before closest approach. The DS-1 mission detected the nucleus of comet 19P/Borrelly at about 3×10^6 km from the nucleus, in good agreement with our model predictions (D. Yeomans, private communication).

The dust coma morphology in the IRAS thermal infrared images appears relatively isotropic, with at most one pronounced jet of moderate strength (Figure 3). There is thus no expected region of minimized coma dust density for the spacecraft to fly through. The trail emission discussed earlier, however, is clearly much brighter "behind" the comet, i.e. along the orbital path in the direction the comet has already traveled, with an expected maximum in the orbital plane of the comet. The trail width from the IRAS images is on the order of 1×10^5–2×10^5 km, similar to the width of the Encke trail found by Reach *et al.* (2000) using the Infrared Space Observatory (ISO). Thus the Deep Impact flyby spacecraft will be traversing the trail at closest approach. The overall risk to the flyby spacecraft can thus be minimized by choosing an encounter geometry that crosses through the orbital plane as late as possible, in the direction of the orbit that the comet has just traversed, after the impactor hits the surface and crater formation has occurred.

Mission Hazard Minimization and Expected Dust Burden

The two parts of Deep Impact, the impactor and the flyby spacecraft, will each take separate, close trajectories through the coma of comet P/Tempel 1. The spacecraft will encounter dust and gas emitted from the comet's surface at roughly $10 \, \text{km s}^{-1}$ relative speed of the spacecraft with respect to the nucleus and at distances up to 10^4–10^5 km from the nucleus. In order for the DI mission to succeed in returning useful science data, it is important for the spacecraft design to be able to deal with the effects of the coma environment on the spacecraft. The most critical of these effects arises due to dust impacts on the spacecraft at high relative velocities, which can cause both damage to the spacecraft and its instruments and tumbling of the spacecraft orientation. Here we discuss the expected number of dust grain impacts on the two spacecraft.

   The fundamental problem in estimating the amount of dust hitting the DI spacecraft, or dust flux, is determining the spatial number density (number per meter^3) of dust of radius **a** (cm) in the coma around the comet's nucleus as a function of distance from the comet. We must integrate along the spacecraft trajectory through this spatial distribution to find the total dust "fluence." The impactor will encounter dust along a trajectory through the coma, modeled as a line of sight (LOS) spacecraft-comet distance that ranges from a value of infinity to 3 km, where the impactor encounters the nuclear surface. The flyby spacecraft will encounter dust along a path through the coma with an impact parameter of 500 km as well as a possible enhancement in the size and number density of the grains, when it crosses the comet's orbital plane and encounters the dust trail.

   To estimate the dust impact burden on the Deep Impact spacecraft during the mission, we have constructed a model coma assuming hemispherical sunward driven emission, a dust particle size distribution obtained from the IRAS spectral modeling results, and total amount of dust surface area at perihelion normalized to the *Afρ* curve of Figure 3. We find a total estimated dust particle fluence for the DI flyby spacecraft traveling from infinity to an impact parameter of 500 km as shown in Figure 8. The ordinate represents the total predicted number of particles of each mass encountered along a chord through the coma. The results for the impactor spacecraft will be exactly the same, scaled by a factor of (500 km/3 km) minimum distance X 0.5 (since the impactor traverses only one-half the coma). The total

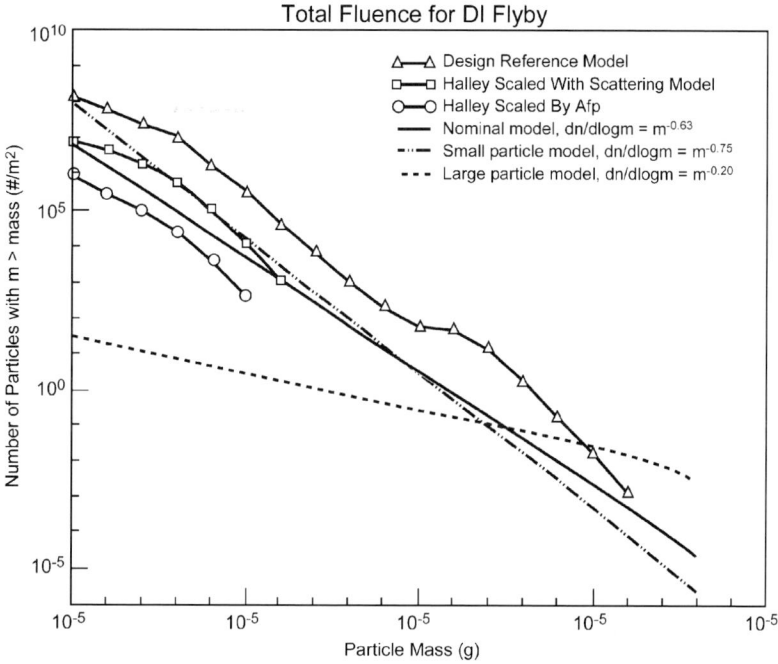

Figure 8. Estimated dust burden on the DI spacecraft during the Tempel 1 coma transit. Maximum $(m^{-0.76})$ and minimum $(m^{-0.63})$ small-particle-loaded particle mass distributions (PMDs) are consistent with the SED of Figure 5. Also shown is an example of a model that fit the mid-IR IRAS data well, but failed to reproduce the optical $Af\rho$ observation by more than an order of magnitude $(m^{-0.20})$. An independent measure of the number of small particles extant has been made using the *in situ* Halley encounter PMD of McDonnell *et al.* (1991), which is then scaled by the relative $Af\rho$ measures of Halley and Tempel 1 and forced to agree with the scattering model. There is good agreement between the two independent calculations. The range of relatively large-particle-dominated best-fit PMDs is consistent with the low dust albedo ($<8\%$) and dust trail observed for the comet.

estimated burden of dust on the Deep Impact flyby spacecraft is $\sim 0.01\,\mathrm{g\,m^{-2}}$, and on the impactor is $\sim 2\,\mathrm{g\,m^{-2}}$.

For comparison, we include the results of Landgraf *et al.*'s (1999) radial model for the Stardust encounter with comet P/Wild 2. This model assumes a Halley size distribution (Lamy *et al.*, 1987), a constant particle density of $1\,\mathrm{g\,cm^{-3}}$, coma structure driven by solar insolation and water and carbon monoxide ice sublimation, and a $70°$ phase angle, as expected for Stardust upon approach to comet P/Wild 2. We have scaled this model to DI at Tempel 1 using the relative activity ratio $Af\rho$ (Tempel 1)/$Af\rho$ (Wild 2) = (150/427), and the relative distance of closest approach of the two spacecraft, impact parameter (DI)/impact parameter (Stardust) = (500 km/100 km). It is interesting to note that while the predicted results for the radial models differ in their predictions for the fluence by a factor of a few in the medium mass ranges, the *total* mass integrated over all grain sizes is quite similar for both, within 30% of 0.01 g m^{-2} at 500 km. The difference is that the two input dust model particle size distributions have very different large particle densities – most of the McDonnell Halley mass is in the abundant, large ($>10\,\mu$m) particles, while the Lamy Halley distribution has more smaller (1–5 μm) particles. The larger particles represent a bigger hazard to the DI spacecraft, however.

As an empirical check, we scaled the Tempel 1 model to the circumstances for the DS-1 flyby of P/Borrelly on September 22, 2001 ($Af\rho = 646$ cm at encounter, flyby distance $= 2000$ km). Our prediction for the number of dust particles detectable by perturbations of the spacecraft attitude control system, i.e. dust particles with mass ≥ 1 mg, was roughly 0.2 particles total burden, a factor of 10 lower than the DS-1 project dust model. For the actual flyby, no particles hits were detected, consistent with our predictions. Our prediction for the number of small ($\sim 1\,\mu$m) particles encountered was also within a factor of 2 of the number reported by Tsurutani et al. (2004) using the on-board 1 m dipole antenna to detect the transient electric field created by the expanding plasma clouds formed during dust particle hits. In a new study, the Tempel 1 dust model, when scaled to the circumstances of P/Wild 2 at the Stardust encounter ($Af\rho = 427$, flyby distance 230 km), was within a factor of 2 of the actual measured cumulative fluences (Tuzzolino et al., 2004) over the mass range 10^{-11} to 10^{-2} g. The Stardust comparison is ultimately of greater use for the DI mission, as a finite number of particles were detected up to a mass of .01 g, whereas the DS-1 measurements only represent an upper limit for particles greater than 1 μg mass.

Encounter Expectations

KIND OF TARGET

From our analysis, Tempel 1 appears to be an excellent target for the Deep Impact mission, as a "canonical" example of short-period cometary bodies. The gas emission from Tempel 1 is similar to other weak-emission-line comets according to Fink and Hicks (1996) and lies at the lower end of the range for "typical" composition comets (A'Hearn et al., 1995), which includes all of the non-Jupiter family comets and one-third of the Jupiter family comets. From the work presented here, the dust emission from Tempel 1 seems entirely normal with respect to amount and kind compared to other short-period comets with many apparitions (Figure 6) like 2P/Encke, 55P/Tempel-Tuttle, and 73P/Schwassmann-Wachmann 3 (Lisse et al., 2002). Similar to these comets, we also estimate that $>90\%$ of the dust is emitted by the comet into bound, stable solar system orbits, and eventually dissipates into the interplanetary dust cloud through perturbations due to dust–dust collisions and Poynting–Robertson drag (Burns et al., 1979). Much of the heaviest bound dust is associated within a "trail," a stream of dust particles extending along the orbital path of the comet, and is also typical of short-period comets on stable orbits (Sykes and Walker, 1992). The Tempel 1 dust trail was detected by IRAS in 1983 (Figure 4c).

Since the IRAS and optical light curves, when combined with respect to time from perihelion, seem to show the same behavior during the 1983, 1994 and 2000 apparitions (Figure 3), we conclude that the release of dust by Tempel 1 is repeatable from orbit to orbit within a factor of order unity. We can use this knowledge to predict the expected coma activity for the encounter phase of the DI mission in July 2005. Two days before perihelion, ($r_h = 1.51$ AU, $\Delta = 0.89$) $Af\rho$ (5500 Å) and $\varepsilon f\rho$ (12 μm) will be about 130 cm, Q_{dust} will be $\sim$250 kg s^{-1}, $Q_{OH} \sim 1.5 \times 10^{28}$ molecules s^{-1} and the Dust-to-Gas ratio will be $\sim$0.5. Given the broad appearance of the weak fan jet in the coma and an estimate of $\sim$8% active area on the surface (Groussin *et al.*, 2004), we surmise that emitted material is flowing from numerous active areas located across the nucleus surface. Thus, Tempel 1 should more closely resemble comet Wild 2, which had multiple weak sources, than comet Borrelly, which had a few very strong jets.

DI Coma Observations

The Deep Impact spacecraft will observe Tempel 1's coma before, during and after the encounter. Observations obtained during the approach will be used to determine the baseline characteristics of the coma, including its composition, structure and any temporal variations in these properties. Any features that are observed will be traced to their origin at the nucleus to help in understanding the link between the nucleus and the coma, a topic that is not well understood, even with several previous in situ comet missions. With constraints on the comet's baseline behavior, observations obtained during and after the impact will reveal changes in the coma that are produced by the impact. Two different instruments will be used to study the coma (Hampton *et al.*, 2005). The medium resolution imaging (MRI) camera, fitted with narrowband filters $\sim$70 Å wide, will obtain images of different gases (OH, CN and C$_2$) in the coma, as well as images of the dust at several different wavelengths. The high resolution imaging infrared spectrometer (HRI-IR) will measure the spectrum of the coma in the wavelength range 1.0–4.8 μm, at a resolving power $700 > R > 200$. Both instruments will provide the highest spatial resolution ever obtained on a cometary coma.

Narrowband imaging and optical spectroscopy, both discussed earlier, have shown that many different species (OH, CN, NH, and NH$_2$) exist in the coma of Tempel 1. Their parent species, H$_2$O, HCN, and NH$_3$ must exist as well. The Vega IKS spectrum of comet 1P/Halley 1986 (Combes *et al.*, 1988) and the ISO PHT-S spectrum of comet C/Hale-Bopp 1995 O1 (Lellouch *et al.*, 1998) represent by far the best knowledge of the neutral species we can expect to find in the cometary coma. The gaseous molecular species H$_2$O, CO, CO$_2$, CH$_3$OH, H$_2$CO (and possibly OCS) were detected from 2–5 μm in these spectra. As the Deep Impact spectrometer will be observing Tempel 1 at higher spatial and spectral resolution and with much better sensitivity than were obtained at either Halley or Hale-Bopp, and since it is of "typical" composition, as were Halley (A'Hearn *et al.*, 1995) and Hale–Bopp (Schleicher, private communication), it is plausible to expect these species to appear in the coma of Tempel 1 as well. Simulations by Sunshine *et al.* (2005) indicate that we should detect H_2O at several wavelengths, CO_2 at 4.2 μm, two peaks of CO at 4.5 μm and possibly another at 2.15 μm. Emission lines in the 3.1–3.6 μm region due to the C–H aliphatic stretching mode will be obtained, mixed in with CH_3OH broad emission lines at 3.4 μm and H_2CO at 3.6 μm. However, as these simulations were based on observations of Oort-cloud comets, whereas Tempel 1 is a Jupiter-family comet with Kuiper belt origin, some differences in the observed chemistry might be expected that could be related to place of formation. Dust should be easily detectable in either case as scattered and thermal continuum IR emission in the spectrum.

Using the multiple gas filters of the DI MRI camera, we expect to have a global overview of the coma with a spatial resolution of a few kilometers per pixel. With the images obtained, the spatial distribution of the majority gas species will be used to help in tracing the sources of the species, as well as providing measurements that can be used to study the photodissociation of their parent molecules (e.g. H_2O and CO). Dust jets will be imaged from the outer to the inner coma and down to their roots on the comet surface. Observations obtained with the Deep Impact infrared spectrometer will allow an accurate determination of the volatile production rates of the coma before and after impact. A large increase in observed CO production due to impact excavation is a likely possibility. The high spatial resolution available with the spectrometer (five times greater than with the MRI camera) also allows us to explore the near nucleus region that is unresolvable from the ground. This should allow DI to detect species with relatively short lifetimes, such as SO_2 and NH_3. These species have very weak emission lines but are located in a very clear region of the spectrum (4.0 and 2.3 μm, respectively) that favors their detection. The high spatial resolution will also allow us to compare the composition of the individual jets to each other and to the general coma composition.

The data return strategy (Klaasen *et al.*, 2005), constrains the observations that are possible, so between impact and the time the DI flyby enters shield mode (800 s after impact) only the infrared spectrometer will be used to study the gas coma. The main objective at this time will be to track changes in the composition of the coma and the potential appearance of new volatiles, and an off-nucleus scan will be performed for that purpose. Because the impact should expose fresh subsurface ices, we expect the coma to be enriched in highly volatile molecules such as CO and CO_2. From 44 min to 2.5 days after impact, after the impact science data has been telemetered to the ground, the lower constraints on data storage and transmission again allow us to observe the coma with the MRI narrowband filters as well as with the infrared spectrometer. During this phase, changes in the coma morphology and composition will be monitored to study the longer-term effects of the impact as the comet returns to a steady-state condition. In particular, we will focus on any new jets to determine their number (one or more), locations (relative to the impact site), and structure (collimated or not). The MRI images will allow us to measure any enrichment in OH and/or carbon-bearing molecules after impact and to determine if the variations of composition in the coma are global or are restricted to the impact region. The production rates of H_2O, CO, CO_2, CH_3OH, H_2CO (and possibly other molecules) will be determined from the infrared spectrometer. The ultimate goal is to compare the production rates before and after impact, to study the effect of the impact on the gas coma and the relation between the coma and the nucleus. If the impact has no effect at all on the gas coma, which would be very surprising, it will still be very interesting, as it will oblige us to review our current view of comets.

LONG-TERM EFFECTS OF THE IMPACT ON THE COMA

Does the Deep Impact experiment effectively end 2.5 days after impact, when the flyby stops observing the comet? Given our derived estimate for the coma dust surface brightness, we can make a conservative estimate of how long the ejecta from the Deep Impact encounter will be detectable. (Note that this analysis ignores the possibility that a new, long-lived, active region will be created, which is a distinctly possible outcome. If a new source is created, this will only increase the duration of the impact-derived effects.)

We derive an estimate of the "impact duration" based on the time at which the spatial density of the impact ejecta has dropped to the level of the density of the nominal coma. The current best crater model predicts a crater some 100 m in diameter and 20 m deep. Assuming an ellipsoidal final crater, a bulk density for the nuclear material near the surface of $0.4\,\mathrm{g\,cm^{-3}}$, a dust-to-gas ratio of 2, and $\sim10\%$ of the excavated material moving at speeds higher than the local escape velocity (Richardson, 2003, private communication), we can expect some $M_{\mathrm{expl}} = 2\times10^7\,\mathrm{kg}$ of dust ejected into the coma due to the impact. Assuming that one-half of the crater material is ejected into a sold angle of $2/3\pi$ (Richardson, 2004, private communication; Schultz et al., 2005), expanding at velocity v_{expl} with the same particle size distribution as the nominal coma dust, we have

$$n(r)_{\mathrm{expl}} = \frac{(1/2) \times M_{\mathrm{expl}}}{(2/9\pi(v_{\mathrm{expl}} \times t)^3)}$$

where t is the time since impact. Setting this quantity equal to the coma dust density from the comet's normal outgassing

$$n(r)_{\mathrm{dust}} = \frac{Q_{\mathrm{dust}}}{4\pi r^2 v_{\mathrm{dust}}}$$

we find that the time after impact when the ejecta shell is reduced to the density of the normal coma is given by

$$t_{expl} = 9 \times \frac{M_{expl}}{Q_{dust}} \times \frac{v_{dust}}{v_{expl}}.$$

Clearly, the more dust that is ejected and the slower the ejection rate, the longer the time the impact perturbation is significant. We find an ejecta velocity of $v_{expl} \sim 45\,\mathrm{m\,s^{-1}}$ by assuming all the kinetic energy of the impactor is converted into kinetic energy of the ejected dust (the recoil energy of the $\sim 10^{14}\,\mathrm{kg}$ comet nucleus is negligible):

$$E_{impactor} = 0.5 \times 350\,\mathrm{kg} \times (10\,\mathrm{km\,s^{-1}})^2 = 0.5 \times 2 \times 10^7\,\mathrm{kg} \times v_{expl}^2.$$

where $v_{expl} \gg v_{escape} \sim 2\,\mathrm{m\,s^{-1}}$, as required for material ejected onto hyperbolic escape velocities into the coma. From the results of experimental cratering studies, Schultz *et al.* (2005) argue for a lower speed for the bulk of the ejecta, $\sim 10\,\mathrm{m\,s^{-1}}$, which is emitted late in the crater formation process, but still in the same order of magnitude as the velocity derived from the simplistic energy balance equation above.

Assuming a compromise average value of $v_{dust} = 20\,\mathrm{m\,s^{-1}}$ and a normal coma dust emission rate of $Q_{dust} = 250\,\mathrm{kg\,s^{-1}}$ (the average value from our modeling of the IRAS data), we find $t_{expl} \sim 27$ days for the density equilibration time. The persistence of the Tempel 1 ejecta signal is due to the large amount of material thrown into the coma at very low speeds above the escape velocity from the impact. How does this compare to other timescales that are observed in the coma? It is short compared to the 115 days estimated time required to cross the observable coma (2×10^5 km; Figure 4) at $20\,\mathrm{m\,s^{-1}}$, and the ~ 100 day lifetime of particles lofted into temporary, near-nucleus orbits (Scheeres and Mazari, 2000). It is long compared to the 5–10-day lifetimes of coma "shells" seen in other comets (e.g. Halley in 1986; Larson *et al.*, 1987; Hale-Bopp in April 1996; Lisse *et al.*, 1999). The 27-day timescale is also long compared to the ~ 1-day timescale for the detection of the large, actively outgassing fragments emitted by comets C/Hyakutake 1996 B2 and C/1999 S4 (LINEAR).

If the pieces of ejected nucleus outgas rapidly, we will see a brighter ejecta structure of shorter duration, as the fragments disperse as they evaporate and accelerate under jet forces. Our estimate of how long the impact ejecta will be visible is also uncertain to $\sim 50\%$ in the ratio of v_{dust}/Q_{dust}, and to a factor of a few in the ratio of M_{expl}/v_{expl}. Given these uncertainties, we believe a conservative range of $5 < t_{expl} < 100$ days is best adopted for planning purposes. Observations of the progress of the ejecta shell through the coma should be designed using this time range.

To answer the question posed at the beginning of this section: The effects of the DI experiment will be observable for at least 5 days, and for as long as 100 days, after the time of the last DI flyby spacecraft observation. Extremely useful information on the ejected particle sizes, compositions, and velocities will be obtained from observations of the progression of the ejecta through the coma, both from the DI flyby spacecraft and from ground-based telescopes. The role of other observing platforms will be critical at this point in time to understanding the nature of the coma and its origin in the material comprising the nucleus of the comet.

Conclusions

Previous observations of the gas emission from comet 9P/Tempel 1 have found its composition to be at the lower end of the range of "typical" comets in carbon-chain molecules relative to water. As the gas emission behavior seems relatively stable from apparition to apparition, we expect a total outgassing rate $\sim 1 \times 10^{28}$ molecules per second and a gas composition of $\log(CN/OH) \sim -2.83$, $\log(C_2/OH) \sim -2.90$, $\log(C_3/OH) \sim -4.06$, and $\log(NH/OH) \sim -2.55$ at encounter.

The dust emission from Tempel 1 is very similar to that seen from other short-period comets that have had a large number of perihelion passages. At encounter, the comet will be emitting ~ 250 kg/s of dust into the coma, with the majority of the particle surface area and mass in particles with radius $\mathbf{a} > 20$ μm. The pre-encounter Dust-to-Gas rated will be ~ 0.5.

The long-term activity of the comet, as depicted in the continuum lightcurve, shows a peak in the activity around 60 days before perihelion. Furthermore, the onset of activity is much steeper than the decrease after the peak, causing an additional asymmetry around the peak itself. This behavior is likely due to the competing effects of seasonal variations and the increasing solar irradiance as the nucleus gets closer to the sun. The seasonal effect produces the early peak, while the increased amount of sunlight at perihelion causes the overall emission to be higher than it was before the peak. It is likely that the emission from the observed jet is related to the seasonal effect and is contributing to the increased activity level. This would indicate that the sun is at its highest elevation, as seen from the source, at -60 days. Although no jets have been detected at other times, this may simply reflect the fact that there are only a limited number of high-quality, high-resolution images available.

The presence of a jet indicates that there is one or more isolated sources on the nucleus, even though the coma generally has an isotropic appearance. This result is consistent with thermophysical models of the nucleus surface (Groussin et al., 2004) that indicate only 8% of the surface area is active. This small active fraction further hints that the isotropic component of the coma may be produced by a number of small, diffuse jets that mimic an isotropic shape (e.g., comet Wild 2, Sekanina et al., 2004).

The nucleus will be detectable against the coma background from DI at ~ 9 days before impact in July 2005. A total dust fluence of ~ 2 g at 10 km s^{-1} per m^2 of projected spacecraft surface area will be experienced by the DI impactor, and

0.02 g per m^2 by the DI flyby (with minimum approach distance of 500 km to the nucleus). The ejecta should be detectable against the coma background for some time after the impact. From simple arguments, we estimate the time for the ejecta to fade into the existing coma will be as little as 5 days and as much as 100 days after impact.

Acknowledgments

The authors wish to thank A. Cochran, P. Feldman, and R. Walker for kindly supplying their archival observations of Tempel 1, Stephanie McLaughlin for helping to produce the long term lightcurves, and S. Kido for mastering the figure graphics. Comments by A. Cochran and an anonymous reviewer were very helpful in improving the manuscript. In pursuing this work, the authors were partially supported by the Deep Impact project, and Planetary Astronomy Grants NAG5-9006 and NAG-11468.

References

A'Hearn, M. F., Schleicher, D. G., Feldman, P. D., Millis, R. L., and Thompson, D. T.: 1984, *Astron. J.* **89**, 579.
A'Hearn, M. F., Millis, R. L., Schleicher, D. G., Osip, D. J., and Birch, P. V.: 1995, *Icarus* **118**, 223.
Belton, M. J. S., Meech, K. J., A'Hearn, M. F., Groussin, O., McFadden, L. A., Lisse, C. M., *et al.*: 2005, *Space Sci. Rev., ibid.*
Bockelee-Morvan, D., Crovisier, J., Mumma, M. J., and Weaver, H. A.: 2005, in Festou, M., Keller, U., and Weaver, H. (eds.), *Comets II*, University of Arizona Press, Tucson.
Boice, D. and Huebner, W.: 2005, in Festou, M., Keller, U., and Weaver, H. (eds.), *Comets II*, University of Arizona Press, Tucson.
Burns, J. A., Lamy, P. L., and Soter, S.: 1979, *Icarus* **40**, 1.
Campins, H., Walker, R. G., and Lien, D. J.: 1990, *Icarus* **86**, 228.
Cochran, A. L., Barker, E. S., Ramseyer, T. F., and Storrs, A. D.: 1992, *Icarus* **98**, 151.
Combes, M., *et al.*: 1988, *Icarus* **76**, 404.
Combi, M. R., Harris, W. R., and Smyth, W. H.: 2005, in Festou, M., Keller, U., and Weaver, H. (eds.), *Comets II*, University of Arizona Press, Tucson.
Crifo, J. F., Fulle, K. M., and Szego: 2005, in Festou, M., Keller, U., and Weaver, H. (eds.), *Comets II*, University of Arizona Press, Tucson.
Farnham, T. L. and Schleicher, D. G.: 2005, *Icarus* **173**, 533.
Farnham, T. L., Schleicher, D. G., and A'Hearn, M. F.: 2000, *Icarus* **147**, 180.
Fernandez, Y. R., Lisse, C. M., Ulrich, K. H., Peschke, S. B., Weaver, H. A., A'Hearn, M. F., *et al.*: 2000, *Icarus* **147**, 145.
Fernandez, Y. R., Meech, K. J., Lisse, C. M., A'Hearn, M. F., Pittichova, J., and Belton, M. J. S.: 2003, *Icarus* **164**, 481.
Feldman, P. and Festou, M.: 1991, in Harris, A. W. and Bowell, E. (eds.), *Proceedings of ACM 1991 on IUE Observations of Periodic Comets Tempel 2, Kopff, and Tempel 1*, Lunar and Planetary Institute, Houston, pp. 171.
Feldman, P., Combi, M., and Cochran, A.: 2005, in Festou, M., Keller, U., and Weaver, H. (eds.), *Comets II*, University of Arizona Press, Tucson.

Fink, U. and Combi, M. R.: 2004, *Plan. Space Sci.* **52**, 573.

Fink, U. and Hicks, M.: 1996, *ApJ* **459**, 729.

Finson, M. L. and Probstein, R. F.: 1968, *Astron. J.* **154**, 327.

Groussin, O., Lisse, C. M., A'Hearn, M. F., Belton, M. J., Fernandez, Y. R., van Cleve, J. E., *et al.*: 2004, Thermal Properties of Deep Impact Target Comet 9P/Tempel 1, COSPAR Meeting 35, abstract B1.1 002204.

Hampton, D. L., Baer, J. W., Huisjen, M. A., Varner, C. C., Delamere, A., Wellnitz, D. D., *et al.*: 2005, *Space Sci. Rev., ibid.*

Harker, D., Woodward, C. E., Wooden, D. H., Witteborn, F. C., and Meyer, A. W.: 1999, *Astron. J.* **118**, 1423.

Jessberger, E. K., Christofordis, A., and Kissel, J.: 1988, *Nature* **322**, 691.

Jessberger, E. K. and Kissel, J.: 1991, in Newburn, Neugebauer, and Rahe (eds.), *Comets in the Post-Halley Era*, Kluwer Academic Publishers, Dordrecht, pp. 1075.

Klaasen, K. P., Carcich, B., Carcich, G., Grayzeck, E. J., and McLaughlin, S.: 2005, *Space Sci. Rev., ibid.*

Krasnopolsky, V. A., Moroz, V. I., Krysko, A. A., Tkachuk, A. Y., and Moreels, G.: 1987, *Astron. Astrophys.* **187**, 707.

Lamy, P. L., Grün, E., and Perrin, J. M.: 1987, *Astron. Astrophys.* **187**, 767.

Landgraf, M., Müller, M., and Grün, E.: 1999, *Plan. Space Sci.* **47**, 1029.

Larson, S., Sekanina, Z., Levy, D., Tapia, S., and Senay, M.: 1987, *Astron. Astrophys.* **187**, 639.

Lebofsky, L. A. and Spencer, J. R.: 1989, *Asteroids II*, University of Arizona Press, Tucson, pp. 128.

Lellouch, E., Crovisier, J., Lim, T., Bockelee-Morvan, D., Leech, K., Hanner, M. S., *et al.*: 1998, *Astron. Astrophys.* **339**, L9.

Lien, D. J.: 1990, *Astron. J.* **355**, 680.

Lisse, C. M., Freudenreich, H. T., Hauser, M. G., Kelsall, T., Moseley, S. H., Reach, W. T., *et al.*: 1994, *ApJ* **432**, L71.

Lisse, C. M., A'Hearn, M. F., Hauser, M. G., Kelsall, T., Lien, D. J., Moseley, S. H., *et al.*: 1998, *ApJ* **496**, 971.

Lisse, C. M., *et al.*: 1999, *Earth, Moon, Planets* **78**(1/3), 251.

Lisse, C. M., Fernandez, Y. R., A'Hearn, M. F., and Peschke, S. B.: 2002, in Green, S. F., Williams, I. P., McDonnell, J. A. M. (eds.), *COSPAR Colloquia Series Vol. 15, Dust in the Solar System and Other Planetary Systems*, pp. 259.

Lisse, C. M., A'Hearn, M. F., Fernandez, Y. R., Gruen, E., Kaufl, H. U., Kostiuk, T., *et al.*: 2004, *Icarus* **171**, 444.

Lynch, D. K., *et al.*: 1995, *Icarus* **114**, 197.

Marsden, B. G. and Williams, G. V.: 2003, *Catalogue of Cometary Orbits*, XVth edn. Central Bureau for Astronomical Telegrams and Minot Planet Center, Cambridge, Massachusetts.

Mason, C. G., Gehrz, R. D., Ney, E. P., Williams, D. M., and Woodward, C. E.: 1998. *Astron. J.* **507**, 398.

McDonnell, J. A. M., Lamy, P. L., and Pankiewicz, G. S.: 1991, in Newburn, Neugebauer, and Rahe (eds.), *Comets in the Post-Halley Era (Bamberg Meeting Proceedings)*, Kluwer Academic Publishers, Dordrecht, pp. 1043.

Meech, K. J., A'Hearn, M. F., Lisse, C. M., Weaver, H. A., and Biver, N.: 2005, *Space Sci. Rev., ibid.*

Ney, E. P. and Merrill, K. M.: 1976, *Science* **194**, 1051.

Osip, D. J., Schleicher, D. J., and Millis, R. L.: 1992, *Icarus* **98**, 115.

Reach, W. T., *et al.*: 1999, *Icarus* **148**, 80.

Reach, W. T., Sykes, M. V., Lien, D., and Davies, J. K.: 2000, *Icarus* **148**, 80.

Scheeres, D. J. and Mazari, F.: 2000, *Astron. Astrophys.* **356**, 747.

Schleicher, D. G. and Farnham, T.: 2005, in Festou, M., Keller, U., and Weaver, H (eds.), *Comets II*, University of Arizona Press, Tucson.

Schultz, P. H., Ernst, C. M., and Anderson, J. L. B.: 2005, *Space Sci. Rev.*, *ibid.*

Sekanina, Z.: 1981, *Astron. J.* **86**, 1741.

Sekanina, Z., Brownlee, D. E., Economou, T. E., Tuzzolini, A. J., and Green, S. F.: 2004, *Science* **304**, 1769.

Storrs, A. D., Cochran, A. L., and Barker, E. S.: 1992, *Icarus* **98**, 163.

Sunshine, J. M., A'Hearn, M. F., Groussin, O., McFadden, L., Klaasen, K. P., Schultz, P. H., *et al.*: 2005, *Space Sci. Rev.*, *ibid.*

Sykes, M. V., Lien, D. J., and Walker, R. G.: 1990, *Icarus* **86**, 236.

Sykes, M. V. and Walker, R. G.: 1992, *Icarus* **95**, 180.

Tokunaga, A. T., Hanner, M. S., Veeder, G. J., and A'Hearn, M. F.: 1984, *Astron. J.* **89**, 162.

Tsurutani, B. T.: 2004, *Icarus* **167**, 89.

Tuzzolino, A. J., *et al.*, 2004, *Science* **304**, 1776.

Yeomans, D. K., Giorgini, J. D., and Chesley, S. R.: 2005, *Space Sci. Rev.*, *ibid.*

COMET GEOLOGY WITH DEEP IMPACT REMOTE SENSING

PETER C. THOMAS[1,*], JOSEPH VEVERKA[1], MICHAEL F. A'HEARN[2],
LUCY MCFADDEN[2], MICHAEL J. S. BELTON[3] and JESSICA M. SUNSHINE[4]

[1]*Center for Radiophysics and Space Research, Cornell University, Ithaca NY, U.S.A.*
[2]*University of Maryland, College Park, MD, U.S.A.*
[3]*Belton Space Exploration Initiatives, LLC, Tucson, AZ, U.S.A.*
[4]*Science Applications International Corporation, Chantilly, VA, U.S.A.*
(*Author for correspondence; e-mail: thomas@baritone.astro.cornell.edu)

(Received 20 August 2004; Accepted in final form 3 December 2004)

Abstract. The Deep Impact mission will provide the highest resolution images yet of a comet nucleus. Our knowledge of the makeup and structure of cometary nuclei, and the processes shaping their surfaces, is extremely limited, thus use of the Deep Impact data to show the geological context of the cratering experiment is crucial. This article briefly discusses some of the geological issues of cometary nuclei.

Keywords: comets, geology, impacts, structure

1. Introduction

The long history of comet observations with telescopes contrasts dramatically with the very brief, recent period of close-up observations of comet nuclei from spacecraft. Fleeting views of comets Halley, Borrelly and Wild-2 have been obtained by flybys, but only the latter data have sufficient resolution to address geological questions concerning the morphology and surface processes of comet nuclei. Deep Impact seeks to explore actively a comet nucleus by impact excavation (A'Hearn *et al.*, this issue). The dynamic and compositional data obtained by observing the impact should provide new information about the properties of comet nuclei. An equally important goal is a detailed remote sensing survey of the nucleus that will put the impact in context, and provide by far the most detailed views of a comet to date. Here we review the investigation strategy and its application to questions concerning the nature and evolution of cometary nuclei. The global properties (including likely rotation state) of the nucleus of Tempel 1, inferred from accumulated telescopic data, are summarized by Belton *et al.* (this issue). The nucleus size, considerably elongated, with a mean radius of approximately 3.4 km, and a spin period of about 2 h, allows nearly half the surface to be imaged at a wide variety of viewing angles and in multiple colors. Here we focus on questions of surface process, morphology and related geological issues.

Space Science Reviews (2005) 117: 193–205
DOI: 10.1007/s11214-005-3391-7

2. Investigation Scheme

The cratering experiment is described by Schultz and Ernst (this issue). The flyby spacecraft, the impactor and the instruments are described by Hampton *et al.* (this issue). The flyby spacecraft carries two imagers, the High Resolution Imager (HRI) and a Medium Resolution Imager (MRI), each provided with a nine-position filter wheel. The respective fields of view and spatial resolution at 1000 km are 2 km, 2 m/pxl, and 10 km, 10 m/pixel. The impactor carries a copy of the MRI, the Impactor Targeting Sensor (ITS) but without color filters, for active targeting and high-resolution images of the impact site. The data taking scheme at encounter (Klaasen *et al.*, this issue) allows for stereo imaging and rapid imaging of the nucleus and of the expanding plume and ejecta from the impact event, which occurs before closest approach of the flyby spacecraft. In addition, it provides for near-IR spectral mapping of the comet surface.

A day before impact, at a range of over 800000 km, the nucleus, about 14 × 5 × 5 km (Belton *et al.*, this issue), is about ~3 HRI pixels across the minimum dimension; see Klaasen *et al.*, in this issue for a detailed presentation of the expected data from Deep Impact. Earlier images yielded data for navigation and coma science. Use of the filter sets in the day before impact will produce a color lightcurve of the comet as it rotates. These data allow a gross comparison of the side seen at high resolution to the rest of the nucleus. The phase angle increases on approach from 28°, reaching 63° close to impact.

On approach only partial images are taken until 13 min before impact. At this time the nucleus, at 30 m/HRI pixel, is about 200 pixels across, and will present almost as much detail as the recent images of Wild-2. However, beginning at 12 min before impact 7-color data will be taken, providing the first ever high-resolution color stereo coverage of a comet nucleus. Color imaging resumes 240 s after impact when the spacecraft has closed to a pixel scale of 12 m/pixel with the HRI, and continues to a highest color scale of 3 m/pixel; here the nucleus should more than fill the HRI frame.

In the few seconds around impact, imaging by both the MRI and HRI is rapid, partial frames to capture impact phenomena at sub-second intervals. The HRI has a pixel scale of 17 m, the MRI, 85 m, at the predicted impact time, which has an uncertainty of about 6 s, and nominally occurs about 850 s before closest approach. Twenty-four seconds after impact full frames alternate with partial frames, as the range to the comet decreases. The expected ~100 m crater (Schultz and Ernst, this issue) and its ejecta plume will be well characterized by the HRI data. The ITS on the impactor spacecraft will give rapid, partial frames of less than 1 m/pixel just before impact, although cometary dust impacts on the optics may degrade the quality of imaging at some point before impact. Partial frames are used to increase the time sampling available, and still cover virtually the entire visible area of the nucleus with the HRI; ITS partial images close to impact have much smaller fields of view.

The changing geometry during the flyby provides the basis for stereo data, with the usual flyby conditions that while lighting conditions remain essentially fixed, resolution changes. Thus, while the highest resolution HRI images will be about 1.4 m/pixel, stereo convergence angles of ~25° are achieved with images of about 3 m/pixel. Four-color stereo of 3 to 6 m/pixel is achieved after impact. Before impact color stereo (7-colors) is achieved between images of about 17 and 30 m/pixel. The highest resolution stereo will include the crater and a significant fraction of the illuminated disk (how much depends on the orientation of the nucleus), the lower resolution stereo will have all the visible disk (less than half the nucleus area). Combination of the ITS images with HRI images provides an additional, higher convergence stereo set, that will at highest resolution cover a small area around the target point (Hampton *et al.*, this issue).

Filters for the MRI (Hampton *et al.*, this issue) are chosen to detect coma components; five of the seven filters in the MRI camera set are narrow bands optimized for coma observations, leaving two wider bands centered at 750 and 850 nm matching those on the HRI. The HRI filters are set at 100 nm intervals between 350 and 950 nm, to determine visible color characteristics without detailed spectroscopy, which investigation is done by the infrared instrument. At visible (.4 to 1 μm) wavelengths, cometary nuclei exhibit a wide range of linear slopes (Jewitt, 2002). Reasons for this diversity remain unclear.

At scales of meters, we will be able to resolve individual morphologic features and units, which may differ in spectral slope. The HRI filters could detect a change in slope toward the ultra violet as often seen in dark (some C, F- and G-type) asteroids. An increased slope toward the infrared would presumably be due to organics. Silicates, if present, would cause a decrease at near 0.9 μm. Finally, a neutral slope with high albedo may be indicative of ice. Any variations in color will also be related to the HRI-IR near-infrared spectra, which while at lower resolution, can provide more detailed constraints on composition (see Sunshine *et al.*, this volume). Correlating color variations with specific features and their physical origins may allow us to better understand the significance of slope diversity observed among the comet population. Moreover, Deep Impact's color images will also, provide an opportunity to compare variations observed on the surface, with those exposed in the interior after our impact event.

The near-infrared, long-slit spectrometer has a slit 10 μrad by 2.5 mrad and spatial pixels of 10 μrad (assuming 2 × 2 binning). It covers the spectral range from 1.05 to 4.8 microns. For details, see Hampton *et al.*, this issue. Spectroscopy on approach is primarily of the coma and is discussed in the article by Sunshine *et al.*, in this issue. The nucleus is an unresolved point source at the center of the slit until well after release of the impactor. Although not spatially resolved, the nucleus will be photometrically resolved at this time, i.e., it will dominate the flux in a single pixel, yielding spectra of the entire illuminated surface as a function of rotational phase. Models of the nucleus, as well as experience with ground-based observations of comets, indicate that the short-wavelength portion of the spectra (1–2.8 μm) will be dominated by reflected sunlight while the long-wavelength (2.8–4.8 μm) portion will be dominated by thermal emission.

After the nucleus becomes spatially resolved, spatial scans are used to map the nucleus with the spectrometer, providing both reflection spectra and thermal emission with varying spatial resolutions. The last pre-impact spectra have a spatial resolution of approximately 100 m on the nucleus; the final spectrum of the impact crater will have a spatial resolution of about 10 m/pixel. With the planned geometry of the encounter, the slit will be oriented parallel to the comet-sun line, with orthogonal spatial scans. Spatial coverage will depend on the orientation of the elongated nucleus, something that can not yet be predicted. The mapping of the reflection spectrum will be used to look for composition variation associated with surface features, while the thermal portion of the spectrum will be used to determine the temperature distribution of the surface layers. Multiple spatial scans will allow a measurement of the phase variation of the spectral reflectance between phase angles of 28–63°.

For later reference we summarize here the best image scales achieved by previous comet flybys. These are about 100 m/pxl for Halley in 1986; 60 m/pxl for Borrelly in 2001; and 15 m/pxl for Wild 2 in 2004. While stereo coverage was obtained both at Borrelly and Wild 2, making it possible to infer surface topography, neither flyby produced high-spatial-resolution color imaging of surface morphology.

3. Surface Morphology of Comets

Spacecraft imaging during the past decade has documented a wide variety of processes that determine the evolution of asteroid surfaces. As noted by Malin (1985), comets, thanks to their activity, hold the potential for even greater diversity in terms of surface processes and morphological forms.

Although considerable effort has been devoted to modeling the activity of comets, most such models do not address surface processes in a comprehensive manner, nor make specific predictions of surface morphology. Unfortunately, images of cometary surfaces obtained to date lack the resolution necessary to test such hypotheses in detail. It is noteworthy, however, that the prediction of sublimation producing a "pit and mesa topography" made by Malin and Zimbleman (1986) seems to have been substantiated by the results obtained by Deep Space 1 at Borrelly (Britt et al., 2004) and by Stardust at Wild 2 (Brownlee et al., 2004).

Expectations for the kinds of topographic forms that may occur on the surfaces of comets have concentrated on forms generated by loss of volatiles from within a matrix of silicates, organics, oxides, or other non-volatile materials (Malin, 1985; Colwell et al., 1990). The loci, longevity, and strength of active areas of sublimation are expected to change, and this has provided the basis for explanations of changing cometary activity (Sekanina, 1990). The sublimation may be in heterogeneous and very loose materials, including voids subject to sudden collapse (Whipple, 1950). Internal structures, by providing pathways for evolved gasses, may be important in non-tidal breakups as well as possibly revealing of cometary assembly (Samarasinha, 2001). The great interest in detecting internal structures has to contend with the complexities of the sublimation effects at the surface and their likely overprinting of internal features.

3.1. ABLATIONAL FEATURES

Ablation of volatile materials and formation of lag deposits should dominate comet nucleus morphology. Only small fractions of nucleus surfaces are active at any one time (Keller, Kramm, and Thomas, 1988; A'Hearn et al., 1995); typically over 90% of the nucleus is observed or inferred to be covered with the very dark (albedo <0.05) mantling materials. As the surface evolves it is expected that the thickness of the mantling lag deposit will vary across the nucleus. While thick deposits will have an insulating effect, thin layers may enhance sublimation (Malin and Zimbelman, 1986; Colwell et al., 1990).

Sublimation of ices on comets with lag deposits of variable thickness may produce topography analogous to that formed in some terrestrial glacial environments (Malin and Zimbelman, 1986; Colwell et al., 1990). These forms include pit and mesa topography and dust mantled cones and mounds that form by inversion of relief (Malin and Zimbelman, 1986). Dust mantles affect the heating of the volatiles at depth (Brin and Mendis, 1979; Grun et al., 1993); they also affect how the gas diffuses to the surface. Porous dust mantles may provide the best insulation, but also may allow the most diffusion of gas from below. It has long been suspected that sections of dust mantles can be blown off, exposing fresh materials and accounting for some cometary outbursts.

Interesting topography may result from inversion of relief and feedback effects on sublimation forms. Inversion of relief occurs when relatively low areas are filled with material that is, or becomes, more resistant to erosional forces than the surrounding materials. On the earth the relevant analogy is covering of snow and ice in low areas by sediment, volcanic ash, or debris flows. The non-volatile covering protects the underlying material from insolation and ablation, and subsequent lowering of surrounding areas yields an inversion of relief: former low areas become relatively high. Inversion can also occur simply due to different amounts of cover. Thin covers can speed up sublimation by lowering the albedo but providing little insulation; thick covers can inhibit sublimation entirely; cover free areas may show an intermediate response (Driedger, 1980; Malin and Zimbelman, 1986).

Feedback effects of topography, most notably focusing of reflected insolation in concavities, can also produce distinctive topography (Colwell et al., 1990). Insolation focusing can deepen pits and other depressions, and may even be accompanied by deposition to form a rim. Consistent alignments of conical

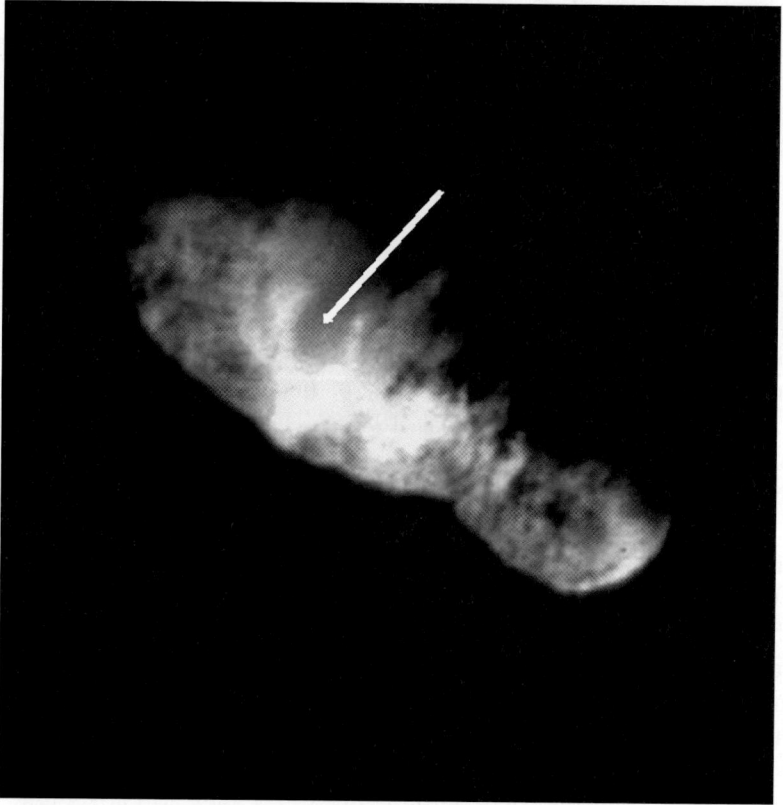

Figure 1. Borrelly from Deep Space 1. Nucleus is about 8 km long in this view. Arrow points to middle of largest of the possible mesas.

forms, pits, or asymmetries of mesas or other hills may be useful in limiting the illumination (rotation) conditions under which forms developed, because their formation is a strong function of specific incidence and azimuth angles of insolation.

Images of Borrelly (Figure 1) marginally resolve mesa-like forms (Britt *et al.*, 2004). These may be remnants of a lag deposit formed by backwasting from sublimation of lower, more volatile materials. They could even be examples of inverted relief, where the lag was once in low areas. However, the forms are not sufficiently resolved to allow more than a speculative interpretation.

Images of Wild-2 show many forms that suggest sublimation. (Figure 2). The steep spires, visible on the limb and in stereo, are certainly consistent with, and likely indicative of, sublimation. These reach heights of over 100 m. Brownlee *et al.* (2004) also note the existence of an overhang, almost certainly a form requiring erosional (sublimation in this environment) formation. Much of the topography of Wild-2 seems determined by arcuate ridges and slopes into the depressions,

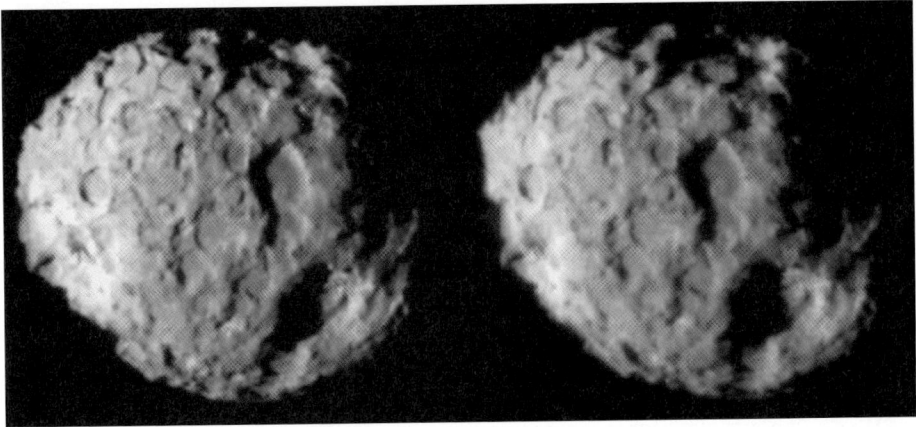

Figure 2. Stereo pair of Wild-2 from Stardust spacecraft. Note the large number of round depressions, many with flat floors, of wide variety of sizes; the arcuate scarps, the pinnacles, and the small patches of relatively high, flat surface. Object is about 5 km across in this view.

strongly indicative of removal of material. More enigmatic is the initiation of the depressions, discussed briefly below.

A major goal of DI high-resolution imaging is to examine ablational landforms in detail and to map their distribution on the nucleus of Tempel 1. We seek to correlate the locations of such features with observed locales of jet activity, with surface expressions of internal structure, and to seek evidence of preferential alignments that might provide clues to past insolation conditions.

3.2. REGOLITH

Regolith (loose material of whatever origin) is found on small asteroids, and may exist in some form on comets. Asteroids with surface accelerations in the range of 0.2 to 0.5 cm/s^2 easily retain sufficient fractions of crater ejecta to produce morphologically noticeable regolith. An object of dimensions $\sim$2 km, and having a density of about 1 g cm^{-3}, has surface gravity $<$0.1 cm/s^2, a value not inherently prohibitive of retention of regolith. However, in view of the importance of sublimation and possible compressional influences of cratering in porous targets, regolith on comets may not be developed by the loose particle sedimentation on asteroids and larger rocky objects. Indeed, most speculations about cometary regoliths involve sublimational lag deposits or redeposition of volatiles. A major unresolved question concerns the degree of cohesion of putative cometary regoliths compared to those found on asteroids and the Moon. The mesas on Borrelly suggest some level of cohesion (Soderblom *et al.*, 2004; Britt *et al.*, 2004), and steep slopes and even overhangs on Wild-2 also indicate cohesion (Brownlee *et al.*, 2004).

Although gravity of substantially less than $1 \, cm \, s^{-2}$ is effective at transporting materials across asteroidal surfaces, it is not obvious how well the process will work on rough objects with less than 1/10th as much gravity. Additional complications are that surface topography can change rapidly (geologically speaking) as ablation proceeds. On small satellites and asteroids, downslope motion is most evident where the surface is smoothed by deposition (Thomas *et al.*, 1996), although Eros shows some large, complex slumps (Veverka *et al.*, 2001). The manifestation of downslope motion on comets might be as material ponded in depressions (Robinson *et al.*, 2001), or as streamers down slopes. Detection of material as a separate deposit in the floor of a depression is sometimes difficult. The ponded materials on Eros are easily discriminated because of the high image resolutions, sharp boundaries of the deposits, color variations, and distinctively flat surfaces.

Wild-2 shows flat floors on many depressions, the larger ones of which are attributed by Brownlee *et al.* (2004) to impacts. Color imaging, such as on DI, would be useful in determining if the floors are distinct materials, and if so, might provide evidence for sedimentation, such as in Eros's ponds, or for impact phenomena. Color was used by Robinson *et al.* (2002) to show that the Eros pond material was slightly different from surrounding areas and possibly size sorted from the average Eros regolith.

The much higher resolution of the DI images compared to those obtained by previous missions will make it possible to search for morphological evidence of comet regolith. The data will also be used to map albedo variations across the surface, correlate those with surface morphology, and to delineate the extent of sublimating active areas by high-resolution albedo and morphology information with data on jet location obtained from distant imaging.

Because of the expectation that sublimation plays a dominant role in the evolution of comet surfaces, even though at any one time only a small fraction is "active," a search for asymmetry of features as a function of latitude may prove fruitful. Spin precession may smooth out latitudinal effects; nonetheless, rates may be rapid enough to produce asymmetries of pole- or equator-facing slopes that can be detected by stereoscopic measurements.

3.3. FAULTS, FRACTURES, STRUCTURES

At least three distinct sources of structure can be expected on comet nuclei: those reflecting the initial agglomeration of the paleonucleus, those caused by later impacts and possible reaccumulation of fragments, and those caused by the incompletely understood phenomenon of comet splitting (e.g. Sekanina, 1990). Primary structures resulting from assembly of smaller bodies to form the nucleus (Keller *et al.*, 1986; Weidenschilling, 1997) might comprise variations in porosity, ice/dust ratios, or ice composition. Their visible manifestations might be ridges or troughs that form as a result of spatially varying rates of sublimation. Reassembly structures following catastrophic breakup would fall in the same category, but being more recent might leave better preserved evidence. Whether the topography associated with such remnants would be prominent enough to be discerned is uncertain. Large, nearly planar features in cometary nuclei might represent metamorphic boundaries (Prialnik and Podolak, 1995) in a larger precursor body, or fracturing in a large body.

Modeling of tidal and other disruptions shows comets can be very weak (Asphaug and Benz, 1994), at least at km scales and probably at 100 m scales (see also the article by Belton et al., this issue). Fractures caused by impacts, tidal encounters, thermal changes, or by other stresses may be young relative to the age of the comet or even compared to the time that the comet has been close to the Sun, and thus may be more likely to be detected than formation structures. In certain cases the prominence of fracturing can be enhanced by subsequent sublimation (Colwell et al., 1990). Stereo imaging from DI will permit three-dimensional mapping of prominent structures. Depending on the length or angular extent around the body of a surface linear feature, the feature can be mapped in three dimensions to test whether it may be fracture related (is it somewhat planar?; does it show preferred crossing angles?). The stereo imaging a >3 m pixel covering most of the visible disk should allow good characterization of linear features more than a few 100 m in extent. Detection of displacements along fractures would be a major piece of evidence in evaluating comet history and mechanical properties.

The single solar orientation during encounters will require specific care to avoid mapping artifacts of lighting as surface lineations (Howard and Larsen, 1972).

3.4. CRATERS

No unambiguous craters are seen in either the Halley or Borrelly images, and there are substantial reasons to question if they are common landforms on comets. The geologically rapid rate of sublimation expected (tens of cm to meters/perihelion passage) is greater than resurfacing rates inferred for Io, which lacks craters of any sort, and may have a crater production rate well above that expected for comets. However, the applicable surface modification rate may be that on the inactive lag areas. Such areas could be modified by slow erosion from diffusing gases or by the fallback of dust ejected at very low speeds. Probably their lifetime is controlled by the migration of centers of activity. The rate of such migration remains poorly known. On Halley, their positions may have remained stable for months (Belton et al., 1991), and possible for decades (Schleicher and Bus, 1991). However, even fairly slow migration of small active areas, if changing at a rate of 1 m/perihelion passage, would produce a rapid average turnover compared to average cratering turnover.

Images of Wild-2 by the Stardust spacecraft (Brownlee et al., 2004) show nearly circular features, of a wide range of sizes (up to 2 km across), many with flat floors. Bownlee et al. attribute those larger than 0.5 km to impact structures, and suggest that their survival is due to recent perturbation of the comet orbit to the inner solar system. Part of the association with craters was made on the basis of comparison to laboratory impact experiments designed to be scalable to low gravity conditions. If the comparison is valid, the features on Wild-2 have undergone fractionally little degredation, although the spires and other forms on Wild-2 demand substantial erosion, at least 100 m vertically in places, as noted by Brownlee *et al.* (2004). We are of the opinion that an impact origin for the circular depressions observed on Wild 2 remains to be established, and that the smaller depressions are not clearly distinguished from the larger ones on a morphological basis. The basic topography of Wild-2: ridges, pinnacles, and intersecting depressions, may not need impacts for an explanation. The impact experiment on DI will provide an important test by demonstrating the expected morphology of an impact crater on a cometary target.

Craters formed in a thick lag material may have different shapes from those formed through a thin lag into a more porous or volatile substrate. If comet nuclei are low-density porous bodies, the effects of impacts may be different from those on many asteroids or planetary surfaces. Asteroid Mathilde is probably very porous, and displays large craters that leave few signs of damage outside the crater bowl and little evidence of ejecta (Davis, 1999). These craters are likely formed dominantly by compression (Housen, Holsapple, and Voss, 1999), and such might be the norm on cometary nuclei.

The best DI resolution expected, about 1.4 m/pxl from the flyby and 0.5 m/pxl from the impactor, should allow the detection of 5 m craters, and the characterization of 10 m craters, if any exist on the surface of Tempel 1. The recent finding of a deficiency of craters <200 m in diameter on asteroid 433 Eros (Chapman *et al.*, 2002) has been explained in two divergent ways. First, that even on small objects, which are not affected by ablation, small craters are erased very efficiently by processes such as seismic shaking induced by impacts (Cheng *et al.*, 2002; Richardson *et al.*, 2004). Second, that our understanding of the population of small impactors far from the Earth-Moon system remains imprecise. The detection of very small craters on the surface of Tempel 1 would serve to constrain the latter possibility. Of course, the spacecraft itself will produce a definitive crater, though we will observe its evolution only for a few minutes.

4. Compositional Units

Ground based and spacecraft remote sensing and *in situ* studies have shown that the nuclei of comets contain ices, carbon/hydrogen/oxygen/nitrogen-rich refractory organic solids, and silicate minerals. Many of these materials have characteristic absorption features that are detectable via remote sensing spectroscopic techniques at the near-infrared wavelengths to be measured by the DI spectrometer (Sunshine, this issue). They include rock-forming silicate minerals such as olivine and pyroxene commonly identified in spectra of asteroids and meteorites (e.g., Gaffey *et al.*, 1989), volatiles like H_2O, CO_2, CO, HCN, NH_3, OCS, SO_2, H_2S, CH_3OH, CH_4, C_2H_2, C_2H_6, and many of the more complex compounds detected in dense molecular clouds. Beyond simple detection of these phases, however, modern radiative transfer and spectral mixture modeling techniques permit quantitative analyses of reflectance and emission spectra for constraining the relative abundances of the components making up the observed surface, and for estimating the dust/ice ratio in different areas.

Recent observations have shown a wide compositional diversity among outer solar system asteroids (Centaurs), planetary satellites, and Kuiper Belt Objects (KBOs). For example, the Centaur object 5145 Pholus shows spectral features indicating the presence of water ice, olivine, and organic molecules–possibly frozen methanol or a photolytic product of methanol (e.g., Cruikshank *et al.*, 1998). On the other hand, Chiron shows a much greyer spectrum dominated by water-ice absorption bands. The DS1 spacecraft acquired a small set of spatially resolved near-IR spectra of parts of the nucleus of Borrelly in the 1.3 to 2.6 μm region (Soderblom *et al.*, 2002). These spectra show a strong thermal emission signature consistent with nucleus surface temperatures in the 300–350 K range. Perhaps not surprisingly (due to these high temperatures), there is no evidence of water or any other ices on the surface of the nucleus. However, there is evidence for at least one spatially varying absorption feature that may be due to the same kinds of relatively simple hydrocarbons inferred for objects like Pholus and some KBOs.

On DI spectral information on scales of 100 m will be correlated with broadband color imaging at 10 m scales to define spectrally distinct units on the nucleus. The composition of the materials in these units, if it can be identified uniquely, will provide important clues to the processes responsible for the formation of the units. Even if such identifications cannot be made uniquely, the patterns observed can provide important clues, for example allowing identification of crater ejecta with a source crater (Geissler *et al.*, 1996), or highlighting the presence of layers, or areas of preferred collection of loose materials of distinctive color/spectra.

5. Conclusions

Deep Impact data promise to advance our understanding of the processes that control the evolution of comet surface in fundamental ways. First, DI will produce the highest spatial resolution views of any comet surface to date. In addition, it will provide the first high-resolution color imaging in stereo, information which will make it easier to distinguish subtle shading effects due to small scale topography from albedo variations. The extended spectral range of the DI spectrometer enhances the chances of identifying spectrally (and hence, compositionally) distinct units on the nucleus. Finally, thanks to its impact experiments, DI will constrain the currently wide ranging speculations as to what impact craters formed on comet nuclei look like.

References

A'Hearn, M. F.: 1998, *Ann. Rev. Earth Planet. Sci.* **16**, 273.

A'Hearn, M. F., Millis, R. L., Schleicher, D. G., Osip, D. J., and Birch, P. V.: 1995, *Icarus* **118**, 223.

A'Hearn, M. F., *et al.*: *Space Sci. Rev.* this issue.

Asphaug, E. and Benz, W.: 1994, *Nature* **370**, 120.

Belton, M. J. S.: 2001, *BAAS* **32**, 1062.

Belton, M. J. S., Julian, H. J., Anderson, A. J., and Mueller, B. E. A.: 1991, *Icarus* **93**, 183.

Bleton, M. J., *et al.*: *Space Sci. Rev.* this issue.

Boenhardt, H., Rainer, N., Birkle, K., and Schwehm, G.: 1999, *Astron. Astrophys.* **341**, 912.

Brin, G. D. and Mendis, D. M.: 1979, *Astrophys. J.* **229**, 402.

Britt, D. T., *et al.*: 2004, *Icarus* **167**, 45.

Brownlee, *et al.*: 2004, *Science* **304**, 1764.

Chapman, C. R., Merline, W. J., Thomas, P. C., Joseph, J., Cheng, A. F., and Izenberg, N.: 2002, *Icarus* **155**, 104.

Chapman, C. R., Veverka, J., Belton, M. J. S., Neukum, G., and Morrison, D.: 1996, *Icarus* **120**, 231.

Cheng, A. F., Izenberg, N., Chapman, C. R., and Zuber, M. T.: 2002, *Meteor. Planetary Sci.* **37**, 1095.

Colwell, J. E., Jakosky, B. M., Sandor, B. J., and Stern, A.: 1990, *Icarus* **85**, 205.

Davies, M. E., *et al.*: 1996, *Celest. Mech. Dynam. Astron.* **63**, 127.

Davis, D. R.: 1999, *Icarus* **140**, 49.

Driedger, C. L.: 1980, in Lipman and Mullineaux D. (eds.), *U. S. Geol. Survey Prof. Paper* **1250**, p. 757.

Fanale, F. P., and Salvail, J. R.: 1984, *Icarus* **60**, 476.

Geissler, P., *et al.*: 1996, *Icarus* **120**, 140.

Grun, E., *et al.*: 1993, *J. Geophys. Res.* **98**, 15091.

Hampton, D. L., *et al.*: *Space Sci. Rev.* this issue.

Helfenstein, P. and Shepard, M. K.: 1999, *Icarus* **141**, 107.

Housen, K. R. and Holsapple, K. A.: 1999, *Lunar Planet. Sci.* **30**, Abstract 1228.

Housen, K. R., Holsapple, K. A., and Voss, M. E.: 1999, *Nature* **402**, 155.

Howard, K. A. and Larsen, B. R.: 1972, *Apollo 15 Preliminary Science Report*, 25–58 to 25–62, NASA, Washington DC.

Jewitt, D.: 2002, *In Proceedings of Asteroids, Comets, and Meteors, 2002*, Abstract [01-00i].

Keller, H. U., *et al.*: 1986, *Nature* **321**, 320.

Keller, H. U., Kramm, R., and Thomas, N.: 1988, *Nature* **331**, 227.

Klaasen, K. P., *et al.*: *Space Sci. Rev.* this issue.

Malin, M. C.: 1985, in *NASA Tech, Mem.* 87563, p. 83.

Malin, M. C. and Zimbelman, J. R.: 1986, *Lunar Planet. Sci. Conf.* **17**, 512.

Meech, K. J., Fernandez, Y., and Pittichova, J.: 2001, *Bull. Am. Astron. Soc.* **33**, 1075.

Miller, J. K., *et al.*: 2002, *Icarus* **155**, 3.

Prialnik, D. and Podolak, M.: 1995, *Icarus* **117**, 420.

Prockter, L. M., Thomas, P. C., Robinson, M., Joseph, J., Milne, A., Bussey, B., *et al.*: 2002, *Icarus* **155**, 75.

Reitsema, H. J., Delamere, W. A., and Whipple, F. L.: 1989, *Science* **243**, 198.

Richardson, J. E., Melosh, H. J., and Greenberg, R.: 2004, *Lunar Planet. Sci. Conf.* **35**, abstract 1864.

Samarasinha, N.: 2001, *Icarus* **154**, 540.

Schleicher, D. G., Bus, S. J.: 1991. *Ast. J.* **101**, 706.

Schultz, P. H. and Ernst, C.: *Space Sci. Rev.* this issue.

Sekanina, Z.: 1990, *Ast. Jour.* **100**, 1293.

Soderblom, L., *et al.*: 2002, *Lunar and Planetary Science Conference* **33**, March 11–15, 2002, Houston, Texas, Abstract no.1256.

Soderblom, L. A., *et al.*: 2004, *Icarus* **167**, 4.
Sunshine, J. M., *et al.*: *Space Sci. Rev.* this issue.
Thomas, P. C., *et al.*: 2002, *Icarus* **155**, 18.
Thomas, P. C., *et al.*: 1994, *Icarus* **107**, 23.
Thomas, P. C., *et al.*: 1996, *Icarus* **120**, 20.
Veverka, J., *et al.*: 2001, *Science* **292**, 484.
Weidenschilling, S. J.: 1994, *Nature* **368**, 721.
Weidenschilling, S. J.: 1997, *Icarus* **127**, 290.
Whipple, F.: 1950, *Astrophys. J.* **111**, 375.

EXPECTATIONS FOR CRATER SIZE AND PHOTOMETRIC EVOLUTION FROM THE DEEP IMPACT COLLISION

PETER H. SCHULTZ*, CAROLYN M. ERNST and JENNIFER L. B. ANDERSON

Department of Geological Sciences, Brown University, Providence, RI 02912-1846, U.S.A.

(*Author for correspondence; E-mail: peter_schultz@brown.edu*)

(Received 20 August 2004; Accepted in final form 10 January 2005)

Abstract. The NASA Discovery Deep Impact mission involves a unique experiment designed to excavate pristine materials from below the surface of comet. In July 2005, the Deep Impact (DI) spacecraft, will release a 360 kg probe that will collide with comet 9P/Tempel 1. This collision will excavate pristine materials from depth and produce a crater whose size and appearance will provide fundamental insights into the nature and physical properties of the upper 20 to 40 m. Laboratory impact experiments performed at the NASA Ames Vertical Gun Range at NASA Ames Research Center were designed to assess the range of possible outcomes for a wide range of target types and impact angles. Although all experiments were performed under terrestrial gravity, key scaling relations and processes allow first-order extrapolations to Tempel 1. If gravity-scaling relations apply (weakly bonded particulate near-surface), the DI impact could create a crater 70 m to 140 m in diameter, depending on the scaling relation applied. Smaller than expected craters can be attributed either to the effect of strength limiting crater growth or to collapse of an unstable (deep) transient crater as a result of very high porosity and compressibility. Larger then expected craters could indicate unusually low density (<0.3 g cm^{-3}) or backpressures from expanding vapor. Consequently, final crater size or depth may not uniquely establish the physical nature of the upper 20 m of the comet. But the observed ejecta curtain angles and crater morphology will help resolve this ambiguity. Moreover, the intensity and decay of the impact "flash" as observed from Earth, space probes, or the accompanying DI flyby instruments should provide critical data that will further resolve ambiguities.

Keywords: Deep Impact, porous targets, comets, oblique impact, NASA Ames Vertical Gun Range, laboratory experiments, Comet 9P/Tempel 1

1. Introduction

The NASA Discovery mission "Deep Impact" (DI) will guide a 360 kg (excluding propellant) probe into the path of Comet 9P/Tempel 1 in July 2005 with a net velocity of $\sim$10.2 km s^{-1}. A companion spacecraft will capture the details of the collision process and resolve the resulting crater during a close approach of about 500 km (see A'Hearn *et al.*, this volume). One of the primary goals of this mission is to analyze spectroscopically pristine cometary materials excavated from below the surface. These materials will be observed in the ejecta plume during excavation and in the annulus of ejecta deposited around the final crater. Such a strategy may seem straightforward, but uncertainties in the actual nature of the upper 10 meters of the surface make specific *a priori* predictions difficult. The outcome of the collision will provide new constraints on the physical nature of the surface and subsurface while establishing a new benchmark for the cratering process.

The scientific basis for using an active probe to characterize subsurface materials of a planetary body dates back to the Ranger missions in the 1960's. In addition to the successful on-board cameras, attempts were made at the Jet Propulsion Laboratory (JPL) to view and calibrate these collisions on the Moon using earth-based telescopes. This concept received more serious study at JPL in 1993 when Marc Adams initiated an internally funded effort to assess possible measurements that could be made by hypervelocity impacts into planetary surfaces, e.g., a multiple asteroid mission. This effort resulted in a Director's Research and Development Fund proposed to JPL in 1995 with promising initial results (Schultz et al., 1996; Adams et al., 1997). Concurrently, JPL program managers excited by the concept actively solicited several possible mission scenarios from the community, including hypervelocity probes as part of possible missions to Europa and Pluto and multiple asteroid encounters. The Department of Defense (and several national laboratories) developed similar mission concepts, including the Clementine II mission, which would have involved impacting an asteroid. But in 1996, Alan Delamere and Mike Belton (with JPL, NOAO, and Ball Aerospace) developed the basis for the current Deep Impact mission concept with emphasis on exposing pristine materials from below the surface layer of a comet using a hypervelocity probe (see A'Hearn et al., this volume). Consequently, the concept of kinetic probes for a planetary exploration has a rich scientific and engineering history, but NASA Discovery's Deep Impact mission is the first to be fully developed to flight status.

The effect of the DI impact on Tempel 1 can be assessed through theoretical calculations and laboratory experiments scaled up to conditions for the DI experiment. In both cases, the low surface gravity ($0.04 \, \text{cm} \, \text{s}^{-2}$) or unusual surface properties (low density) on Tempel 1 may result in surprises. At one extreme, the crater may be much smaller than expected due either to energy losses during the compression stage or to the effects of strength, both processes potentially contributing to little ejecta or surface expression (e.g., Housen et al., 1999). At the other extreme, a low-strength, porous particulate upper surface (>25 m) may result in a large crater limited in its lateral dimensions only by its meager surface gravity or the effects of deep penetration (e.g., Schultz, 2001; 2003a; Schultz et al., 2002). Because imaging sequences for DI require planning for the maximum crater size, initial estimates focused on direct extrapolations of gravity-controlled scaling relations for a loose particulate target (e.g., Schultz, 2001).

In order to be prepared for the event, a series of laboratory impact experiments are being performed at the NASA Ames Vertical Gun Range (AVGR). These experiments are designed to explore the effect of a wide range of target variables on crater evolution, final crater dimensions, and the photometry of the brief "flash" induced by the thermal plume. A range of possible target properties can be imagined. First, physical reworking (due to volatile escape or regolith development) may result in a loose, porous surface of silicate/organic particulates. Second, cometary processes could produce a weakly bonded, highly porous surface layer. Third, volatile losses could create an indurated surface lag. These three extremes then could be extended to additional scenarios where the underlying substrates (including pristine volatiles) exist at depths from 1 m to 20 m.

The first three scenarios provide end-member properties that could dramatically affect the size of the crater and the amount of material exposed for remote analysis. For the purposes of discussion here, the first scenario will be called the porous regolith (PR) model; the second, the "under-dense regolith" (UR) model; and the third, the "strength-controlled" (SC) model. The fourth scenario will be collectively termed the "composite and layered" (CL) model.

The targets used in the experiments were not chosen to provide direct simulations of the DI crater. Rather, each target type provided different properties that would aid in relevant extrapolations. For example, carbonates do not represent cometary materials but impacts into carbonates result in strong atomic/molecular emissions with little thermal radiation. These spectra can be used to characterize the vapor plume at impact velocities available in laboratory experiments (Schultz, 1996; Sugita et al., 2003). Similarly, a thick, purely silicate surface layer is unlikely, but use of such a target in laboratory experiments establishes scaling relations and partitioning of kinetic into radiant energy for heated particulates. Moreover, the use of very different physical properties of the target material allows exploring a wide range of cratering outcomes.

Recent results from encounters with Comets Borelly (Soderblom et al., 2002) and Wild 2 (Brownlee et al., 2004) have provided unprecedented views of cometary surfaces at scales of 100 m's. It is the response of the surface and substrate at meter scales to the Deep Impact collision, however, that will affect what is observed. Consequently, the approach here is to consider a range of scenarios in order to assist in rapid interpretations of the observations soon after the event. Discussion first places the various crater-scaling regimes in a conceptual framework. It then considers the contrasting evolution of the ejecta plume. Next, expectations for the impact "flash" are reviewed. These various observations are then combined in order to anticipate diagnostic observations for understanding the nature of the upper surface layers of the comet.

The goal of this contribution is to clarify the range of possible outcomes from the DI collision. While the impactor properties are known, some impact parameters (impact angle, comet properties) are not. Impact angle with respect to the comet surface may range from 90° (vertical) to as low as 10° depending on the targeted region and local slope effects. As will be shown, interpretations will need to depend not only on the appearance of the final crater but also on the evolution of the initial radiant energy, early-time plume evolution, ejecta curtain appearance, and coma brightening created by the total ejected mass.

2. Crater Scaling

2.1. GENERAL CONSIDERATIONS

Various studies have assessed the effect of critical independent variables on crater size, particularly in particulate targets (e.g., Gault *et al.*, 1975; Gault and Wedekind, 1977; Schmidt, 1977; Schmidt and Holsapple, 1982; Holsapple and Schmidt, 1982; Schultz and Gault, 1985a, b; Housen *et al.*, 1999). Additionally, new diagnostic tools are being applied to the evolution of the ejecta and their relation to the cratering flow field (e.g., Cintala *et al.*, 1999; Anderson *et al.*, 2003, 2004).

Impacts into competent targets result in craters with diameter limited by material strength; hence, this is termed the strength-controlled crater-scaling regime (Gault and Wedekind, 1977; Schmidt, 1977). Impacts into loose particulate targets (e.g., PR models), however, grow "freely" until gravity prevents material from escaping the cavity. This is called the gravity-controlled crater-scaling regime (Post, 1974; Chabai, 1977; Gault and Wedekind, 1977; Schmidt, 1977).

A third regime has been proposed for under dense particulate targets (UR models). Experiments using highly compressible, porous targets (e.g., pumice dust) resulting in reduced cratering efficiencies (relative to sand) due to internal energy losses (e.g., Schultz and Gault, 1985a). Nevertheless, crater-scaling relations indicated gravity-controlled growth over a wide range of projectile sizes and velocities. Housen *et al.* (1999) found that energy losses in highly porous targets mixed with fine fly ash result in smaller craters due to significant compaction with minimal ejecta. Other studies showed, however, that hypervelocity impacts into highly porous targets result in deep penetration prior to complete transfer of energy and momentum (O'Keefe *et al.*, 2002; Schultz *et al.*, 2002; Schultz *et al.*, 2003a). Computational codes (O'Keefe *et al.*, 2002) demonstrated the deep penetration in under-dense targets until R-T instabilities disrupt the impactor during the earliest stages of cratering. Projectile disruption during penetration in porous targets is the rationale for using brittle Pyrex projectiles in laboratory experiments since this process has been shown to affect key crater-scaling exponents (e.g., Schultz and Gault, 1983, 1985b; Schultz, 1988). Laboratory experiments, however, also allow tracking the process beyond the penetration stage described by O'Keefe *et al.* (2002) and reveal that crater formation in loose particulates resembles a deeply buried explosion at late stages with significant amounts of ejecta launched at high angles (Schultz *et al.*, 2002; Schultz, 2003a). The high-angle ejecta were observed to return to the crater, which subsequently collapsed to produce a much smaller final crater.

In the following discussion empirical scaling relations widely used to estimate crater dimensions are first reviewed. Second, the various cratering regimes are considered in the context of possible conditions for the DI collision, e.g., very low gravity and high porosity targets. Third, experimental design is used to illustrate some of the processes controlling phenomena associated with highly porous and compressible targets. Our focus here is on the size of the crater and what observations could be used to interpret the observed crater size and plume evolution. Additional details concerning the evolution of the plume over much longer time scales (days) are deferred to other contributions and future work.

2.2. SCALING RELATIONS

The following relation (see Schmidt and Holsapple, 1982; Holsapple and Schmidt, 1982) describes strength-controlled crater scaling:

$$\frac{M_c}{m_p} = k\pi_Y^{-\beta} \tag{1}$$

where k is an empirical constant; M_c, the total displaced mass (deformation and ejecta) at the end of crater formation; and m_p, the original impactor mass. The term π_Y is given by the following:

$$\pi_Y = \frac{\delta_t v^2}{Y} \tag{2}$$

where δ_t is the target density; v the impactor velocity; and Y the target strength. The exponent β depends on the controlling cratering process. For momentum-controlled cratering $\beta = 2$; for energy-controlled scaling, $\beta = 1$.

By contrast, gravity-controlled cratering is controlled by outward material flow field accelerated behind the shock front and is redirected by rarefactions off the free surface (Gault et al., 1968; Gault et al., 1975). This flow continues provided that it has sufficient velocity to escape the cavity. When ejecta velocities no longer allow escape from the cavity, then the following relation applies for the transient crater (maximum displacement prior to collapse):

$$\frac{M_c}{m_p} = k\pi_2^{-\alpha} \tag{3}$$

where π_2 is an inverse Froude number that scales inertial to gravitational forces (Chabai, 1977; Schmidt, 1977; Gault and Wedekind, 1977):

$$\pi_2 = 3.22\left(\frac{gr}{v^2}\right) \tag{4}$$

where r is the projectile radius; g the gravitational acceleration at the surface; and v the vertical velocity component represented by $v \sin \theta$ (Gault and Wedekind, 1977; Chapman and McKinnon, 1986; Schultz and Gault, 1990).

As recognized in experiments (Stöffler et al., 1975) and computations (e.g., Schultz et al., 1981), the total displaced mass is about twice the mass of the ejecta actually launched out of the crater. The amount of ejecta is important for the DI mission since it will control the over-all coma brightening following the impact as the total ejected mass (gas and dust) greatly exceeds the nominal daily flux. The total displaced mass is important because it determines the final crater dimensions to be observed during the flyby.

The final diameter and depth for simple (un-collapsed) craters are commonly assumed to form a constant ratio, i.e., the crater diameter and depth are proportional. If the crater diameter is referenced to the pre-impact surface, it is termed the "apparent crater," and this aspect ratio ranges from 3:1 for strength-controlled metals (matched projectile/targets) to 4:1 for gravity-controlled particulate targets. Projectile aspect ratio, strain-rate effects, grain size (relative to the projectile), projectile failure, and projectile/target density also appear to affect scaling (e.g., Schultz, 1988). The transient crater diameter and depth, however, do not grow proportionally with each other. Laboratory and computational experiments demonstrate that crater depth reaches its maximum prior to crater diameter (Schultz et al., 1981; Schultz, 1988). Consequently, non-proportional crater growth can result in pre-collapse crater aspect ratios very different from the nominal values depending on the effect of gravity, density contrast, and compressibility as will be discussed.

The independent variables for the impactor (the DI probe) are known, excepting the impact angle. The independent variables for the comet, however, are unknown or poorly constrained, including the strength, physical structure (layering), and the density. Consequently, these are the important variables to be explored through experiments. For the DI impact, the relative encounter velocity will ensure eventual catastrophic failure of the projectile at impact, even for the most extreme scenario of an under-dense surface layer. Consequently, Pyrex spheres are used in the laboratory experiments here as a more relevant analog for higher velocity impacts, including the DI impactor. Although composed of copper, holes in the DI impactor and the attached structures reduce the effective bulk density to about $1 \, g \, cm^{-3}$; consequently, Pyrex also provides reasonable density match.

The maximum transient crater diameter, D, for a gravity-controlled crater scaled to the projectile diameter, a, is given by (again following Holsapple and Schmidt, 1982):

$$\frac{D}{a} = k' \left(\frac{\delta_p}{\delta_t} \right)^{1/3} \left(3.22 \frac{gr}{v^2} \right)^{-\alpha/3}. \tag{5}$$

The exponent α for particulate targets typically ranges from 0.48 to 0.56, depending on specific material properties and whether energy or momentum transfer dominates. Consequently, Equation (5) shows that direct extrapolations can become problematic when considering large sizes or extreme gravitational accelerations at the surface (Schmidt and Holsapple, 1982; Gault and Wedekind, 1977; Schultz, 1992).

Crater depth can be simply calculated by assuming a fraction of the final crater diameter. Experiments, however, reveal that crater depth can be arrested by a competent substrate at depth without significantly affecting the diameter, provided that the depth of the layer is greater than about three times the impactor diameter in the case of vertical impacts (Schultz, 2003b). Conversely, an under-dense target will allow the projectile to penetrate deeply before termination. In this case, sufficient strength near the surface (or a weakly coupled shock at first contact) can result in an initially small diameter but deep crater. Unless allowed to grow to large diameter, this profile is unstable and collapses. Consequently, the strategy to watch the crater grow during the DI encounter (and the evolution of the ejecta plume and curtain) may prove as important as measurements of the final crater dimensions.

The density of the comet plays three roles. First, the bulk density and radius affect gravitational acceleration forces at the surface. Second, the near-surface density affects the amount of material (mass) displaced, i.e., crater size. Third, an under-dense target may result in a weakly coupled shock near the surface, thereby resulting either in deep penetration of the projectile prior to transferring its momentum and energy or in compression effects with little excavation. The first two roles are readily incorporated into Equation (5). The last role is best appreciated in the context of the decay of peak pressure in relation to strength, gravity, and the initial stages of coupling.

Figure 1 illustrates the various cratering regimes expressed in terms of the peak pressure decay in the target. This graphical representation is based on theoretical considerations consistent with observations from experiments and computational

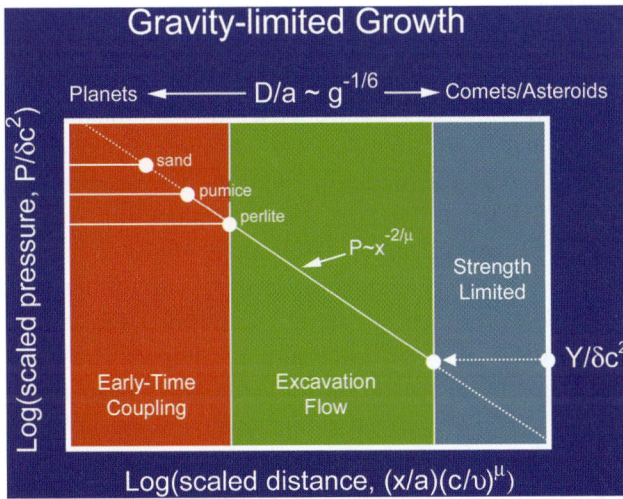

Figure 1. Schematic plot showing the stages of cratering in terms of scaled peak pressure decay as a function of scaled distance from impact (following Holsapple, 1987). Peak pressure is scaled to target density (δ) and sound speed (c); distance (x) is scaled by impactor diameter (a), velocity (v), and target sound speed. Different initial conditions during the compression stage eventually converge on a common pressure-decay relation at different scaled distances. The "excavation stage" occurs in the rarefaction zone behind the shock front. The limit of excavation is represented by two conditions. Strength-controlled craters are prevented from growing further when the peak pressures approach the strength limit (Y) as indicated by the lower axis. Gravity-controlled craters are limited by the ballistic excavation (upper axis) as described in the text.

codes (Holsapple, 1987). Immediately after impact, the impactor transfers its energy and momentum to the target. The transfer of energy and momentum is generally complete (coupled to the target) after the shock front has passed through the projectile to its back surface (Gault *et al.*, 1968). As a result, the stage of crater growth (i.e., distance, x, or depth, d at a given time) needs to be scaled to either a final crater dimension (e.g., radius, R) or to the projectile diameter. Alternatively, stage of growth can be represented in terms of time (t) relative to the total crater formation time or the length of time for the projectile to travel its diameter (i.e., the penetration time scale, a/v, along its trajectory):

$$\tau = \frac{t}{(a/v)} \tag{6}$$

where v is the impact velocity.

As recognized early in the study of cratering (e.g., Dienes and Walsh, 1970; O'Keefe and Ahrens, 1977), peak pressures eventually decay to a common material-dependent relationship after a sufficient time or distance from the impact. This far-field equivalence can be expressed in a useful description based on dimensional analysis for peak pressure (P) scaled to the target density (δ_t) and ambient sound speed (c) in the target:

$$\frac{P}{\delta_t c^2} = \left[\left(\frac{x}{a} \right) \left(\frac{c}{v} \right)^\mu \right]^{-2/\mu} \tag{7}$$

where μ is a pressure-decay exponent that depends on which coupling process dominates cratering. If cratering is controlled by momentum, then $\mu = 2/3$; if controlled by energy, $\mu = 1/3$ (Holsapple, 1987). At large distances from the source, however, the exponent $(2/\mu)$ approaches 2 in order to conserve momentum in the shock (see Dahl and Schultz, 1999, 2001). Lower values of μ (momentum scaling) also apply to porous particulate targets (Holsapple, 1987).

Equation (7) is useful for visualizing the effect of controlling variables on crater size relevant to the DI experiment. If strength-controlled, the crater can grow until the material strength exceeds the peak pressure. In competent targets, this represents the yield strength (ductile) or tensile (brittle) strength. Consequently (x/a) simply approaches (D/a) at the end of excavation (the lower abscissa in Figure 1). In reality, the decay slope also depends on target porosity, but this is ignored for purposes of illustration. At the other extreme, a completely strengthless target will permit the crater to grow until gravity no longer allows material to escape the cavity (upper abscissa). Under high gravity, the cavity becomes smaller for a given projectile size. Under very low gravity, the cavity grows until ejecta velocities are insufficient to achieve a ballistic range beyond the crater rim.

Because most materials exhibit some form of strength (e.g., cohesion), crater growth ceases when the peak pressure in the shock front no longer exceeds the scaled material strength ($Y/\delta_t c^2$ in Figure 1). Powdered pumice exhibits a high angle of friction under static conditions as demonstrated by its ability to form a vertical face due to the irregular shapes of the constituent grains. Hypervelocity impact experiments into pumice powder demonstrate, however, that gravity controls excavation. This paradox is readily explained by the fact that excavation occurs under extension in the rarefaction zone behind the shock front (Gault *et al.*, 1968). For example, a post-shock zone of distended powder ("bulking") is observed well beyond the rim of craters produced in pumice. Consequently, the relevant strength term for particulate targets is not simply the static strength but the resistance (frictional shear) during dynamic flow in the rarefaction wave behind the shock front. In laboratory experiments, the gravity-controlled crater is much smaller than the disturbed annulus beyond the rim. But under very low gravity, it is likely that the resistance to flow will arrest the crater before it reaches its gravity limit. In Figure 1, the dashed horizontal line represents $Y/\delta_t c^2$ corresponding to this limit.

Compression primarily affects the initial stages of cratering rather than the final stages, as in gravity- and strength-controlled growth (Figure 1). As shown by O'Keefe and Ahrens (1977), different initial conditions eventually follow a single decay curve reflecting material properties behind the shock front. They converge at a distance approximately where the projectile detaches from the shock to the left in Figure 1. Within this zone, early-stage coupling (such as compression) is thought to be unimportant for later stage excavation flow, according to point-source theory (Holsapple and Schmidt, 1982, 1987). If experiments are performed under high gravity for a given target material (or for larger impactor size at a given gravity), then the gravity-controlled limit in the excavation flow regime will move to the left in this diagram (abscissa above) and shut down growth. For highly compressible targets under very high gravity, the cratering limit may overlap with the early-stage coupling.

The DI crater will form under very low gravity (from $0.04\,\mathrm{cm\,s^{-2}}$ to $0.08\,\mathrm{cm\,s^{-2}}$). Consequently, the relevant issue here is the possible effect of unusual cometary materials resulting in significant internal energy losses at the outset (compression effects) such that the early-time compression zone consumes more of the excavation flow (to the right). The excavation limit is ultimately controlled either by the peak pressure relative to the material strength or by gravitational acceleration.

This general discussion relating input to output allows understanding the wide range of possible outcomes due to material properties of Tempel 1, including substrates with very low density, high porosity, and high compressibility as considered next.

2.3. UNDER-DENSE, POROUS TARGETS (PR MODELS)

There is a physical difference between the bulk target density (porosity) and compressibility. A highly porous target with a low bulk density can be relatively incompressible under static conditions. For example, particulate targets composed of hollow ceramic microspheres can be compressed to maximum packing with

considerable bulk compressive strength. Particulate pumice, vermiculite, and perlite targets represent the other end member where each constituent particle is highly
compressible even after achieving maximum packing.

Experimental data reveal that even large projectile/target density ratios (>10) for
incompressible, highly porous targets do not significantly affect crater scaling for
gravity-controlled growth (Schultz and Gault, 1985a). Experiments using targets
composed of hollow microspheres (bulk $\delta_t = 0.09 \, \text{g cm}^{-3}$; porosity $\sim 90\%$) and
vermiculite (bulk $\delta_t = 0.09 \, \text{g cm}^{-3}$; porosity $\sim 90\%$) resulted in enormous transient
craters consistent with expectations from Equation (5). Craters produced in targets
composed of microspheres collapse due to the unstable crater profile, whereas
craters in vermiculite retain their profile (see Figure 2).

Target density (porosity), however, does affect the peak pressure and its decay
rate through the target. Consequently, experiments also were performed with different degrees of compaction of pumice powder: compressed ($1.5 \, \text{g cm}^{-3}$; porosity of
35%); uncompressed ($1.3 \, \text{g cm}^{-3}$; porosity of 43%); and lightly fluffed ($1.1 \, \text{g cm}^{-3}$;
porosity $\sim 50\%$). The compressibility of these targets can be illustrated by low-
velocity, free-fall experiments. A solid rod dropped from 0.5 m will penetrate completely through to the bottom of the target container (below the crater floor) for the
fluffed pumice target. Hypervelocity impacts ($>5 \, \text{km s}^{-1}$) by Pyrex or aluminum

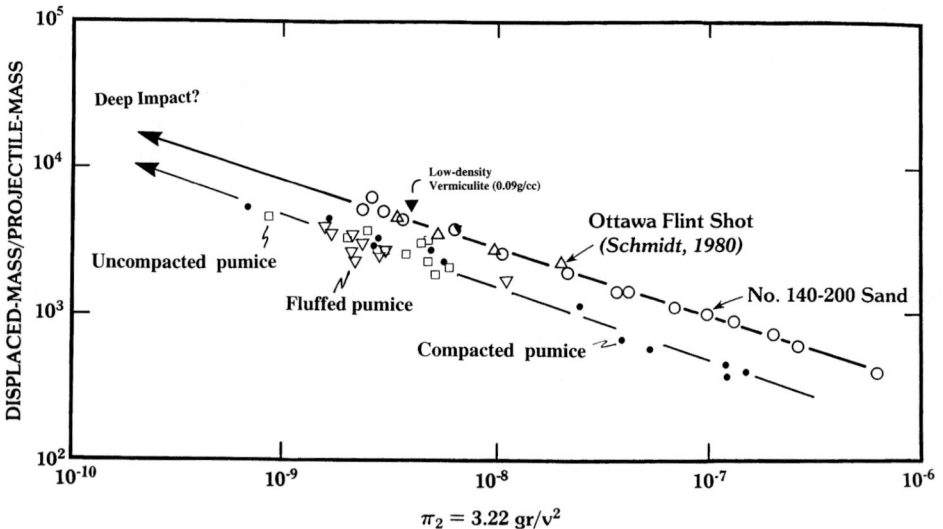

Figure 2. Cratering efficiencies (total mass displaced relative to the impactor mass) plotted against the
gravity-scaling parameter (π_2) expressed in terms of the impactor radius (r), gravitational acceleration
(g), and impactor velocity (v). Data include craters produced in No. 140–200 μm sand (Schultz and
Gault, 1985a), Ottawa flint shot (Schmidt, 1980), and compressible targets (pumice, vermiculite).
Three types of pumice with different bulk densities are shown: compacted, uncompacted, and fluffed.
The lower cratering efficiency for pumice is attributed to internal energy losses during the early stages
of compression.

(or even copper) into sand or pumice, however, do not penetrate below the crater floor throughout crater growth due to the role of shock disruption of the projectile. Figure 2 summarizes cratering efficiencies for a wide range of targets with different densities. These experiments indicate that impactor/target density ratios, target porosity ($<50\%$), and cohesion all have little effect on crater scaling of loose particulates at 1 g over the range of variables in use.

2.4. UNDER-DENSE, COMPRESSIBLE TARGETS (UR MODELS)

Perlite targets provide a useful surrogate to assess the consequences of both high bulk porosity and high compressibility of constituent particles (Housen *et al.*, 1999). The experiments for the present study included targets of perlite granules (composed of a highly porous silicate) sieved to retain grains smaller than 1 mm (bulk $\delta_t = 0.20\,\mathrm{g\,cm^{-3}}$). In contrast with the experiments by Housen *et al.* (1999), additives such as fly ash were not included because such fine-grained components can introduce viscous drag that can dominate the process, especially at low impact velocities ($<2\,\mathrm{km\,s^{-1}}$). Such experimental conditions also can complicate the distinction between early-stage compression effects (and poor shock coupling) and late-stage strength limits (Figure 1).

Cratering efficiency for impacts into sieved perlite increases with decreasing impact angle (referenced to the horizontal) as shown in Figure 3, directly opposite to well-coupled impacts into sand and pumice (Gault and Wedekind, 1978). "Quarter-space" experiments reveal the evolution of crater growth and expose the underlying cause for this counter-intuitive result. In this approach, the projectile is aimed at the surface directly adjacent to a plexiglass sheet, thereby allowing observations of crater growth using high-speed imaging. Such an approach has been used in numerous investigations (e.g., Piekutowski, 1977; Schmidt and Housen, 1987). The evolution of the transient crater for the sieved-perlite experiments is shown in Figure 4. In vertical impacts, the projectile penetrates deeply before outward (lateral) growth occurs (Figure 4a). Because of the unstable transient crater shape under 1 g, the final crater collapses and becomes much smaller. At slightly oblique impact angles (Figure 4b) the initial penetration funnel eventually opens to produce a large transient crater below the surface. As impact angle decreases still further (15–30°), the crater becomes stable and cratering efficiency actually increases relative to expectations (Figure 4c) as demonstrated in Figure 3.

The difference between the transient and final crater shape is shown in Figure 5 for different particulate targets. The diameter-to-depth aspect ratio (x/d) evolves from 0.5 within the first 4 ms to 1.5 at the end of excavation in the quarter-space experiments. In contrast, the final D/d value for nominal (half-space) experiments with perlite is almost 3. Consequently, crater growth for both vertical and oblique impacts into sieved perlite targets is non-proportional throughout most of crater growth and results in an unusual aspect ratio for the pre-collapse transient crater.

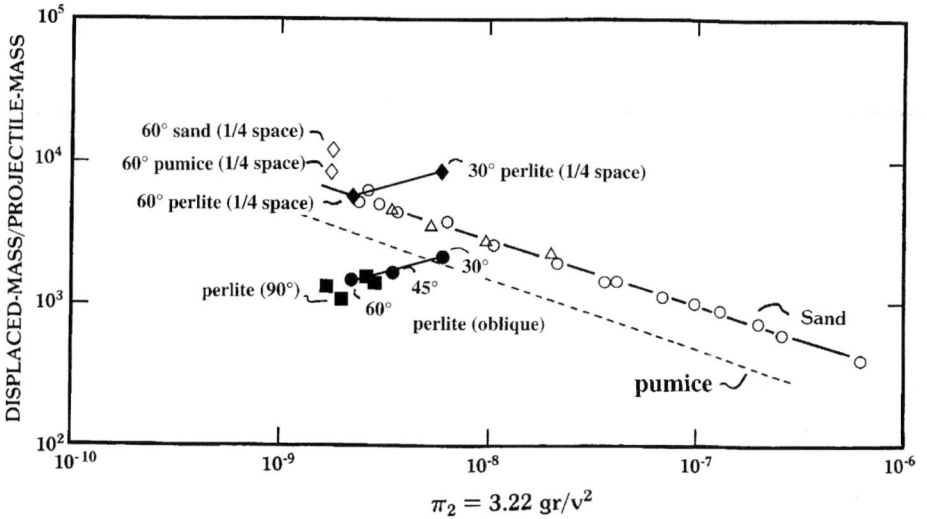

Figure 3. Cratering efficiencies for under-dense (sieved perlite) targets in contrast with results for sand and pumice targets. Final craters produced in perlite by vertical impacts (solid squares) are much less efficient than the final craters produced in pumice (dashed line, see Figure 2) and sand. Oblique impacts into perlite (solid circles) become progressively more efficient with decreasing impact angle, in contrast with impacts into denser particulate targets (pumice and sand). Transient craters measured from quarter-space experiments (open diamonds for 60° impacts into pumice and sand) reveal that cratering efficiencies prior to collapse of the cavity are generally consistent with expectations for the appropriate target (but slightly enhanced due to the quarter-space design). Oblique impacts (30°) into perlite using the quarter-space design (solid diamond), however, resulted in greater cratering efficiencies due to shallower depths of energy/momentum transfer.

Such results indicate that the underlying controlling process for under-dense, highly compressible targets is the effective penetration depth where the impactor energy and momentum are fully transferred to the target. This process is illustrated for the earliest stages of cratering using quarter-space experiments (Figure 6). Figure 6a provides a reference for a 60° impact into sand. Vertical (Figure 6b) and near-vertical (Figure 6c) impactors into perlite disappear in a long penetration funnel beneath the surface similar to early-time computational results of O'Keefe *et al.* (2002). The deep penetration in both cases results in a crater formed by a process resembling a deeply buried explosion. In contrast, a low-angle impact (30°) into perlite (Figure 6d) couples at a more optimum depth for both maximum cratering efficiency and stable final crater shape. The late-stage transient craters are shown in Figure 7.

The laboratory experiments used weak impactors (Pyrex) that totally disrupt at impact, even in the low-density perlite. Disruption also should occur for the 10.2 km s^{-1} DI collision, even for extremely low densities of the comet surface materials. This is a major difference between impact experiments that do and do not have sufficient velocity to exceed the sound speed and failure limit in both target and projectile, especially for under-dense targets.

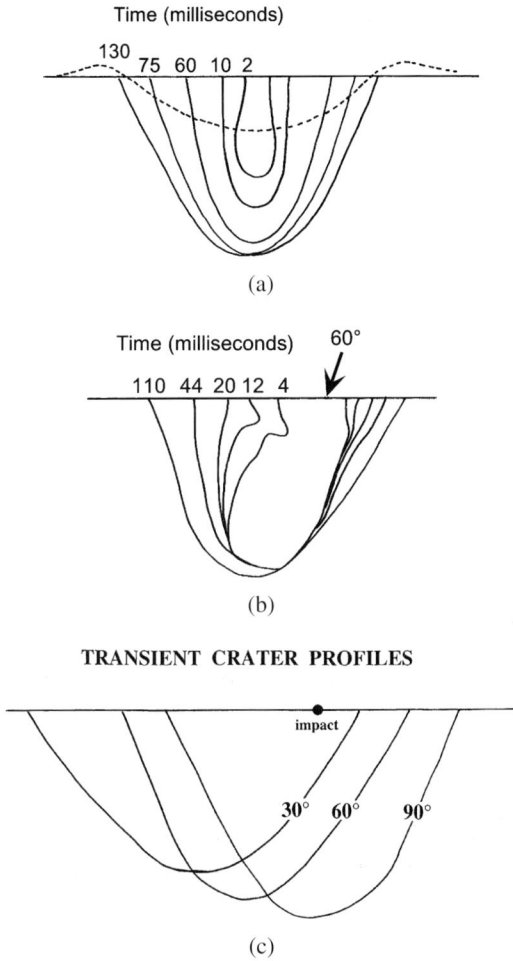

Figure 4. (a) Crater evolution for vertical hypervelocity ($\sim$6 km s^{-1}) impact by a 0.318 cm diameter pyrex sphere into sieved perlite using the quarter-space experimental design. An early-stage penetration funnel eventually expands into a deep paraboloid. The final crater (dashed line) is much smaller than the transient crater (at 130 ms) due to crater collapse and the return of high-angle ballistic ejecta to the cavity. (b) Crater evolution for an oblique (60°) hypervelocity (5.6 km s^{-1}) impact by a 0.318 cm diameter Pyrex sphere into sieved perlite revealed by quarter-space design. The initial penetration funnel eventually opens ("blooms") to produce a large transient crater before collapsing. (c) Comparison of maximum transient crater profiles for different-angle impacts into perlite for the same impact velocity and projectile size. Immediately after formation, transient craters collapse.

The quarter-space experiments specifically reveal four material displacement regimes that evolve during hypervelocity impacts (e.g., see Figures 4–7). A penetration funnel characterizes the first regime at early times as target material is compressed in front of the fragmenting projectile. This funnel expands cylindrically, similar to the mach tube creating during hypervelocity atmospheric entry (e.g.,

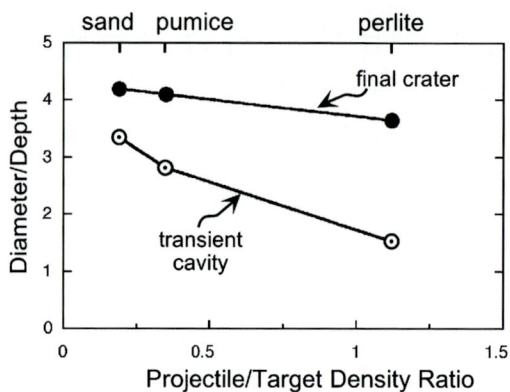

Figure 5. Effect of projectile/target density ratio on the crater aspect ratio (diameter/depth) for final (half-space target) and transient (quarter-space target) craters. Data are for hypervelocity (6 km s^{-1}) impacts at 60° from the horizontal.

Figure 6. Comparison of initial energy/momentum-transfer process revealed in quarter-space experiments using high-speed imaging (0.17 ms inter-frame time). High-angle (60°) impact into sand (a) produces an exposed flash and growing cavity lined with incandescent material over the first 0.5 ms. Vertical (90°) impact into sieved perlite (b) produces a hidden flash within a deep penetration funnel that later opens into a large transient crater (see Figure 4a). High-angle (60°) impact into sieved perlite catastrophically disrupts the pyrex sphere into fragments that disperse ahead of the penetration funnel and the subsequent cavity (c). Low-angle impact (30°) disperses melt target/projectile fragments at a shallow depth and produces a much larger maximum transient crater (d) as documented in Figure 3.

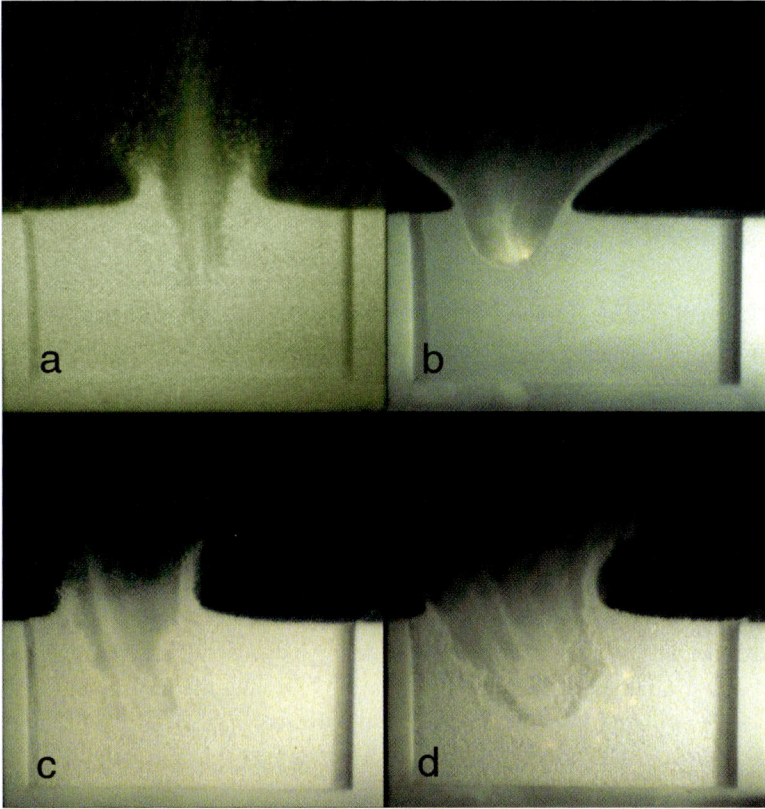

Figure 7. Late-stage (100 ms) transient craters for a vertical impact into perlite (a); high-angle (60°) impact into sand (b) and sieved perlite (c); and low-angle (30°) impact into sieved perlite (d).

Boslough *et al.*, 1994). In under-dense targets, the explosively disrupted/melted fragments continue into the target ahead of the funnel. In sand targets, these fragments line the growing cavity. The second stage of displacement is represented by high-angle (>80°) ejecta. Such high ejection angles develop as the result of an expanding cavity at depth, analogous to a deeply buried explosive charge. A pillar-like plume of high-angle ballistic ejecta characterizes the third stage and lasts throughout crater formation. This stage represents the combined effects of cavitation (inward flow) and escape. Highly directed vapor plumes also have been observed during hypervelocity impacts into dry ice targets (Schultz, 1996). The fourth stage is an outward-moving ejecta curtain in response to rarefaction-deflected excavation flow from the free surface as witnessed in experiments using sand targets (Figure 7b).

The collision of Comet Shoemaker-Levy with the atmosphere of Jupiter may provide an end-member case for a low-density projectile into a very low-density target at a much higher velocity (e.g., see Boslough *et al.*, 1994). This was not a solid-surface impact, and a final crater (and ejecta from the surface) did not form.

Nevertheless, ejecta (from the impactor) did produce a high-angle reverse plume similar to the early-time results shown in Figure 6 due to the deep penetration prior to fully coupling its energy.

3. Possible Crater Dimensions for the Deep Impact Crater

Can gravity scaling observed in the laboratory be extended to Tempel 1? This question may not be answered until 2005 when such an experiment finally will be made during the DI collision. Nevertheless, small bowl-shaped craters exist on Mathilde and generally resemble lunar craters formed in a thick regolith. Blocks and boulders around these craters also could be breccias created by compressed regolith materials. Impact experiments using sand do not generate such products, whereas impacts into pumice do create compressed sediments and melt products comparable in size to the projectile due to the lower melting temperature of the constituent grains.

One of the more critical unknowns will be the nature of the transmission of the compression wave through a highly porous regolith. If an AVGR experiment could be done under the same gravity as on Tempel 1 ($0.04 \, \mathrm{cm \, s^{-2}}$), the crater rim would extend almost 7 times farther with the peak pressure reduced by a factor of 50 (depending on the value of $2/\mu$ in Equation (7)) with the terminal ballistic ejection velocity approaching $10 \, \mathrm{cm \, s^{-1}}$ (versus $50 \, \mathrm{cm \, s^{-1}}$ at 1 g). Late-stage material motions observed in quarter-space experiments approach this velocity within the target but only contribute to rim uplift prior to collapse (rather than ejecta beyond the rim).

Table I provides summaries of expectations for gravity-controlled growth (Equation (5)) for different target densities extrapolated to a value of π_2 for the Deep Impact collision. Table II provides empirically derived values for the corresponding

TABLE I

Different materials (Extrapolated to Deep Impact).

	Target density 0.3 g cm^{-3}		Target density 1.0 g cm^{-3}	
	Size	Formation times	Size	Formation times
Pumice	$D_a = 89$ m	225	62 m	190
	$d_a = 22$ m		15 m	
Sand #140–200	$D_a = 141$ m	280	94 m	230
	$d_a = 35$ m		24 m	
Sand #140–200 (compression)	$D_a = 111$ m	250	74 m	205
	$d_a = 28$ m		19 m	
Sand (energy scaling)	$D_a = 238$ m	370	160 m	300
	$d_a = 60$ m		40 m	

TABLE II

Values of empirically derived constants and exponents.

	k	k'	α
Pumice	0.0963	0.756	0.518
Loose sand	0.240	1.30	0.51
Sand (compressible)	0.120	1.03	0.51
Sand (energy scaling)	0.0029	0.273	0.75

constants and exponents. These expectations provide maximum diameters and depths for proportional growth based on experiments as benchmarks. If the bulk density of the comet (and the impacted surface) is overestimated by a factor of two, the predicted crater diameter would be decreased by about 15%. The predicted crater-formation time (in seconds) for the craters in Table I could range from 200 to ~400 s if this regime applies. Three results for sand targets are shown. The first estimate directly extrapolates the experimental results. The second assumes that energy losses due to compaction reduce cratering efficiency at the outset but not the exponent α. The third estimate considers the possibility that energy scaling controls cratering efficiency with an exponent of $\alpha \sim 0.75$ (pure energy scaling). Energy scaling may apply when impact velocities greatly exceed the sound speed in the projectile, as well as in the target (Gault and Wedekind, 1977; Schultz and Gault, 1983; Schultz, 1988). Under these conditions, energy coupling occurs soon after first contact.

Both strength-controlled (late stage) and compression-dominated (early stage) effects may produce small craters close to the resolution limit of the DI high-resolution instrument (HRI). As discussed below, however, those two extremes can be distinguished by the nature of the ejecta plume (photometry and shape), in addition to the final crater morphology. The extrapolations shown in Table I assume proportional growth, but various late-stage hydrocode models (e.g., Orphal, 1977; Schultz et al., 1981) and experiments (e.g., Figures 4–7) indicate that craters reach their maximum crater depth before growing laterally. Consequently, a purely gravity-controlled crater produced on Tempel 1 should be shallower than craters observed in 1-g laboratory experiments, even without the effects of a stratified target. But if a regolith-covered substrate exists (CL model), it likely would produce one of the distinctive morphologies observed in small lunar craters (e.g., Quaide and Oberbeck, 1968). Experiments indicate that the gravity-controlled crater diameter is not appreciably affected for a surface layer depth three times the projectile diameter since the lateral shock would have been fully coupled (Schultz, 2000). In the inverse case (indurated surface over a porous substrate), experiments demonstrate the formation of large spall plates (Gault et al., 1968).

In a completely strengthless target, crater scaling should follow predictions, but any residual post-shock target strength at very large distances from the impact point would reduce these predictions. Conversely, rapid vapor expansion by volatiles in Tempel 1 also could result in backpressures (or exothermic release from chemical reactions) that might offset the effects of strength. Hypervelocity impact experiments using dry ice at the impact point have been observed to augment cratering efficiency at 1 g. In this case, vapor driven downward into the cavity creates a backpressure that boosts crater growth in strengthless or low-strength targets.

4. Crater and Ejecta Evolution

Witnessing crater formation on comet Tempel 1 will help constrain conflicting interpretations based only on the final morphology. Even though the crater will not be resolved at impact, the ejecta curtain profile will be imaged either directly or indirectly through its shadow crossing the cometary surface. Four observables will assist in interpretations: initial photometric evolution, shape and expansion of the ejecta curtain, opacity of the curtain during formation, and the total coma brightening due to the sudden addition of gas and dust to the coma.

First, photometry of the initial thermal plume will depend on the composition, porosity and structure of the target. Ongoing experiments are assessing these effects (Ernst and Schultz, 2002–2004) and initial results are described in the following section. Thermal radiation, however, will only partly control the light output. Strong aluminum oxide molecular bands due to reactions between the aluminum within the impactor ($\sim$6%) and cometary oxygen also may enhance peak intensity in the visible (as noted in laboratory experiments).

Second, the shape and expansion rate of the ejecta curtain (directly or through its shadow on the surface) will provide an indirect measure of crater growth (Figure 8). The advance of the ejecta curtain (speed and diameter at a given time) across the

---

Figure 8. Sequence showing ejecta evolution for a 60° impact at $\sim$5 km s^{-1} (0.635 cm diameter Pyrex sphere) into pumice covered by a thin layer of red dry paint powder. First frame (at impact) shows the rapidly expanding vapor and plasma plume directed downrange (to the left), followed by a conical curtain of incandescent ejecta (next 5 frames or $\sim$0.93 ms). After the fifth frame, the sequence is shown every 50 ms. A high-angle plume (directed back toward the initial trajectory) develops during the earliest stages but is surrounded by an annulus that expands across the surface after crater formation as ballistic ejecta within the curtain strike the surface. After crater formation, the curtain speeds up since it is composed of progressively faster ballistic ejecta.

Figure 11. Ejecta velocity vectors within a thin laser sheet cutting through the ejecta curtain $\sim$8 cm above the target surface at a given time. The laser sheet illuminates individual ejecta particles and an algorithm allows deriving the velocity vectors shown. Vertical impacts into loose sand (a) show that the flow-field is very well defined and symmetrical: ejecta crossing the laser plane occur in a narrow ring ejecta velocities are all very similar at a given time. A vertical impact into sieved perlite (b) shows a similar pattern (with slightly higher ejecta angle) but appears to be split. Very fine-grained powder in a vertical plume (see Figure 7a) blocks the laser used in the technique (see Anderson *et al.*, 2004) and produces the gap in velocity vectors in Figure 11b.

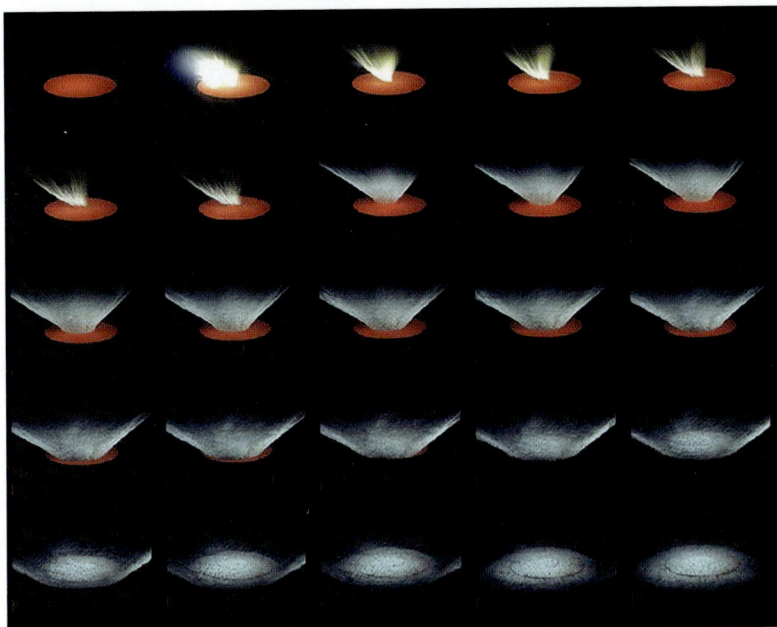

Figure 8.

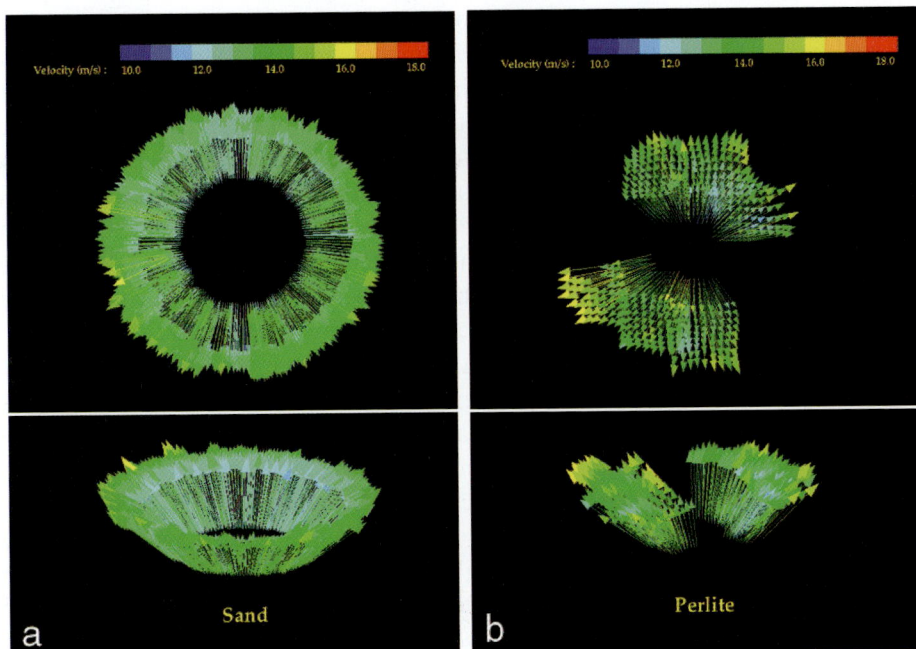

Figure 11.

surface provides a measure of the appropriate controlling variable, i.e., gravity or strength. If gravity constrains the maximum crater diameter, then empirically derived scaling equations can be used (e.g., Housen et al., 1983; Anderson et al., 2003, 2004) for extrapolation to the DI impact. More recent analyses reveal that such scaling equations depend on projectile/target density contrast and impact angle (e.g., Anderson et al., 2003, 2004). At late stages of growth (>50%) the following scaling relation applies for fine-grained sand targets:

$$v_e = \kappa (gR)^{1/2} \left(\frac{x}{R} \right)^{-w}$$

(8)

In Equation (8), κ and w are empirically derived values; R is the final apparent crater radius; and x is the radius of crater at time of ejection. Point source models (Housen et al., 1983) predict that the exponent will depend the coupling process (energy or momentum) at impact with values of w ranging from 1.5 (momentum) to 3.0 (energy). Experimental data for low-velocity impacts (aluminum at 1 km s^{-1}) into medium-grained sand yield a value of $w = 2.53$ (Anderson et al., 2004).

Large spall fragments or incandescent ejecta clumps may allow tracking trajectories through time. The advance of the ejecta curtain (Figure 8) across the surface, however, will provide additional clues not just for the value of gravitational acceleration on the surface but also for inferences about target properties. Laser sheets used in 3D-PIV studies slice the ejection curtain and provide a unique view of the advancing wall of particles. Figure 9a shows the diameter of the ring of ejecta illuminated by the laser (8.9 cm above the pre-impact target surface) as a function of time after impact (t) scaled by the total time of crater formation (T). The diameter of the curtain (C_d) at any given time reflects where the ejecta left the surface (stage of growth), ejection angle, and ejection velocity. Consequently, C_d will not follow a simple relationship for different impact angles very early in crater growth but will converge at later times. For oblique impacts (30° and 60°), curtain diameter is taken transverse to the trajectory. Laser-illuminated ejecta in the curtain actually left the surface at an earlier stage of crater growth even when scaled time indicates that the crater has finished forming ($t/T \sim 1.0$) as described in Anderson et al. (2003).

The expansion velocity of the curtain with scaled time is shown in Figure 9b. Curtain velocity is now normalized to gravity-scaled ejecta velocity and reveals that this strategy accommodates a wide range of impact speeds and impact angles. The common relationship for very different experimental conditions reflects the horizontal velocity component of ballistic ejecta. At launch, ejecta speeds decrease with scaled time raised to an exponent of $-\gamma$ where $\gamma = 0.70$ for low-velocity impacts into sand. This exponent is higher for hypervelocity impacts when the projectile completely fragments at impact and approaches energy scaling (Schultz and Gault, 1983; Schultz, 1988). Curtain velocities progressively decrease with time but then increase after the crater finishes forming, as higher speed ejecta comprise the curtain at a given height above the surface. Values for a gravity-controlled 141 m-diameter crater produced by the DI collision are also shown in Figure 9b. Curtain velocities decrease to about 30 cm s^{-1} at about one crater radius above the surface of the comet (in this case ~70 m) for the case of sand-like targets with gravity-controlled growth. Specific values on the ordinate to the right will increase by $2^{1/2}$ for a surface gravity twice as high. If gravity limits crater growth, the minimum outward curtain velocity will be greater when $t/T_c = 1$. Unusual target properties such as underdense materials (Figures 4–6) will affect specific values by increasing ejection angles and changing crater growth rates.

In gravity-controlled cratering, the expansion velocity of the ejecta curtain at the base initially decreases until the crater has finished forming since the curtain is tied to the growing cavity. Crater formation can be inferred when the ejecta curtain begins to make a sharp angle with the surface. At this time the curtain diameter at the base is about 25% larger than the final transient crater diameter. After crater

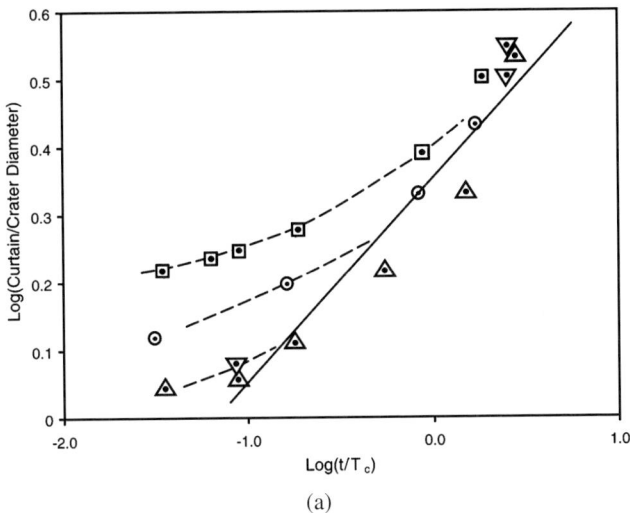

(a)

Figure 9. (a) Outward expansion of the ejecta curtain as a function of time (scaled by the total time for crater formation). Late-time expansion of the curtain for different impact conditions converge on a slope of 0.29 as expected for gravity-scaled growth. Early-time expansion of the curtain departs from expectations due to departure from a point source assumption and the effects of crater growth (ejecta launch angle from the surface). The following symbols apply for different impact conditions: 0.318 cm diameter aluminum spheres at ~5.5 km s^{-1} (circles for 90°; squares for 30°); 0.635 cm diameter spheres at ~1 km s^{-1} (upside-down triangle for 90°; triangles for 60°). (b) Outward curtain expansion speed observed in laboratory experiments for gravity-controlled growth. Speeds (ordinate to the right) for specific times (upper abscissa) are shown for the DI collision with the assumption of gravity-controlled crater scaling. Increase in growth after the crater has formed ($t/T = 1$) is due to the curtain being composed of faster ejecta launched from earlier times. If the comet is composed of weakly bonded grains, then the minimum observed velocity will be truncated. In both cases, height of the curtain is about 0.5 crater diameter (final) above the surface. Symbols are the same as in Figure 9a.

(Continued on next page)

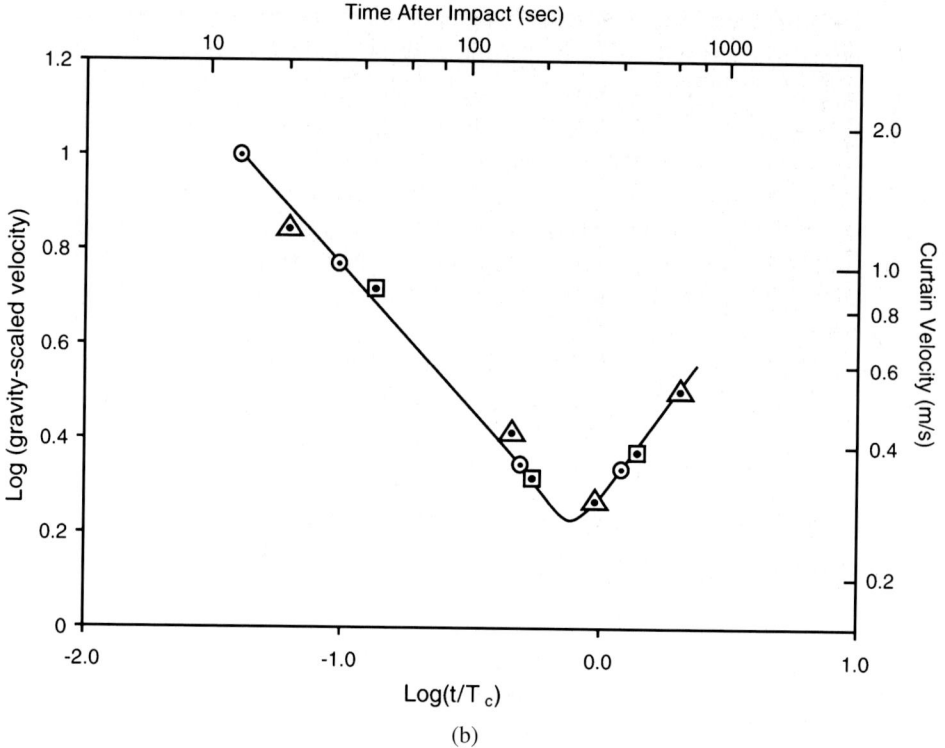

Figure 9. (Continued)

formation, the curtain diameter at the base progressively increases in velocity since the constituent ejecta were launched at higher velocities (e.g., Schultz and Gault, 1979). If strength limits crater growth, the ejecta curtain may detach from the crater (and perhaps even the comet).

Very high porosity targets will produce high ejection angles (60°) at late times in contrast with sand and pumice targets (45°–55°). Under-dense, compressible targets result in a two component, late-stage ejecta curtain (Figure 10): first, a vertical plume due to cavitation (temporary containment and redirection of vapor by the transient crater below the surface); second, an outward-moving curtain due to rarefaction-controlled excavation off the free surface. Figure 11a illustrates the vector-velocity field of particles within the curtain during ejection using the 3D-PIV technique for a loose sand target (see Anderson *et al.*, 2003, 2004). Figure 11b illustrates this velocity field for the ejecta curtain from a sieved perlite target. The gap in Figure 11b is due to fine (micron-size) debris within an opaque vertical plume that occults the laser beam used to image the slice through the ejecta curtain. In both examples, the ejecta have not been extrapolated back to the target surface. Figure 12 compares the final craters and the image slices for impacts into sand,

Figure 10. Impact into sieved perlite at 60° (0.318 cm diameter Pyrex sphere at $\sim$5.6 km s^{-1}) results in a two-component ejecta plume at late stages: a high-angle plume due to deep penetration and an outward-advancing curtain related to shock rarefaction off the free surface. The high-angle plume gradually evolves as the cavity opens.

pumice, and perlite targets. At this stage of crater growth, the laser illuminates only the central, vertical plume of ejecta.

The third observable is the opacity of the curtain through time, which is related to both the evolving number density within the outward-advancing ejecta curtain and crater scaling. It is very unlikely that the ejecta curtain will form a uniform sheet as in experiments using a narrow range of particle sizes for the purposes of scaling relations. Instead, experiments using natural particulates demonstrate that the curtain forms clumps and gaps due to inhomogeneities in the target that evolve through time.

Theoretical and empirical models of ejecta curtain growth can be readily constructed from existing analytical models based on scaling theory or by extrapolating direct observations of the ejecta-mass velocity field, e.g., using data from 3D-PIV observations (see Anderson *et al.*, 2003, 2004). This technique provides simultaneous measurements of the evolution of the ejecta curtain dimensions (size and width), opacity, constituent ejection angles, and ejection velocities for a wide range of projectile and target properties. Even though gravity eventually limits crater growth, measured particle motions in the target beyond the rim allow assessing peak pressure decay and material flow affecting the potential growth at much lower gravity.

The fourth observable is the total brightening of the comet due to the sudden contribution of gas and dust into the coma. For sand targets, the total ejected mass is

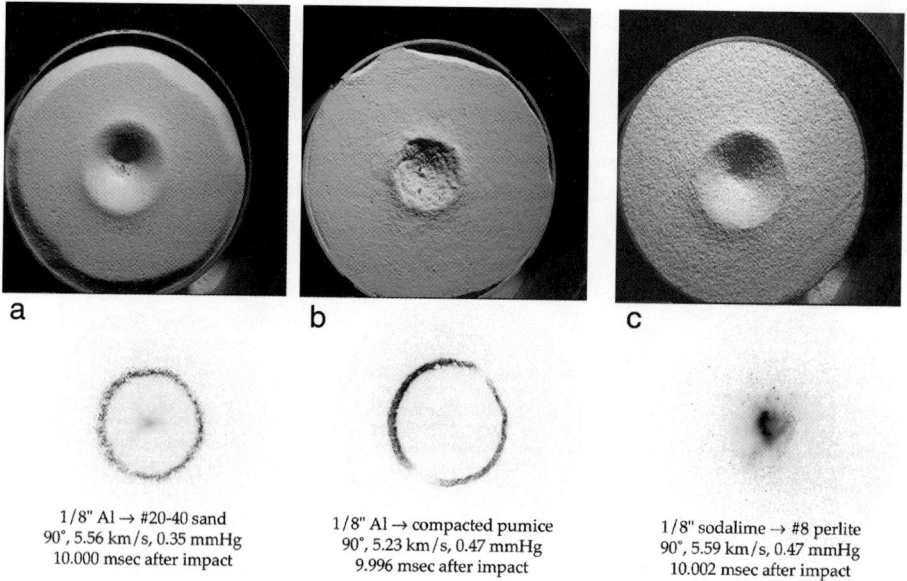

1/8" Al → #20-40 sand	1/8" Al → compacted pumice	1/8" sodalime → #8 perlite
90°, 5.56 km/s, 0.35 mmHg	90°, 5.23 km/s, 0.47 mmHg	90°, 5.59 km/s, 0.47 mmHg
10.000 msec after impact	9.996 msec after impact	10.002 msec after impact

Figure 12. Comparison of final craters and ejecta illuminated by a laser sheet for sand (a), pumice (b), and perlite (c). The laser sheet illuminates individual ejecta particles as they pass through the plane. The illuminated individual grains have been reversed (black) in order to assist visualization. At any given time, ejecta from impacts into sand and pumice are confined within a narrow sheet (curtain) having a narrow range of velocities (see Figure 11a). The times correspond to dimensionless times of $\tau = 17512$ (12a), $\tau = 16,466$ (12b), and $\tau = 17610$ (12c). The central debris plume in Figure 12c is responsible for the split ejecta pattern in Figure 11b, as also shown in Figure 7a in quarter-space experiments. Note that the final crater in perlite appears only slightly larger than the crater in sand. This is the result of high-angle ejecta returning to the crater and the effect of rim collapse. Narrow band-pass filters minimized the effect of thermal radiation in the image.

about 50% of the total displaced mass shown in Figures 2 and 3. With the assumption of gravity-controlled growth as a reference, the DI impact could add 10^7–10^8 kg of dust and gas to the coma over the 3 to 6 min of crater formation. Late stages of ejection contain most of the mass but have low velocities (< 10 m s^{-1}); consequently, the effect of this component will take hours to days to have a visible effect from Earth. For highly compressible targets (Figures 4–7), the total ejected mass may be reduced by a factor of 3 to 4 but may be collimated within a narrow cone (30° to 40° solid angle) in a region between the direction of the trajectory and the local surface normal.

5. Impact Radiation

5.1. IMPACT FLASH

The emitted light energy from impacts is commonly termed the "impact flash", a term carried over from micro-particle experiments (e.g., Eichhorn, 1976) since the duration is less than a microsecond. More recent studies using macroscopic projectiles ($>500\,\mu$m) into porous silicate targets demonstrate, however, that this term is a misnomer because the radiation can last well over a millisecond (e.g., Ernst and Schultz, 2003). When scaled to the penetration time (Equation (6)), this represents a value of τ exceeding 1000. Light emissions can be characterized by two sources: atomic and molecular emissions from impact-generated vapor and blackbody radiation, whether from melt phases or vapor condensates. An extensive series of laboratory experiments are providing important clues for expectations for the Deep Impact collision. Despite the difference in scale, scaling relations for different substrate types (porous, easily volatized, competent) allow first-order extrapolations. The goal here is to estimate the intensity evolution and character (color temperature, peak intensity, decay) as a function of substrate type and impact angle in order to understand the underlying causes (or at least limit the possibilities) for the final crater appearance. The detailed view of light emissions from laboratory experiments will not be resolved fully by either spacecraft or earth-based instruments. But understanding the underlying causes for observed light emission requires such details. Consequently we first review expectations for the total visible and near infrared emissions.

5.2. INTEGRATED LUMINOUS INTENSITY

For a purely kinetic event (kinetic energy transferred to luminous energy), the Stefan-Boltzmann relation requires the following for the radiance, R, at the source for condensed phase radiation:

$$R = \sigma T^4 \tag{9}$$

where total radiance is in $\mathrm{W\,m^{-2}}$; σ is the Stephan-Boltzmann constant ($5.67 \times 10^{-8}\,\mathrm{W\,m^{-2}\,K^{-4}}$); and T is the color temperature (K). The total luminous energy, LE, is then:

$$\mathrm{LE} = (\sigma T^4)(A_s \Delta t) \tag{10}$$

where A_s represents the effective source area and Δt is the duration of the emissions. More accurately, LE should be written as:

$$\mathrm{LE}(\theta) = \int_0^t \mathrm{LE}(\theta, t)\, dt$$
$$= \sigma \int_0^t T_\theta^4 A_\theta\, dt \tag{11}$$

where the radiating source color temperature and area vary with time for a given impact angle (θ) and target.

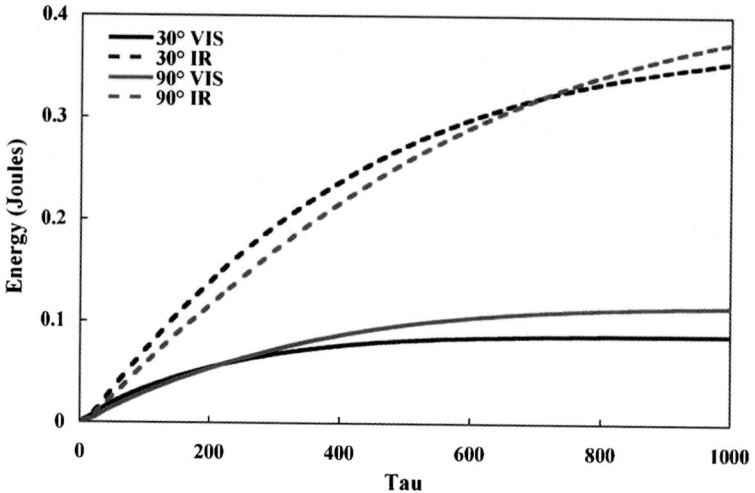

Figure 13. Observed cumulative radiant energy for hypervelocity impacts ($5.5 \, \text{km s}^{-1}$) into pumice at 90° and 30° in laboratory experiments. Radiant energy is calculated from the observed blackbody (color) temperature and source area through time. The evolution of the radiant energy in the visible (0.34 to 1.0 μm) and infrared (1.0 to 4.8 μm) is shown.

Various studies (Melosh *et al.*, 1993; Nemtchinov *et al.*, 1999, 2000; Artm'eva *et al.*, 2000) suggest that LE can be expressed as a fraction of the kinetic energy of the impact (KE_i):

$$LE = \eta \cdot KE_i. \tag{12}$$

Previous estimates suggested that η (the luminous efficiency) might range from 10^{-5} to 10^{-6} depending on the target. These efficiencies are based on gas phase radiation and are consistent with direct measurements of efficiencies in the laboratory (Sugita *et al.*, 2003).

Hypervelocity impacts into silicate targets (velocities of 5 to 6 km s^{-1}), however, indicate a much higher total (all wavelengths) luminous efficiency of 10^{-4} to 10^{-3}. Observations of color temperature and radiating source area through time (see Ernst and Schultz, 2004) allow integrating Equation (11) to give the cumulative radiant energy with dimensionless time, τ, for 90° (vertical) and 30° impacts into pumice at around 5.6 km s^{-1} as shown in Figure 13. Here the radiant energy is shown only for the wavelength ranges of interest to the DI instruments for the visible CCD (0.34 to 1.0 μm) and the infrared spectrometer (1.0 to 4.8 μm). The total integrated radiant flux for all wavelengths represent about 0.7 joules over $\tau = 1000$, which represents a luminous efficiency of about 3.5×10^{-4} largely due to heated silicates that incandesce long after first contact.

If the light energy is a simple fraction of the impact energy for a given target, then Equation (12) requires that $(R \cdot A_s \cdot \Delta t)$ is proportional to impact energy as well. Each of the independent variables (color temperature, area, and duration) can be directly related in laboratory impact experiments in order to test the assumption that η is approximately constant for a given target. Such measurements are part of an ongoing experimental study for the radiation from the thermal (Ernst and Schultz, 2002, 2003, 2004) and atomic/molecular emission lines (Sugita and Schultz, 1999; Sugita et al., 1998; Sugita and Schultz, 2003a, b; Eberhardy and Schultz, 2003, 2004).

For purposes here, it is assumed that the laboratory experimental results can be used to derive the color temperature, source area, and duration as a function of impactor size and velocity. This approach strictly applies only to the silicate targets used in these experiments. The rationale for using silicates is that minimal vaporization occurs at the available impact velocities, thereby minimizing the complicating effect of phase changes and energy expended in atomic/molecular transitions. The possible effect of volatile-rich components is discussed below.

Experiments indicate that the peak radiant flux is proportional to v^ω where $\omega \sim 5.5$ (see Ernst and Schultz, 2002; 2003). Such a strong dependence on impact velocity seems counterintuitive but the same relation is found for atomic-emission line intensity (Sugita et al., 1998). After the initial peak intensity decays, the experimental ω approaches 3, a value consistent with conservation of energy (see Schultz et al., 2004). This also can be expressed as $T \sim v^{0.75}$. Consequently, the integrated radiant flux for DI (LE_{DI}) relative to the laboratory experiments (LE_L) is given by:

$$\frac{\mathrm{LE}_{\mathrm{DI}}}{\mathrm{LE}_L} \sim \left(\frac{T_{\mathrm{DI}}}{T_L}\right)^4 \left(\frac{A_{\mathrm{DI}}}{A_L}\right) \left(\frac{\Delta t_L}{\Delta t_{\mathrm{DI}}}\right) \sim \left(\frac{a_L}{a_{\mathrm{DI}}}\right)^3 \left(\frac{v_{\mathrm{DI}}}{v_L}\right)^2. \tag{13}$$

The projectile ratio of about $(90/0.476)^3$ and velocity ratio of $(10.2/5.5)^2$ would indicate an integrated radiance approximately 2.3×10^7 times that observed in the laboratory experiments. This would scale to a total of about 15 MJ of light energy for the DI collision (luminous efficiency of $\sim 0.08\%$). Such a thermal plume could last more than 60 ms.

Experiments indicate that the luminous efficiency could be almost an order of magnitude greater for a silicate-rich, particulate target. Conceivably, it could increase to 6th magnitude, depending on the duration and visibility (e.g., view angle). Figure 14 provides results of a more complete extrapolation where the temperature and source area are scaled to the DI collision and integrated over time for the visible camera and the infrared spectrometer. The sampling rate of the MRI is about 0.06 s at the time of impact. Consequently, the event might be captured in only one or two frames. It also could extend over 10–50 frames if the comet surface contains a particulate silicate-rich surface.

The estimates for the radiant energy (apparent magnitude) and duration depend on several important factors. First, the porosity of the target dramatically affects the peak intensity, decay, and duration (Ernst and Schultz, 2003). Highly porous particulate targets (such as sieved perlite used in laboratory experiments) produce a peak intensity reduced by half and a duration shortened by more than two orders of magnitude. Second, the nature of the target (e.g., solid pumice vs. particulate

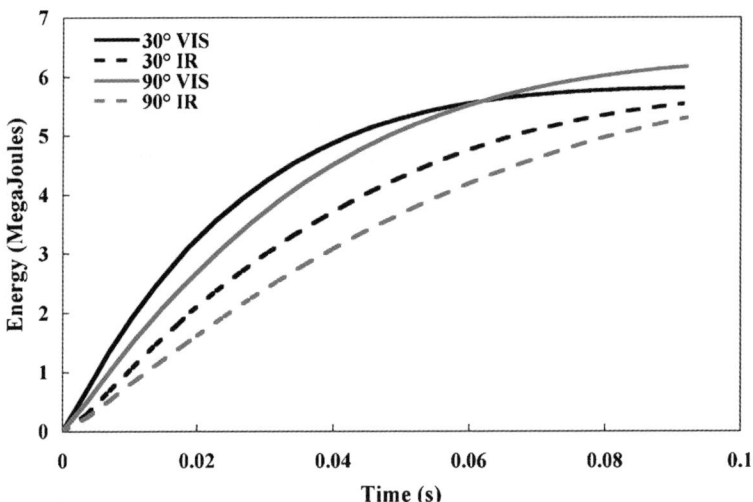

Figure 14. Observed cumulative radiant energy extrapolated from observations of color temperature and radiating source area in laboratory experiments (Figure 13) to the Deep Impact collision as discussed in the text.

pumice) also reduces the peak intensity by half and duration by a factor of two. Third, the target volatile content dramatically affects the blackbody radiation responsible for the longer duration radiation. The greater volatile content results in greater radiant energy represented in emission lines over limited wavelengths. Such sources are subject to the issues of self-absorption, ionization, and optical depth (see Sugita and Schultz, 2003a, b; Sugita *et al.*, 2003), thereby complicating unique interpretations of this component by the available DI instruments. Fourth, interactions between DI ejecta and the surrounding coma gas may create secondary radiation sources, thereby enhancing and prolonging the inferred luminosity and duration greatly (e.g., Sugita and Schultz, 2003a). In spite of such caveats, the ongoing laboratory experiments provide key strategies for using various instruments to interpret the nature of the impacted substrate, particularly if the crater is unexpectedly small.

6. Concluding Remarks

The Deep Impact "hard" encounter represents a unique large-scale experiment. In laboratory experiments, all independent variables are controlled. For the Deep Impact experiment, the composition, structure, density (porosity), gravity, and impact angle are not known or will not be well constrained until after the impact. Nevertheless, results of laboratory experiments allow predictions for the DI event:

1.) Smaller than expected crater (<50 m in diameter)

    a.) If the crater is smaller than predicted due to post-shock surface strength (SC model), then the crater will nevertheless be deep due to the anticipated low density ($<0.5 \, \mathrm{g \, cm^{-3}}$) that allows deep penetration. The ejecta plume will initially expand spherically above the target, followed by a vertical, high-temperature plume (see Figure 15). Spall fragments will dominate late-stage excavation; a well-defined ejecta curtain will not develop. The flash will be very short (several frames in the MRI).

    b.) Large pockets (1 to 10 m) of highly porous (>80%) and compressible material embedded in a matrix of fine particulates could reduce the coupling of the initial shock to the comet. The weak shock front then might not overcome the interparticle binding forces in the matrix or frictional drag could arrest growth before achieving the gravity limit.

    c.) The high encounter velocity still should induce catastrophic disruption of the DI probe during entry and result in a large transient crater. An extremely

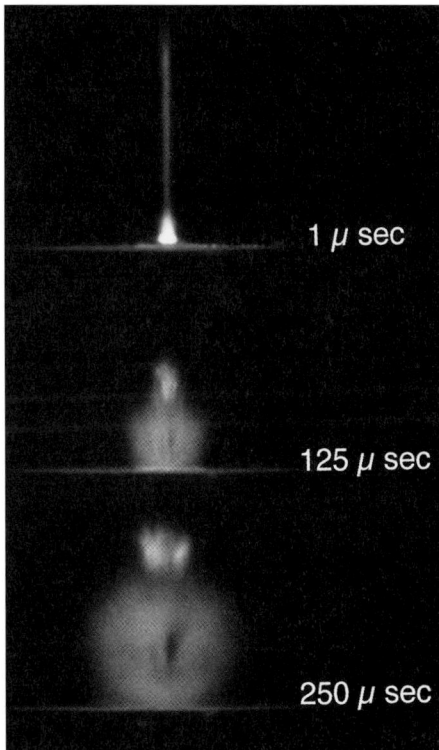

Figure 15. Vertical impact by a 0.635 cm diameter pyrex sphere into dry ice (side view) at $5.4 \, \mathrm{km \, s^{-1}}$ under near-vacuum conditions. In the first frame a high–velocity jet-like spike is directed out of the small penetration opening. This "jet" is actually a parcel of hot gas that is smeared in the first frame due to its high velocity ($\sim 3 \, \mathrm{km \, s^{-1}}$). A spherical cloud of CO and CO_2 vapor expands above the impact followed by a high-temperature plume that "blooms" above the spherical plume after 125 μs.

high porosity (80%) and compressibility, however, may produce a final crater that is unexpectedly smaller primarily due to wall collapse of the transient cavity, even at $g = 0.04\,\mathrm{cm\,s}^{-2}$. In this scenario (and in Scenario 1b), a three-component plume should be observed in the MRI and may allow the identification of this surface property.

   i. $T = 0.06\,\mathrm{s}$ (first frame): very weak flash.
   ii. $T = 1\,\mathrm{s}$ (~17 frames): high-angle jet-like plasma plume due to cavitation and redirection back out the penetration funnel.
   iii. $T = 1$ to $20\,\mathrm{s}$: opaque spherical cloud forming a shadow above the comet surface.
   iv. $T = 20\,\mathrm{s}$ to $100\,\mathrm{s}$: high-angle ejecta plume extending $>10\,\mathrm{km}$ above the surface surrounded by a high-angle (~60°) ejecta curtain.
   v. $T > 200\,\mathrm{s}$: advancing annulus of disturbed cometary surface near the crater as low-velocity components return. High-angle central plume may detach from the crater due to collapse and shut off of deep ejecta.

2.) Nominal crater dimensions (~80 to 130 m)

   a.) A crater produced in a lunar regolith-like material (PR model) should grow with a final diameter:depth (rim-to-floor) ratio of 4:1 to 5:1.
   b.) If the near-surface material is volatile-rich (silicate poor), a well-defined ejecta curtain will be delayed until very late time (~100 s). The flash will be weak and dominated by atomic and molecular emission lines with little blackbody until solid condensates form. Strong absorptions may develop as cooler gases expand in front of the thermal background.
   c.) If the near-surface material is silicate rich (volatile poor), a well-defined ejecta curtain will evolve as in laboratory experiments. The flash will be largely thermal and of long duration (up to 1 s) as ejecta fragments cool. A strong thermal source will remain on the crater floor but cooling melt breccias (1 to 10 cm) may be traced to large distances in the IR.

3.) Larger than expected (>180 m)

   a.) Backpressures created by vapor expansion may augment crater growth, particularly if deep penetration occurs. Exothermic reactions also may contribute if metastable components (C, H, N, O) at cold conditions react under high temperatures or if latent heat is released at depth due to phase changes. The classic ejecta curtain may not develop due to rapid expansion and dissipation of the vapor phases. The crater may continue to "feed" volatiles to the coma as the thermal wave penetrates more deeply into the comet.
   b.) Extremely low density cometary surface ($0.1\,\mathrm{g\,cm}^{-3}$ down to a depth of <50 m depth) could produce a much larger than expected crater since cratering efficiency reflects mass, not volume (UR model). The evolution

of the ejecta plume and curtain (direct or in shadow) would follow Scenario 1b.

4.) Craters in composite or layered surface materials (CL model)

    a.) A layered comet structure with weak, low-density surface (1–10 m deep) over a competent layer would result in a large but shallow crater with a central penetration pit. The diameter of the outer crater would depend on $(\delta_p/\delta_t)^{1/3}$ of the surface layer. A two-component ejecta curtain should be evident: a vertical plume surrounded by a thin (less opaque) advancing curtain.

    b.) A layered comet structure with a strong surface layer over a low-density substrate would result in an irregularly shaped crater with radial and concentric fractures, large ejecta spalls and rayed ejecta. The crater rim would be very ill defined and bulbous due to under-thrusting and upward displacement followed by collapse.

The qualitative predictions above represent a combination of direct observations from laboratory experiments and the morphology of small (<100 m) craters on the lunar surface (e.g., Schultz, 1976). The wide range in alternative scenarios underscores the fact that the results of this mission will provide a new understanding of not only comet properties but also key physical processes related to impact cratering at a much larger scale and higher impact velocities.

Acknowledgments

The authors gratefully acknowledge the professional assistance by the technical crew (D. Holt, D. Bowling, R. Smythe, and C. Cornelison) at the NASA Ames Vertical Gun Range, funded through the Planetary Geophysics Program, Office of Space Science, and NASA Headquarters. The research was made possible by support from NASA's Deep Impact Discovery mission, managed through the University of Maryland and Michael A'Hearn. The authors also gratefully acknowledge the assistance of J. T. Heineck for the 3D-PIV data presented here, as well as fruitful discussions with Seiji Sugita.

References

Adams, M. A., Schultz, P. H., Sugita, S., and Goguen, J. D.: 1997, *Lunar Planet. Sci. XXVIII*, Abstract, p. 3.

A'Hearn, M., Belton, M. J. S., Delamere, A., and Blume, W. H.: 2005, Space Science Reviews (this volume).

Anderson, J. L. B., Schultz, P. H., and Heineck, J. T.: 2003, *J. Geophys. Res.* **108**, E8, doi:10.1029/2003JE002075.

Anderson, J. L. B., Schultz, P. H., and Heineck, J. T.: 2004, *Meteor. Planet. Sci.* **39**, 303.

Artem'eva, N. A., Kosarev, I. B. V., Nemchinov, I., Trubetskaya, I. A., and Shuvalov, V. V.: 2000, *Solar System Res.* **34**, 453.

Boslough, M. B., Crawford, D. A., Robinson, A. C., and Trucano, T. G.: 1994, *Geophys. Res. Lett.* **21**, 1555.

Brownlee, D. E. *et al.*: 2004, *Science* **304**, 1764.

Chabai, A. J.: 1977, in Roddy, D. J., Pepin, R. O., and Merrill, R. B. (eds.), *Impact and Explosion Cratering*, Pergamon Press, New York, p. 1191.

Chapman, C. R. and McKinnon, W. B.: 1986, in Burns, J. A. and Matthews, M. S. (eds.), *Satellites*, University of Arizona Press, Tucson, p. 492.

Cintala, M. J., Berthoud, L., and Hörz, F.: 1999, *Meteor. Planet. Sci.* **34**, 605.

Dahl, J. M. and Schultz, P. H.: 1999, *Lunar Planet Sci. XXX*, LPI, Abstract No. 1854.

Dahl, J. M. and Schultz, P. H.: 2001, *Internat. J. Impact Engin.* **26**, 145.

Dienes, J. K. and Walsh, J. M.: 1970, in Kinslow, R. (ed.), *Hypervelocity Impact Phenomena*, Academic Press, New York, p. 45.

Eichhorn, G.: 1976, *Planet. Space Sci.* **24**, 771.

Eberhardy, C. A. and Schultz, P. H.: 2003, *Lunar Planet. Sci. Conf. XXXIV*, Abstract No. 1855.

Eberhardy, C. A. and Schultz, P. H.: 2004, *Lunar Planet. Sci. Conf. XXXV*, Abstract No. 2039.

Ernst, C. M. and Schultz, P. H.: 2002, *Lunar Planet. Sci. Conf. XXXIII*, Abstract No. 1782.

Ernst, C. M. and Schultz, P. H.: 2003, *Lunar Planet. Sci. Conf. XXXIV*, Abstract No. 2020.

Ernst, C. M. and Schultz, P. H.: 2004, *Lunar Planet. Sci. Conf. XXXV*, Abstract No.1721.

Gault, D. E. and Wedekind, J. A.: 1977, in Roddy, D. J., Pepin, R. O. and Merrill, R. B. (eds.), *Impact and Explosion Cratering*, Pergamon Press, New York, p. 1231

Gault, D. E. and Wedekind, J. A.: 1978, *Proceedings Lunar Planet. Sci. Conf. 9th*, 3843.

Gault, D. E., Quaide, W. L., and Oberbeck, V. R.: 1968, in French, B. M. and Short, N. M. (eds.), *Shock Metamorphism of Natural Materials*, Mono Books, San Francisco, p. 87.

Gault, D. E., Guest, J. E., Murray, J. B., Dzurisin, D., and Malin M. C.: 1975, *J. Geophys. Res.* **80**, 2444.

Holsapple, K. A.: 1987, *Int. J. Impact Engin.* **5**, 343.

Holsapple, K. A. and Schmidt, R. M.: 1982, *J. Geophys. Res.* **87**, 1849.

Holsapple, K. A. and Schmidt, R. M.: 1987, *J. Geophys. Res.* **92**, 6350.

Housen, K. R., Schmidt, R. M., and Holsapple, K. A.: 1983, *J. Geophys. Res.* **88**, 2485.

Housen, K. R., Holsapple, K. A., and Voss, M. E.: 1999, *Nature.* **402**, 155.

Melosh, H. J., Artemieva, N. A., Golub, A. P., Nemtchinov, I. V., Shuvalov, V. V., and Trubetskaya, I. A.: 1993, *Lunar Planet. Sci. Conf. XXIV*, 975.

Nemtchinov, I. V., Shuvalov, V. V., Artemieva, N. A., Kosarev, I. B., and Trubetskaya, I. A.: 1999, *Int. J. Impact Engin.* **23**, 651.

Nemtchinov, I. V., Spalding, R. E., Shuvalov, V. V., Artem'eva, N. A., Kosarev, I. B., and Popel, S. I.: 2000, *Lunar Planet. Sci. Conf.* XXXI, Abstract No. 1334.

O'Keefe, J. D. and Ahrens, T. J.: 1977, *Lunar Planet. Sci. Conf.* VIII, 3357.

O'Keefe, J. D., Stewart, S. T., and Ahrens, T. J.: 2002, *Lunar and Planet Sci. XXXIII*, Abstract No. 2002.

Orphal, D. L.: 1977, in Roddy, D. J., Pepin, R. O., and Merrill, R. B., (eds.), *Impact and Explosion Cratering: Overviews*, Pergamon Press, New York, p. 907.

Piekutowski, A. J.: 1977, in Roddy, D. J., Pepin, R. O. and Merrill, R. B. (eds.), *Impact and Explosion Cratering*, Pergamon Press, New York, p. 67.

Post, R. L.: 1974, *Rep. AFWL-TR-74-51*, Air Force Weapons Lab, Kirtland, AFB, N. M.

Quaide, W. L. and Oberbeck, V. R.: 1968, *J. Geophys. Res.* **73**, 5247.

Schmidt, R. M.: 1980, *Proc Lunar Planetary Science Conference XI*, 2099.

Schmidt, R. M.: 1977, in Roddy, D. J., Pepin, R. O., and Merrill, R. B. (eds.), *Impact and Explosion Cratering: Overviews*, Pergamon Press, New York, p. 1261.

Schmidt, R. M. and Holsapple, K. A.: 1982, in Silver, L. T. and Schultz, P. H. (eds.), *Geological Implications of Impacts of Large Asteroids and Comets on the Earth*, Geological Society of America Special Paper, p. 93.

Schmidt, R. M. and Housen, K. R.: 1987, *Int. J. Impact Eng.* **5**, 543.

Schultz, P. H.: 1976, *Moon Morphology: Interpretations Based on Lunar Orbiter Photography*, University of Texas Press, Austin, 626 pp.

Schultz, P. H.: 1988, in Vilas, F., Chapman, C., and Mathews, M. (eds.), *Mercury*, University of Arizona Press, Tucson, p. 274.

Schultz, P. H.: 1992, *J. Geophys. Res.* **97(E10)**, 16,183.

Schultz, P. H.: 1996, *J. Geophys. Res.***101(E9)**, 21,117.

Schultz, P. H., Adams, M. A., Perry, J. W., and Goguen, J.: 1996, *Lunar Planet. Sci. Conf. XXVII*, Abstract, p. 1149.

Schultz, P. H.: 2000, *Lunar Planet. Sci. Conf.* XXXI, Abstract No. 2071.

Schultz, P. H.: 2001, *Bull. Am. Astron. Soc.* **33**, 1095.

Schultz, P. H.: 2003a, *Lunar Planet. Sci. Conf.* XXXIV, Abstract No. 2067.

Schultz, P. H.: 2003b, *Sixth International Conference on Mars*, Abstract No. 3263.

Schultz, P. H. and Gault, D. E.: 1979, *J. Geophys. Res.* **84**, 7669.

Schultz, P. H. and Gault, D. E.: 1983, *Lunar and Planet. Sci. XV*, Abstract, p. 730.

Schultz, P. H. and Gault, D. E.: 1985a, *J. Geophys. Res.* **90**, 3701.

Schultz, P. H. and Gault, D. E.: 1985b, *Lunar and Planet. Sci. XVI*, Abstract, p. 742.

Schultz, P. H. and Gault, D. E.: 1990, in Silver, L. and Schultz, P. (eds.), *Geological Implications of Impacts of Large Asteroids and Comets on the Earth*, Geol. Society of Amer. Spec. Paper. 247, p. 239.

Schultz, P. H., Orphal, D., Miller, B., Borden, W. F., and Larson, S. A.: 1981, in Merrill, R. B. and Schultz, P. H., (eds.), *Proceedings of the Conference on Multi-Ring Basins: Formation and Evolution*, Pergamon Press, New York, p. 181.

Schultz, P. H., Anderson, J. L. B., and Heineck, J. T.: 2002, *Lunar Planet. Sci. Conf. XXXIII*, Abstract No. 1875.

Schultz, P. H., Sugita, S., Eberhardy, C. A., and Ernst, C. M.: 2004, *Lunar Planet. Sci. Conf. XXXV*, Abstract No. 1946.

Soderblom, L. A., *et al.*: 2002, Science **296**, 1087.

Stöffler, D., Gault, D. E., Wedekind, J., and Polkowski, G.: 1975, *J. Geophys. Res.* **80**, 4062.

Sugita, S., Schultz, P. H., and Adams, M. A.: 1998, *J. Geophys. Res.* **103(E8)**, 19,427.

Sugita, S. and Schultz, P. H.: 1999, *J. Geophys. Res.* **104(E12)**, 30,825.

Sugita, S. and Schultz, P. H.: 2003a, *J. Geophys. Res.* **108(E6)**, 5152, doi: 10.1029/2002JE001960.

Sugita, S. and Schultz, P. H.: 2003b, *J. Geophys. Res.* **108(E6)**, 5151, doi: 10.1029/2002JE001959.

Sugita, S., Schultz, P. H., and Hasegawa, S.: 2003, *J. Geophys. Res.* **108(E12)**, 5140, doi:10.1029/2003JE002156.

IMPACT CRATERING THEORY AND MODELING FOR THE DEEP IMPACT MISSION: FROM MISSION PLANNING TO DATA ANALYSIS

JAMES E. RICHARDSON[1,*], H. JAY MELOSH[1], NATASHA A. ARTEMEIVA[2]
and ELISABETTA PIERAZZO[3]

[1]*Lunar and Planetary Laboratory, University of Arizona, Tucson, AZ, U.S.A.*
[2]*Institute for the Dynamics of the Geospheres, Moscow, Russia*
[3]*Planetary Science Institute, Tucson, AZ, U.S.A.*
(*Author for correspondence; E-mail: jrich@lpl.arizona.edu)

(Received 10 September 2004; Accepted in final form 28 December 2004)

Abstract. The cratering event produced by the Deep Impact mission is a unique experimental opportunity, beyond the capability of Earth-based laboratories with regard to the impacting energy, target material, space environment, and extremely low-gravity field. Consequently, impact cratering theory and modeling play an important role in this mission, from initial inception to final data analysis. Experimentally derived impact cratering scaling laws provide us with our best estimates for the crater diameter, depth, and formation time: critical in the mission planning stage for producing the flight plan and instrument specifications. Cratering theory has strongly influenced the impactor design, producing a probe that should produce the largest possible crater on the surface of Tempel 1 under a wide range of scenarios. Numerical hydrocode modeling allows us to estimate the volume and thermodynamic characteristics of the material vaporized in the early stages of the impact. Hydrocode modeling will also aid us in understanding the observed crater excavation process, especially in the area of impacts into porous materials. Finally, experimentally derived ejecta scaling laws and modeling provide us with a means to predict and analyze the observed behavior of the material launched from the comet during crater excavation, and may provide us with a unique means of estimating the magnitude of the comet's gravity field and by extension the mass and density of comet Tempel 1.

Keywords: impact cratering: theory, modeling, experiments, comets: structure, composition, space missions: deep impact spacecraft design

1. The Inception of Deep Impact

The idea of impacting a space probe into a small solar system body in order to investigate its composition and structure has its beginnings with a 1994 JPL concept study. Shortly thereafter, the Deep Impact mission was conceived in the fall of 1995, when Mike Belton met with Jay Melosh at the Lunar and Planetary Laboratory of the University of Arizona and asked the simple question "how large a crater would be produced by the impact of a 500 kg spacecraft at 10 km/s on a comet?" A quick estimate using the Schmidt–Holsapple scaling law gave a rough estimate of about 100 m diameter (Holsapple and Schmidt, 1982). Belton realized that this would excavate material from a substantial depth below the surface of the comet and that his idea of using an impact to probe the comet's interior made sense. This confirmation expanded into the assembly of a science team and writing assignments for the first round of the Deep Impact proposal (which proposed Phaeton, not Tempel 1, as a target) in of June 1996.

As this project evolved, we were always painfully aware of the difficulty of making exact predictions of crater size on an almost entirely unknown target. No one then knew (nor yet knows) the density of a comet. Estimates for the density of Halley range from 0.03 to 4.9 g/cm^3 (Peale, 1989). Is the surface material of a comet inert like sand or will it release large amounts of volatile gases when it is struck and heated? Is the surface material strong like rock or as weak as the 100 Pa strength inferred for Comet SL9 (Scotti and Melosh, 1993)? Precise estimates of the size of the crater and the course of excavation depend on answers to these unknowns. In the end, we decided that the experiment itself must answer these questions: We would try, for the first time, to probe a comet by direct impact and deduce its mechanical properties from the response.

Although we remain very uncertain about what the Deep Impact experiment will eventually show, we have nevertheless tried to do the best job we could in predicting the outcome. Fortune, after all, favors the prepared mind. This paper represents our current best attempts to understand what we can expect to see when the 360 kg Deep Impact impactor strikes Tempel 1 at 10.2 km/s in early July 2005.

2. Scaling Relations for Crater Diameter

In one respect, our understanding of the Deep Impact cratering event is much better constrained than that of the multi-kilometer scale impact craters observed on the Earth and other moons and planets. Unlike the large craters that form a major part of the landscapes of most airless bodies, the relatively small Deep Impact crater is a good match to our ability to compute or experimentally model such impacts.

The formation of an impact crater does not involve any new physics. The impact and the subsequent growth of the crater are governed by a set of classical differential equations known as the Navier–Stokes equations, supplemented by an equation of state that describes the thermodynamic properties and constitutive equations that describe the strength of materials (see Melosh, 1989 for a review of cratering mechanics). The Navier–Stokes equations express the conservation of mass, energy and momentum. The "equation of state" relates the pressure in all materials and mixtures of materials to their densities and internal energies. The constitutive equations define a material model that links shear stresses and strains. The principal uncertainty in using these equations is the equation of state and material model. These relations are not well known for most natural materials. However, the equations themselves offer some hope for a simple solution. As in many such equations, they posses several "invariances:" changes of some variable that leaves the overall equation unchanged. If gravity or rate-dependent strength is not involved (which, as we will see, may be too drastic a simplification in practice), one of the principal invariances is a coordinated change of size and time. Thus, a 1 mm projectile striking a target at 10 km/s will yield the same result as a 1 m projectile striking at the same speed, provided all distances are scaled by the same ratio of 1,000 = 1 m/1 mm and all times are multiplied by the same factor. Thus, if the 1 mm projectile makes a crater 6 cm in diameter in 300 ms, the 1 m projectile will create a 60 m diameter crater in 300 s. In this scaling, velocities, densities and strengths are unchanged. Thus, the target from which the problem is scaled must be the same material as the actual target.

This simple scaling invariance thus opens the door to detailed experimental study of the Deep Impact crater, providing we can find close matches to the actual material of a comet and achieve velocities similar to that of the Deep Impact collision. In fact, laboratory studies using two-stage light gas guns are limited to about 6–8 km/s, but this is not very far from the actual conditions. Schultz and Ernst (this volume) describe a detailed laboratory simulation approach using just this correspondence. The main factors that spoil this rosy picture are (1) target materials that posses a rate-dependent material strength, and (2) the Earth's gravity. Although many target materials do not have this first problem, rate dependence is observed for carbonates (Larson, 1977) and other materials, especially for tensile failure, where a strong rate dependence is expected (Melosh *et al.*, 1992), so caution is needed here. If gravity is important in limiting the crater's growth (and, depending on the strength of the material in the target, it may or may not be), then this simple invariance does not hold. Gravity is a function of $(distance)/(time)^2$; so, for a strictly correct comparison between the laboratory and the actual event, the acceleration of gravity must be scaled as the inverse of the distance or time ratio. Thus, the 1 mm projectile in a terrestrial gravity field corresponds to a 1 m projectile in a gravity field of 1/1,000 of Earth's surface gravity. This is certainly a step in the right direction for the Deep Impact event, in which the surface gravity on Tempel 1 may be only be $0.0008\ m/s^2$, but it actually goes about a factor 10 too far! To simulate the comet impact correctly under Earth gravity we really need a projectile about $80\ \mu m$ in diameter, made of the same materials as the Deep Impact impactor and striking a target of the same composition as the comet at 10.2 km/s. Even the grain size of the Earth simulant target must be reduced by the same factor of 8×10^{-5} from the grain size in the comet. This is a pretty tall order for experimental studies and may require the numerical methods described later to make serious progress, although any numerical computation must, of course, be checked by experimental findings under all possible circumstances.

Although it is often difficult to satisfy the requirements of the exact space/time/material invariance, an approximate form of invariance has been recognized in impacts and explosions. This invariance ultimately stems from the fact that the final crater is much larger than the projectile, so that projectile-specific properties such as diameter, shape, composition, angle of impact, etc., do not affect the final outcome. Only a single-dimensional parameter, the "coupling parameter" that depends on the projectile's total energy and momentum may affect the size and shape of the end result (Holsapple and Schmidt, 1982). When this is the case, a number of "scaling relations" can be derived that link impacts at different sizes, velocities and gravitational accelerations. Such scaling relations take the form of power laws relating dimensionless combinations of quantities describing the projectile and final crater:

$$\pi_D = C_D \pi_2^{-\beta}, \tag{1}$$

where π_D is a dimensionless measure of crater diameter,

$$\pi_D = D \left(\frac{\rho_t}{m} \right)^{1/3}, \tag{2}$$

in which D is the transient crater diameter measured at the level of the pre-impact surface (the "apparent" diameter), ρ_t the target density, m the projectile mass, and π_2 the inverse of the Froude number,

$$\pi_2 = \frac{1.61 g L}{v_i^2}, \tag{3}$$

where g is the surface gravity, L the projectile diameter and v_i the impact velocity. C_D and β are constants that are determined empirically. These constants depend on the nature of the target material, in particular on its porosity or coefficient of internal friction. Table I lists values of these coefficients determined from a suite of experiments by Schmidt and Housen (1987). Note that such scaling laws apply only to the final state of the crater. Early-time phenomena such as melt or vapor production depend on the details of the impactor and thus cannot be predicted from a scheme of this type.

In most previous planetary studies, the large craters that are of principal interest form at values of π_2 much in excess of what can be measured in the laboratory, making extrapolation to large values of π_2 necessary. In contrast, the Deep Impact crater will occur at a value of π_2 of about 10^{-11}, much *smaller* than has been observed in laboratory experiments, where it usually ranges from 10^{-5} to about 5×10^{-10}. Little data have been reported for high-velocity impacts into dry sand and water in which π_2 ranged down to about 10^{-10} (Schmidt and Housen, 1987). In this case, we must extrapolate π_2 to the opposite extreme from most previous

TABLE I

Experimental parameters for diameter scaling law (Schmidt and Housen, 1987).

Target material	C_D	β
Water	1.88	0.22
Loose sand	1.54	0.165
Competent rock or saturated soil	1.6	0.22

TABLE II

Diameter estimates for the Deep Impact crater from three
different scaling laws (Melosh 1989).

Scaling method	Diameter (m)
Yield scaling	63.5
Pi scaling	60.4 for loose sand
	250 for competent rock
Gault scaling	65.7

Impact conditions: projectile diameter: 1 m; projectile density: $0.46\,g/cm^3$; impact velocity: 10.2 km/s; angle from horizontal: $90°$; target density: $1.0\,g/cm^3$; acceleration of gravity: $0.0008\,m/s^2$. The formation time is 350 s.

studies. Table II illustrates the range in uncertainty in crater diameter predictions using scaling relations from a variety of sources. The expected range in diameter for a gravity-dominated crater is thus 60–250 m, depending on the nature of the target (loose sand versus competent rock). The reader may be surprised that the crater diameter in "competent rock" is larger than that of "loose sand". This is a simple consequence of the more rapid attenuation of shock waves in loose materials. This relationship has been observed in a large number of impact experiments and numerical simulations (see, e.g., Figure 7.3 of Melosh, 1989).

When material strength, not gravity, finally halts the crater growth, similar power-law scaling relations can be constructed in which the important dimensionless variable is not π_2 but a combination depending on some measure of strength, Y:

$$\pi_3 = \frac{Y}{\rho_p v_i^2}, \tag{4}$$

where ρ_p is the density of the projectile (Holsapple and Schmidt, 1982). Although implementation of this sort of relationship appears simple, at the moment it is unclear how to implement strength scaling. The problem is that it is uncertain what the strength measure Y really is: One could use either tensile strength or crushing strength, but the two often differ by an order of magnitude.

The entire concept of what is meant by "strength" in impact cratering is presently somewhat fuzzy. Modern theories of dynamic fracture indicate that the actual failure strength should be strongly rate dependent (Grady and Kipp, 1987), a factor not considered in the derivation of the original strength scaling relations. Furthermore, numerical computations indicate that the strength of the material surrounding an impact is often strongly degraded by the shock wave long before the excavation flow clears the material out of the crater interior (Croft, 1981; Asphaug and Melosh, 1993; Nolan et al., 1996). This pre-excavation fracture appears to depend strongly on crater size as well as intrinsic strength, being more important in small gravitational fields than in large ones. At a much larger scale, the collapse of multi-kilometer craters clearly requires some form of extreme strength degradation mechanism to match observations with theory (Melosh and Ivanov, 1999).

Another area of uncertainty is the effect of the special aspects of a cometary surface on crater growth. We presently have very little experience with high-velocity impacts on highly volatile targets in which a significant amount of material may be melted or vaporized, although a number of low-velocity analog experiments have been performed (Schultz *et al.*, 1992). The expansion of a large quantity of volatile material can only be expected to increase the crater size by pushing more material away from the impact site. Some experiments with highly volatile cadmium projectiles and targets have been performed (Poorman and Piekutowski, 1995), but the relevance of these experiments to the surface of a comet with an unknown suite of ices with unknown abundances is unclear.

Comet surfaces are also expected to be highly porous, with estimates of porosity of the outer crust ranging from essentially zero to more than 50% (Sagdeev *et al.*, 1988). In the extreme limit of low porosity, the impacting spacecraft might simply pass through the comet. We do not really expect this to happen. The best current estimate of cometary density, derived from the tidal breakup of SL9, suggests that is in the neighborhood of 0.5 g/cm^3 (Asphaug and Benz, 1994), which is certainly high enough to prevent the spacecraft from penetrating the entire object.

The presence of a certain amount of porosity is actually advantageous to producing a large crater. It has long been known that the excavation efficiency of a buried explosion is a strong function of the depth of burial (Nordyke, 1962). Experience with both high explosives and nuclear detonations indicates that the scaled crater diameter, $D/W^{1/3}$, where W is the energy release, increases with scaled depth of burial, $h/W^{1/3}$, until the "optimum depth of burial" is reached at about 0.003 m/J$^{1/3}$. Explosions at greater depths produce smaller craters, until at depths in excess of 0.009 m/J$^{1/3}$ crater formation is entirely suppressed. For the Deep Impact spacecraft, which delivers about 1.9×10^{10} J to the comet, this depth is about 8 m. The penetration depth of a projectile is given roughly by $d = L\sqrt{\rho_p/\rho_t}$, which is based on simple momentum conservation (Melosh, 1989). For a projectile 1 m in diameter and average density $\rho_p = 0.46$ g/cm^3 impacting a target of similar density, the penetration depth is only 1 m, well short of the optimum depth. This accords with general experience: Impact craters are generally considered similar to explosions buried at shallow depth (Holsapple, 1980). Thus, within wide limits a lower target density permits the projectile to penetrate deeper and creates a larger crater. Only if the comet density were lower than 0.007 g/cm^3 (which gives a penetration depth greater than 8 m, based on the penetration formula given earlier) would the crater size decrease as a result of excessively deep penetration. Nevertheless, some concern has been expressed about the possibility that a high-density projectile might penetrate so deeply that crater excavation is suppressed. This concern has partly driven the design of the Impactor Spacecraft, as discussed in the next section.

3. Cratering Constraints on the Impactor Spacecraft Design

Some of the unique features of the Deep Impact mission generated strong constraints on the design of the Impactor Spacecraft. The mass and shape of the Impactor were optimized to create the largest possible crater under the widest possible range of circumstances. Moreover, because we hope to determine the comet's composition from spectral emissions from the material vaporized by the impact, constraints arose on the elemental composition of the Impactor.

The Impactor itself is a roughly cylindrical spacecraft about 1 m in diameter and 1 m long (Figure 1). In addition to the impacting mass itself, it contains a small propulsion system, guidance, communications, and imaging systems. Because it must fly through the coma of an active Tempel 1 before impact, a three-plate

Figure 1. Schematic view of the Deep Impact Impactor Spacecraft.

Whipple shield protects its forward end. We expect to be able to survive impacts by 1 g coma dust particles at the approach velocity of 10.2 km/s. Only about 1/3 of the total mass could be devoted to inert impact mass. An early decision was to make this material from the metal copper. Copper also possesses well-defined lines that should not mask emissions of other species of geochemical interest (assuming that it is in the vapor phase, not in the form of incandescent droplets). In addition to the desirable thermal and mechanical properties of copper, it is an element in which few geochemists have expressed a serious interest. The fact that the emissions from the copper vapor, from the projectile will overwhelm the signature of any copper in the target was thus not considered a drawback.

The abundances of elements of more geochemical interest than copper, however, generated a unique set of constraints on the impactor. Assuming that the impact will vaporize a mass of target roughly equal to the mass of the impactor, we estimated the probable mass of a suite of elements likely to be vaporized from the target (Table III), based on abundances of elements in the lunar soil (Heiken et al., 1991), Type I carbonaceous chondrites (Taylor, 1982), and in Comet Halley dust (Jessberger, 1999). We then required that the impactor not contain more than 20% of these strategic elements. This limit was further lowered for elements that have strong emission lines that might mask those of other elements of interest. Table III

TABLE III

Estimated composition of vaporized comet crust vs. Impactor Spacecraft.

Element	Mass in DI vapor plume (kg)	Mass in DI Impactor Spacecraft (kg)
O	123	1.8
Na	2.8	[a]
Mg	49	1.0
Al	4.5	78.2
Si	55	0.4
K	0.3	[a]
Ca	4.9	[a]
Ti	0.23	24.9
Mn	1.2	[a]
Fe	95.2	14.4
Co	0.1	0.0
Ni	0.3	2.0
Cu	0.006	178.1

Masses are based on a devolatilized composition similar to Type I carbonaceous chondrites (Taylor, 1982). Mass vaporized: 350 kg.
[a]These elements were not reported separately. The total mass of unreported elements was 4.4 kg.

also lists the final composition of the projectile (Alice Phinney, 2004, personal communication) for comparison.

The precise distribution of the inert mass generated considerable discussion during the design phase, executed by Alice Phinney and her team at Ball Aerospace, Inc. For structural reasons, the engineers initially wanted to put all of the copper in a thin disc-shaped plate about 1 m in diameter. However, because we are not certain that we will ever know the precise angle of impact, and because we expect that the outcome of the event will depend on whether the plate entered on edge or face-on, the science team wanted to put as much mass as possible into a sphere. In the end, a compromise was reached in which the leading end of the impactor was faced with a 1 m diameter copper disk of mass 15 kg on which is mounted a 0.64 m diameter spherical segment 0.16 m high. The plate and spherical segment are constructed from a stack of copper plates (Figure 2) in which numerous non-overlapping holes are milled, lowering the average density of the mass to 4.0 g/cm^3. The entire copper fore-body has a mass of 113 kg out of a total spacecraft mass (at impact) of 360 kg. This configuration has the advantage that, if the comet turns out to have a very low density, the broad extension of the plate and after-body will couple the spacecraft momentum gradually into the target. On the other hand, if the comet is very dense, the spherical segment alone will penetrate deeply while the remainder of the spacecraft is stripped away near the surface. In both cases, we will couple a large fraction of the impactor's kinetic energy deeply into the target.

The complicated structure of the spacecraft (and probably of the target) make it impossible to be confident about the validity of the simple scaling relations described earlier. In the final analysis, a coupled series of detailed numerical simulations and experimental studies will probably have to be carried out to resolve the details of the impact process.

Figure 2. Forebody copper mass. Diameter 0.64 m, height of stack 0.16 m, mass 113 kg.

4. The Early Stages of Impact and Vapor Plume Formation

One of the many unique features of this mission is the ability to determine the composition of the comet's crust by observing the vaporized material produced by the impact in its earliest stages. As discussed earlier, this vapor plume will consist of a mixture of impactor and comet material, and in addition to avoiding elements of interest and cataloguing exactly what is in the impactor (Table III) it can be useful to model the evolution of the vapor plume to understand how its observed composition and thermodynamics will change over time. The earliest stages of contact of the Deep Impact spacecraft with Tempel 1 can be simulated either experimentally (Shultz and Ernst, this volume), or theoretically using a computer code to solve the equations describing the projectile and target, so far as they are known. The penetration phase of this impact is short: about 0.1 ms for the 1 m projectile to fully contact the target at 10.2 km/s. Nevertheless, we can get a rough idea of the early events using a modern numerical hydrocode (Anderson, 1987).

We performed high-resolution three-dimensional (3D) hydrocode simulations of the early stages of impact cratering events using the hydrocode SOVA (Shuvalov, 1999) coupled to tabular versions of the ANEOS equations of state (Thompson and Lauson, 1972) for the materials of interest. The objective of this study is to investigate impactor penetration, and the development and evolution of the expansion plume at early times.

SOVA (Shuvalov, 1999) is a two-step Eulerian code developed at the Institute for Dynamics of Geospheres (Russia) that can model multidimensional, multi-material, large deformation, strong shock wave physics. It is based on the same principles utilized in the well-known hydrocode CTH (McGlaun et al., 1990), developed at the Sandia National Laboratories (Albuquerque, NM). In particular, care has been used in SOVA to develop a more appropriate formalism for the simultaneous conservation of energy and momentum. We use the code in 3D geometry; in this case, it is common practice to use bilateral symmetry, which will allow us to model only the positive half-space originating at the impact plane (the plane perpendicular to the target defined by the impact direction), thus significantly reducing the rather large mesh size (and accompanying computational requirements) needed for the simulation. Three-dimensional benchmark tests have shown that SOVA produces shock melting and vaporization patterns and volumes comparable to the well-known CTH (Pierazzo et al., 2001; Artemieva and Ivanov, 2001). Tabular equations of state were constructed using a revised version of the ANEOS code (Thompson and Lauson, 1972; Melosh, 2000). The availability of reliable, wide range equations of state relating thermodynamic parameters of materials, such as pressure, density and temperature, is vitally important for numerical simulations. By using different physical treatments in different domains of validity, ANEOS provides a thermodynamically consistent equation of state whose validity extends over a wide range of temperatures and pressures. ANEOS also offers a limited treatment of material's phase changes. As a result, this code can model the thermodynamic evolution of a material well beyond the initial shock stage, accounting for melting and vaporization, and providing consistent estimates of material's energies and entropies.

In the simulations, we tried to simulate the Deep Impact copper projectile as closely as possible. The impactor consists of a copper cylinder 0.4 m in radius and 0.8 m in height, with an average density of 660 kg/m^3, with a leading face made of a combination of two solid copper plates (density of 8,400 kg/m^3), one 0.4 m in radius followed by another 0.2 m in radius, with thickness of 2 cm each. The total mass of the projectile is 370 kg, with 120 kg concentrated on the solid leading face. The impact velocity is 10.2 km/s (as expected for the Deep Impact projectile). Our simulations model impact angles of 90° from the surface. At this early stage in the calculation, the acceleration of gravity is unimportant. The spatial resolution is 40 cells per projectile radius in the initial stage of penetration (i.e., the total thickness of the two solid plates on the leading edge spans four cells). Many believe that comets have a rocky surface crust few meters thick covering an ice-rich interior (Brandt and Chapman, 2004). We used serpentine (hydrated olivine, similar to the composition of carbonaceous chondrites) to model the crustal material. Since the typical density of the surface crust is not known, we carried out exploratory runs for (a) fully dense serpentine (unrealistic case; $\rho_t = 2.55$ g/cm^3); (b) 50% porous serpentine ($\rho_t = 1.275$ g/cm^3); (c) 80% porous serpentine ($\rho_t = 0.51$ g/cm^3). For comparison, we also carried out a simulation of a 45° impact and same setup as case (b). Figure 3 shows the outputs for the various simulations 0.7 ms after impact. The material colors are graded according to density variations, while colors of the Lagrangian tracers in the target represent the maximum shock experienced by the material. Figure 3 shows that the density of the target material affects the evolution of the expansion plume: in the highly porous surface case, Figure 3C, the impactor penetrates deep into the target, allowing target vapor to emerge relatively free of copper contamination. For a denser target, Figure 3A and B, copper vapor emerges early, mixed with target vapor. In the case of the oblique impact Figure 3D, the copper vapor appears to envelope the vapor from the target. This occurs from the very early stages of the impact, as a result of the asymmetry in the impact event, as shown in Figure 4. Further study will investigate if the copper vapor opacity is high enough to significantly affect the identification of target material from the plume.

5. Hydrocode Modeling of Crater Excavation

Once the kinetic energy of the projectile has coupled into the target, the ensuing shock wave spreads out and initiates the excavation stage of cratering (e.g. Melosh, 1989). The shock wave first compresses, then releases the engulfed material to low pressure. In the process, this material is accelerated away from the impact site and eventually opens a crater as comet surface materials are both displaced further downward into the comet or ejected from the surface. The most visible part of this

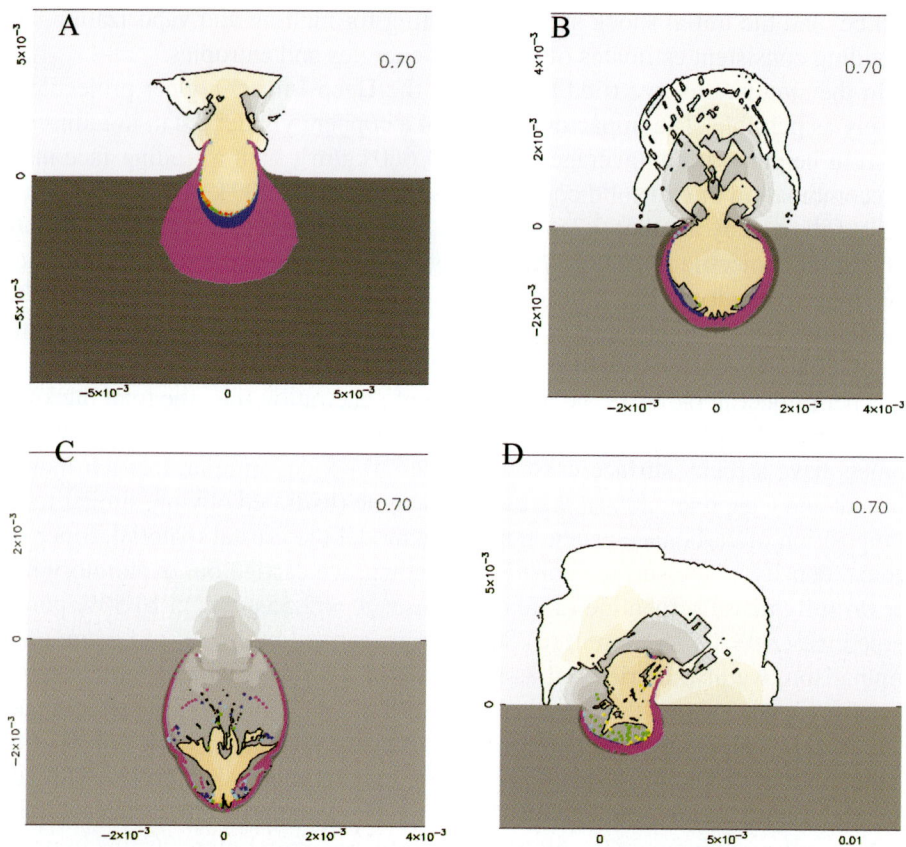

Figure 3. Impact simulation outputs 0.7 ms after impact for the various simulations: (A – *upper left*) fully dense serpentine ($\rho = 2.55$ g/cm^3); (B – *upper right*) 50% porous serpentine ($\rho = 1.275$ g/cm^3); (C – *lower left*) 80% porous serpentine ($\rho = 0.51$ g/cm^3); (D – *lower right*) 45° impact and same setup as (B). Material colors: gray: serpentine; yellow: copper impactor. Tracer colors represent target material shocked at various levels: yellow, 100–150 GPa; green, 50–100 GPa; cyan, 30–50 GPa; blue, 18–30 GPa; magenta, 5–18 GPa.

process is the ejecta plume, described in detail in the next section. The crater cavity itself is a consequence of both displacement and ejection.

Numerical modeling of impact crater excavation has now reached a high degree of sophistication (Collins *et al.*, 2004). Successful models have been created for a number of large terrestrial and extraterrestrial impact craters, and more are currently in progress. However, the Deep Impact crater presents a number of unique challenges. The principal one is the importance of treating porosity (Love *et al.*, 1993), dilatancy and strength in the comet crust. While we are not certain that the crust is highly porous, current best estimates of comet density strongly suggest that the volatile-depleted lag material we expect to find mantling the icy interior is probably an open granular aggregate of some kind, although opinion varies as to whether it is loose or sintered. The Stardust images of Comet Wild 2 showed

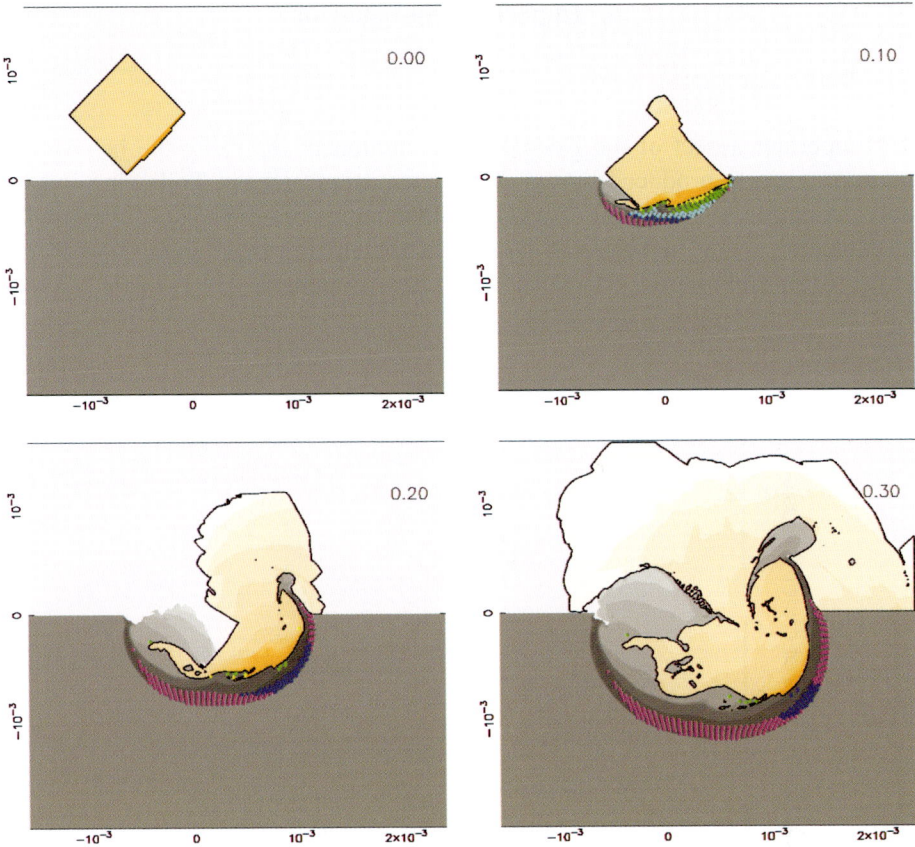

Figure 4. Impact simulation outputs for the 45° impact simulation (case D in Figure 3) in the early stages of impact, and shown for time after impact of: (*upper left*) 0 ms, (*upper right*) 0.1 ms, (*lower left*) 0.2 ms, and (*lower right*) 0.3 ms. Material and tracer colors are the same as in Figure 3.

spectacular 100 m high vertical cliffs that require some cohesion (Brownlee *et al.*, 2004), although in the feeble gravitational field of the comet this strength may be no larger than 100 Pa. Although we have recently made substantial progress in introducing realistic strength models into hydrocodes (Collins *et al.*, 2004; Wünnemann and Ivanov, 2003), the addition of porosity is less advanced (Rubin *et al.*, 2000) and dilatancy still needs a great deal of development (Wroth and Bassett, 1983).

To date, there are no complete numerical simulations of the Deep Impact crater from initial contact to final excavation, in spite of attempts by workers such as the O'Keefe and Ahrens team at Caltech and David Crawford of Sandia National Laboratory. Work on this topic is presently progressing and some results may be obtained by the time of impact on Tempel 1. Because laboratory models cannot adequately simulate every aspect of the comet's surface, research of this type is a prime necessity for establishing a link between the observed crater and the mechanical properties of Tempel 1's surface.

6. Impact Ejecta Behavior

6.1. INTRODUCTION

One important aspect of the Deep Impact mission is analyzing the behavior of
the impact ejecta produced by the crater excavation process. Following impact,
individual ejecta particles are launched ballistically from the edge of the bowl of
the expanding crater, and collectively these particles form an inverted, cone-shaped
plume (or curtain) which also expands over time (Figure 5). The ballistic behavior

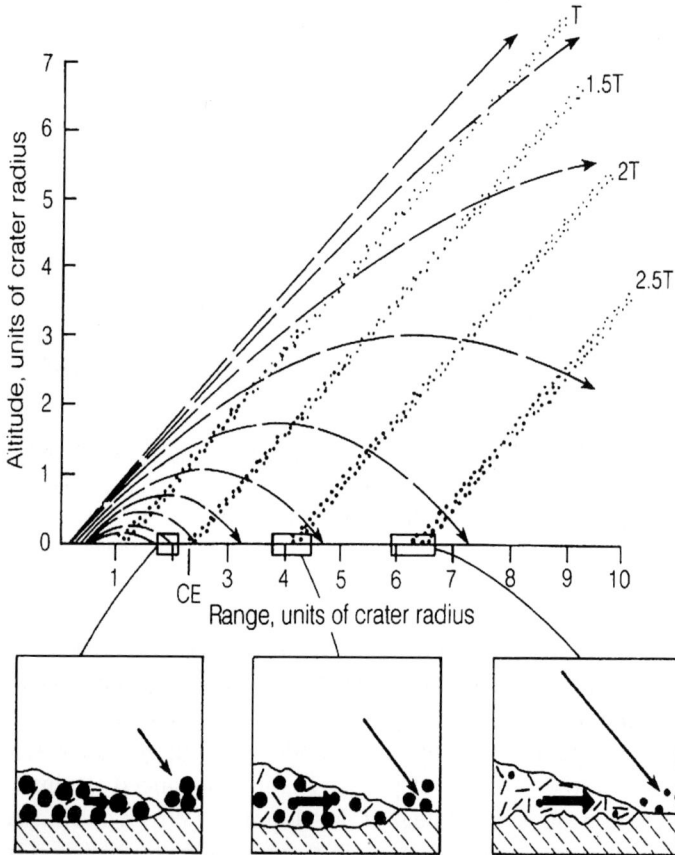

Figure 5. The debris ejected from an impact crater follows ballistic trajectories from its launch
position within the transient crater (horizontal scale is in units of *R*). The innermost ejecta are launched
first and travel fastest, following the steepest trajectories shown in the figure. Ejecta originating farther
from the center are launched later and move more slowly, and fall nearer to the crater rim. Because of
the relationship between the position, time, and velocity of ejection, the debris forms an inverted cone
that sweeps outward across the target. This debris curtain (plume) is shown at four separate times
during its flight, at 1, 1.5, 2, and 2.5 t_f, where t_f is the crater formation time. Figure reproduced from
Melosh, 1989.

of the individual particles and the collective behavior of the ejecta plume are both heavily affected by the velocity and characteristics of the impactor (which we control); and target parameters such as the density, strength, porosity, and gravity field. By modeling the observed behavior of individual ejecta particles and the collective ejecta plume resulting from the mission, we hope to place constraints on these parameters.

6.2. IMPACT EJECTA SCALING LAWS

To model the impact ejecta behavior, we develop a revised set of crater growth, ejection time, and ejecta velocity scaling laws, based on the scaling laws described by Housen *et al.* (1983). The two relationships used here are based on gravity-dominated cratering in an experimental environment and are given by:
Crater formation time:

$$t_f = \frac{1}{2}\sqrt{\frac{2R}{g}}, \tag{5}$$

Ejecta velocity as a function of crater radius:

$$v_{ej} = C_e\left(\frac{r}{R}\right)^{-\varepsilon}, \tag{6}$$

where R is the gravity-dominated transient crater radius $(D/2)$, r the crater radius as a function of time, g the surface gravitational acceleration, and ε is a material constant ranging from 1.8 for competent rock to 2.6 for quartz sand (Melosh, 1989). The coefficient of 1/2 in front of the crater formation time equation (Equation (5)) comes from an empirically derived value of 0.54 given by Melosh (1989).

We find the constant C_e by assuming that the crater rim advancement velocity must be equal to the horizontal component of the particle ejection velocity, such that the ejecta plume base and crater rim advance at the same rate. We take advantage of this by setting the particle ejection angle to a mean of $\theta = 45°$ above the horizon, which gives $v_{horizontal} = \sqrt{2}/2v_{ej}$. Letting $v_{horizontal}$ equal the rim advancement speed produces:

$$\frac{\partial r}{\partial t} = \frac{\sqrt{2}}{2}C_e\left(\frac{r}{R}\right)^{-\varepsilon}. \tag{7}$$

Solving this differential equation such that r is allowed to move from $0 \rightarrow r$ while t moves from $0 \rightarrow t$, yields:

$$r = \left(\frac{\sqrt{2}}{2}(1+\varepsilon)C_e R^\varepsilon t\right)^{1/1+\varepsilon} \tag{8}$$

Letting $r = R$ and $t = t_f$ produces a crater formation time of:

$$t_f = \frac{\sqrt{2}R}{C_e(1 + \varepsilon)}.$$

(9)

However, t_f is also given by Equation (2), allowing a solution for C_e to be found:

$$C_e = \frac{2\sqrt{Rg}}{1 + \varepsilon}.$$

(10)

This gives the following two model equations, in addition to Equation (5):
Ejecta velocity as function of crater radius (replaces Equation (6))

$$v_{ej} = \frac{2\sqrt{Rg}}{1 + \varepsilon}\left(\frac{r}{R}\right)^{-\varepsilon},$$

(11)

Particle ejection time as a function of rim position:

$$t_{ej} = \frac{\sqrt{2}r^{1+\varepsilon}}{2\sqrt{g}R^{(\varepsilon+1/2)}}.$$

(12)

We further modify Equation (11) to simulate late-stage ejection velocities more properly, when the crater radius is approaching its final value. In gravity-dominated cratering, the particle ejection (ballistic) velocity should go to zero as r goes to R, while Equation (11) instead goes to a constant (a weakness also described in Housen *et al.*, 1983). We correct this by subtracting a higher-order term, which has negligible effect throughout most of the excavation process, but which ramps the velocity expression to zero as the final (transient crater) rim is approached – essentially applying a mathematical bridge between known good behaviors. This gives the following equations.
Ejecta velocity as function of crater radius (replaces Equations (6) and (11)):

$$v_{ej} = \frac{2\sqrt{Rg}}{1 + \varepsilon}\left(\frac{r}{R}\right)^{-\varepsilon} - \frac{2\sqrt{Rg}}{1 + \varepsilon}\left(\frac{r}{R}\right)^{\lambda},$$

(13)

where the power λ is selected by the model user ($\lambda \approx 6–10$). Figure 6 shows a plot of ejection velocities produced from Equation (13), compared to experimentally derived values.

6.3. EJECTA PLUME TRACER MODEL

These scaling law equations are then applied to a dynamical simulation which models – *via* thousands of point tracer particles – the ejecta plume behavior, ejecta blanket placement, and impact crater area resulting from a specified impact on an irregularly shaped target body (similar to Geissler *et al.*, 1996). Figure 7 shows an example of one impact simulation, visualized in 3D polygon fashion. Placing the target body (shape-model) into a simple rotation state about one of its principal axes, the user then inputs an impact site and a set of projectile/target parameters.

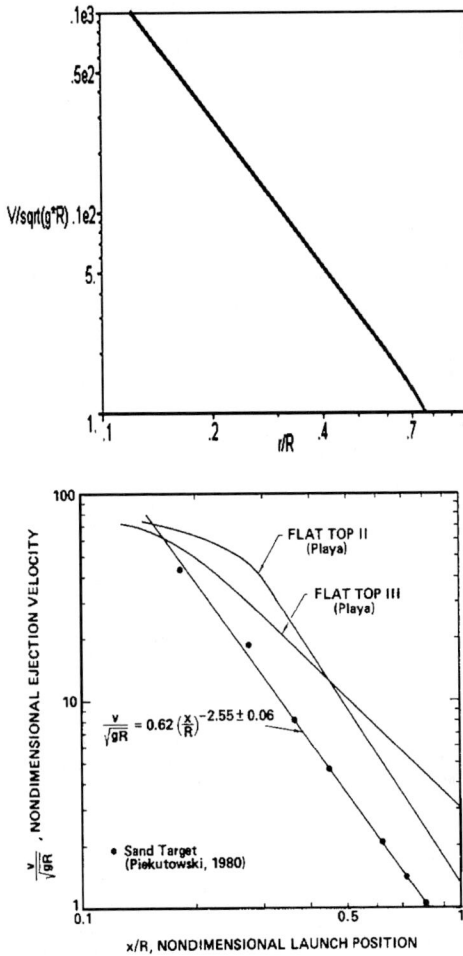

Figure 6. (A) Normalized (non-dimensional) ejecta velocities produced using Equation (9) as a function of the normalized radial position r within the transient crater of radius R. The values computed are for loose sand, with $\varepsilon = 2.44$ and a normalized equation coefficient of 0.58. (B) Experimentally measured ejecta velocities produced from large explosion craters and published in Figure 4 of Housen *et al.* (1983). The best model-experiment agreement corresponds to the lower sand target values.

From this information, the program places a circular transient crater area on the surface and populates this area with random tracer particles that have a spatial distribution such that each particle represents a roughly equal volume of ejecta. Once positioned, each particle is assigned an ejection time (Equation (12)), velocity (Equation (13)), and direction (radially outward at ejection angle θ above the horizon, discussed later), after which the simulation clock begins. While in flight, the gravitational acceleration from the irregular target body on each tracer particle is computed using the polygonized surface (polyhedron) gravity technique developed

J. E. RICHARDSON ET AL.

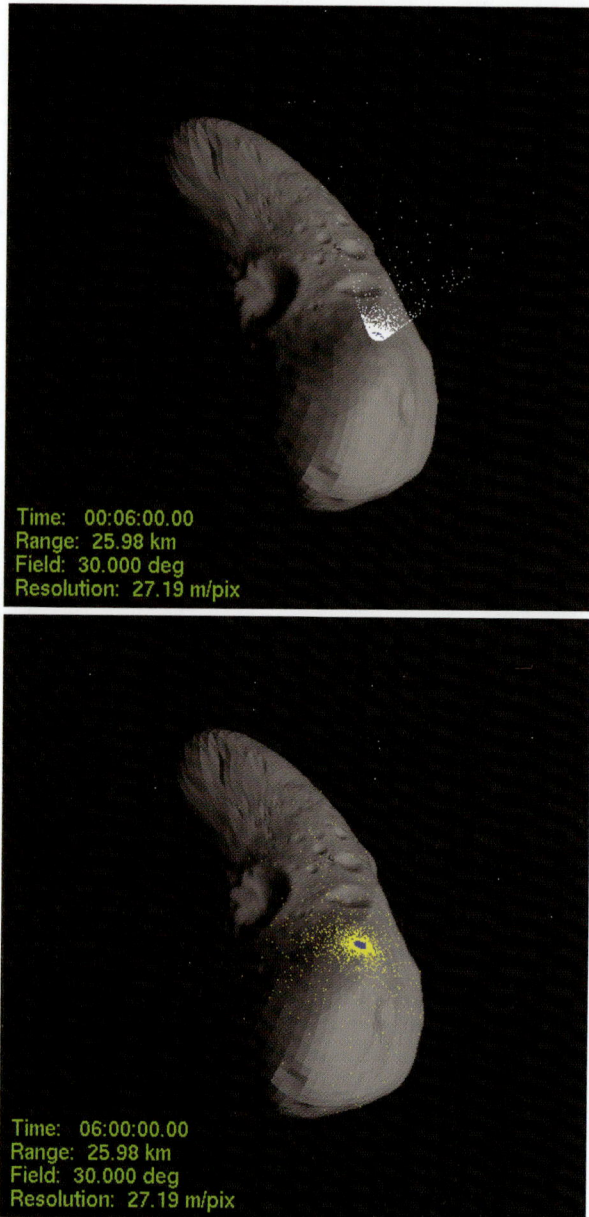

Figure 7. A simulation showing the ejecta plume (*white tracer particles*), ejecta blanket (*yellow tracer particles*), and impact crater surface area (*shown in blue*) resulting from a small impact on an Eros-shaped target body having a 6-h rotation period about its principal z-axis. The top panel shows the state of the ejecta 6 min after the impact, using 2000 tracer particles to map its behavior. The ejecta plume is fully formed at this stage, with the slowest particles beginning to fall out near the crater rim. The bottom panel shows the state of the ejecta 6 h after the impact (one rotation), with most of the tracer particles landed again on the surface to form the ejecta blanket. This blanket is slightly asymmetrical, with more ejecta in the trailing direction (to the right) than in the leading direction.

by Werner (1994). The model tracks all tracer particles until they have either left the gravitational sphere of influence of the body (escaped) or landed again on the surface.

To properly model the ejection angle variations that occur over time in impact cratering experiments (Cintala *et al.*, 1999; Anderson *et al.*, 2003), we mimic the empirical data by allowing the particle ejection angle to drop from $\theta = 60$ to $30°$ as the crater rim (r) moves from 1 projectile diameter (its starting point) to the transient crater radius (R). This feature also causes the ejecta plume shape to change as a function of time, demonstrated in the Cintala *et al.* (1999) experiments and compared to the model in Figure 8. If a gravity-dominated cratering event occurs as a result of our impact on Tempel 1, the shape of the ejecta plume will provide a means for marking the end of the crater formation process (end of excavation flow) as the ejecta plume changes shape from concave during excavation, to straight at the transient crater rim (end of excavation), to convex during the post excavation (fall-out) stage.

In the event that the crater excavation is dominated by strength, we have added a target strength parameter (R_s) to the model, which cuts off crater growth and excavation flow when the inertial stress on the material reaches a user-assigned material yield stress (Y). This is determined by the equation (Melosh, 1989):

$$R_s = \frac{\rho_t v_{ej}^2}{Y}. \tag{14}$$

where ρ_t is the surface density. Unlike gravity-dominated excavation, in which the ejecta plume remains attached to the target surface throughout its formation and fall-out stages, in strength-dominated excavation the ejecta plume detaches completely from the target. In this case, the plume's bottom edge will follow a radial (with respect to the crater) ballistic path away from the edge of the truncated impact crater. While this form of cratering will give us a smaller final impact crater (and perhaps less chance of looking inside of the cavity), the ballistic path followed by the bottom edge of the ejecta plume (and perhaps some large late-ejected fragments) may provide us with our best opportunity for the determining magnitude of the comet's gravity field.

Collectively, the expansion rate of the ejecta plume (especially in a gravity-dominated event) can itself be used to gain a measure of the surface gravity field g, in part due to a g dependence in the particle ejection velocities (Equation (13)), but primarily due to the effect of gravity on the ballistic paths followed by the individual particles. Figure 9 shows the effect of varying the gravitational force (by varying the density of a constant volume model) on the ejecta plume base position and velocity as a function of time. If this form of plume behavior can be observed at high enough resolution for several minutes following the impact, then a surface gravity and comet mass can be estimated (albeit roughly), as well as obtaining an approximation for the comet's density by using the volume obtained from shape-modeling (Thomas, this volume).

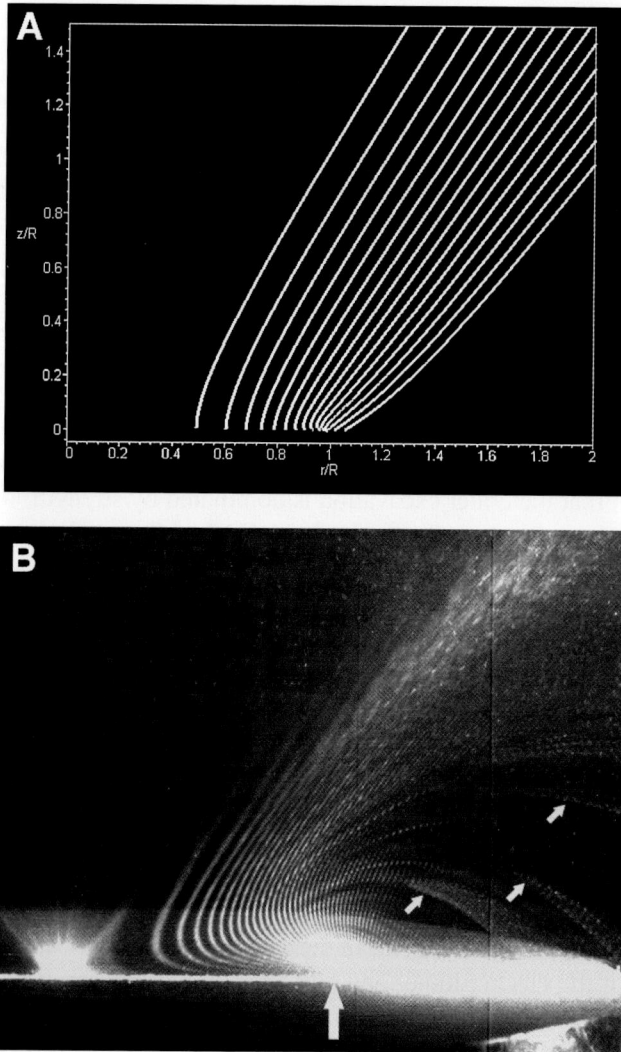

Figure 8. Ejecta plume profile comparisons between the ejecta plume model (A) and a small-scale cratering experiment (B) performed by Cintala *et al.* (1999) and based on their Figure 11, in which a 3.18 mm glass sphere was shot into fine-grained sand at 1.24 m/s to form an ∼4 cm diameter crater. The top panel shows both horizontal and vertical scales normalized to the transient crater radius R, while the bottom panel shows a large vertical arrow at the transient crater rim (small arrows point out the ballistic paths of three individual ejecta particles). The plume profiles in the top panel are shown in 4 ms increments, with the transient crater formed at about 44 ms (11 time steps). The plume profiles in the bottom panel are shown in 2 ms increments, with the transient crater formed at 45 ms (23 time steps). Note the change in plume profile from concave to convex as the final crater rim is passed, along with a noticeable change in velocity with position – rapidly slowing as the final rim is approached and gaining speed again as the slower particles fall out first. Compare these figures to the more basic Figure 5.

Plume Base Position vs. Time

Plume Base Velocity vs. Time

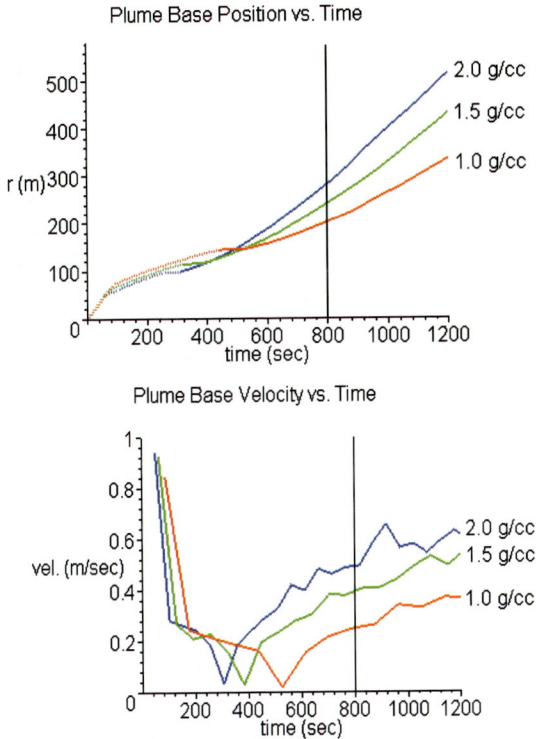

Figure 9. Plots of model ejecta plume position as a function of time (A) and ejecta plume velocity as a function of time (B) for three different assigned densities to a homogeneous, ellipsoidal comet shape-model: $1.0\,g/cm^3$ *red*, $1.5\,g/cm^3$ *green*, and $2.0\,g/cm^3$ *blue*. Dotted portions of the lines in the upper figure indicate the excavation (plume formation) stage, while the solid portions show the plume fall-out stage. The vertical black line at 800 s indicates the limit of observation time for the Deep Impact mission. Note the different expansion rates as a function of comet gravity, particularly evident in the velocity curves. The velocity curves also display a noticeable decrease in velocity during crater formation, an inflection point at the crater formation time t_f, and increasing velocity as slower particles fall out. With good ejecta plume resolution, we plan to use this type of plume behavior to estimate the gravity and mass the comet Tempel 1.

6.4. EJECTA PLUME POLYGON MODEL

A more realistic method for simulating the physical properties of an ejecta plume and eventual blanket resulting from an impact on a small, irregular target body is to model the ejecta plume as a 3D polygon object rather than randomly generated tracer particles. At each time step, the surface area and opacity of each polygon of the ejecta plume is calculated and rendered appropriately (assuming a user-specified particle distribution). Figure 10 shows an example of this model variant, for both a gravity-dominated and strength-dominated cratering event.

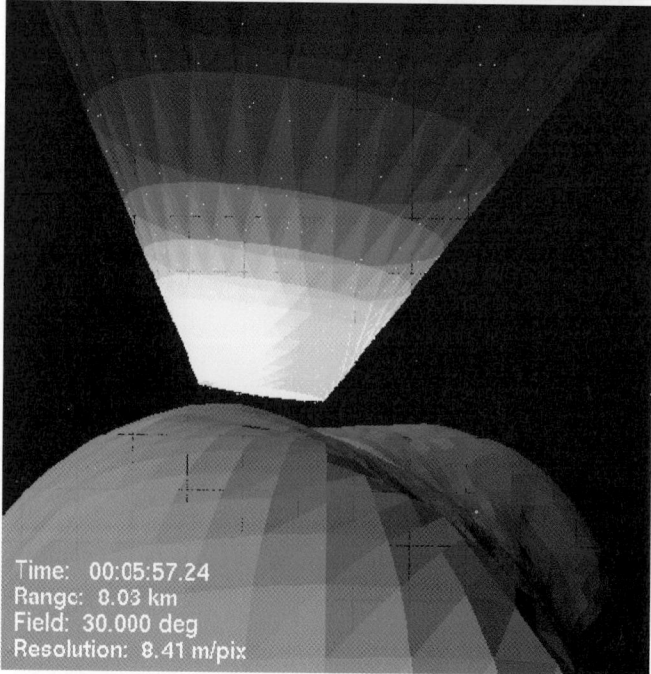

The polygon ejecta plume is initially formed by placing a mesh of 1,800 regularly spaced tracer particles on the starting surface area of the crater, such that 3,540 roughly equal-area triangular polygons are formed (each formed by connecting three tracer particles). This creates 59 rings of 60 polygons each, ranging from Sqrt((1/60) * R^2) to R in radius – the hole in the center is intentional, in order to avoid the region of very fast ejecta particles which produce extremely deformed polygons and do not contribute significantly to the visible ejecta plume.

We currently calculate the mass of impact ejecta that each polygon represents by dividing the excavated portion of the crater into a series of simple paraboloid shells. The mass of material injected into each ring of 60 polygons is given by:

$$m_i = \frac{\rho_t \pi}{8} \left(r_{i+1}^3 - r_i^3 \right), \tag{15}$$

where m_i is the mass injected in ring i. The mass per polygon in this ring is thus $m_0 = (1/60)m_i$. This estimate is based on the assumption that the excavation depth of the transient crater is about $D/8$ or $R/4$ (Melosh, 1989). The initial mass loading per polygon will remain constant throughout the simulation, while the surface area of the polygon A will change dramatically throughout ejection and flight. Along with the target surface density ρ_t (which is also used as a particle density), the user must supply a mass distribution description, consisting of the smallest and largest particle diameters – which are converted to a minimum and maximum particle mass – and a cumulative distribution power-law exponent. This is described by:

$$dN = K m_p^{-b} dm_p, \tag{16}$$

where N is the cumulative number of particles, K is a constant, m_p is the particle mass, and b the supplied power-law exponent.

This gives us a means to derive the optical scattering properties of each ejecta plume polygon, which is a function of the scattering properties of each individual particle's surface area and albedo. The surface area of an individual particle is given by:

$$a = \pi \left(\frac{3m_p}{4\pi \rho_t} \right)^{2/3}. \tag{17}$$

Figure 10. A simulation showing the ejecta plume as a 3D shape-model for a small gravity-dominated cratering event (*top*) and a small strength-dominated cratering event (*bottom*) on an Eros-shaped target body. An ejecta particle size distribution has been assumed (maximum particle size, minimum particle size, and power-law distribution) with the resulting ejecta plume opacity calculated and rendered. Note that the ejecta plume detaches from the target body in the case of (*bottom*) strength-dominated cratering, with the first particles landing on the surface again at some distance from the impact crater site (if they do at all). In both simulations, a few random ejecta blocks are also included as discrete points. In the (*top*) gravity-dominated event, the convex shape of the ejecta plume indicates that the transient crater has finished forming and that the ejecta plume is now in the fall-out stage.

The total surface area per unit volume σ_v is found by solving $d\sigma_v = a\,dN$, which yields:

$$\sigma_v = \pi \left(\frac{3}{4\pi\rho_t}\right)^{2/3} \left(\frac{K}{(5/3) - b}\right) \left(m_1^{(5/3)-b} - m_s^{(5/3)-b}\right), \tag{18}$$

where m_1 and m_s are the mass of the largest and smallest particles, respectively. We place the constant K in terms of the mass density within the plume ρ_e, by solving $d\rho_e = m_p\,dN$:

$$\rho_e = \left(\frac{K}{2 - b}\right) \left(m_1^{2-b} - m_s^{2-b}\right), \tag{19}$$

Solving this expression for K, substituting back into the expression for σ_v, and re-arranging to find the surface area per unit mass σ_m gives:

$$\sigma_m = \frac{\sigma_v}{\rho_e} = \pi \left(\frac{3}{4\pi\rho_t}\right)^{2/3} \left(\frac{2 - b}{(5/3) - b}\right) \left(\frac{m_1^{(5/3)-b} - m_s^{(5/3)-b}}{m_1^{2-b} - m_s^{2-b}}\right), \tag{20}$$

which is an intrinsic property of the ejecta plume, based on the user supplied mass distribution of particles.

To determine the opacity of individual ejecta plume polygons as a function of their changing surface areas A, we make use of the *Lambart Exponential Absorption Law* (Chamberlain and Hunten, 1987):

$$I_f = I_0\, e^{-\sigma_m \psi}, \tag{21}$$

where I_0 is the initial light intensity, I_f the final light intensity, and ψ the mass loading per unit area within the plume polygon ($\psi = m_0/A$). Note that $\sigma_m \psi$ is equivalent to the optical depth of the plume. Normalizing the light intensity and bringing in the change in polygon area over the course of the simulation gives an opacity O equation for each polygon:

$$O = 1 - e^{-\sigma_m \psi_0 (A_0/A)}, \tag{22}$$

where ψ_0 and A_0 are the initial mass per unit area and initial polygon area, respectively. Note that this opacity applies to viewing the plume surface from a normal (perpendicular) direction, and does not yet take into account the variable albedo of the particles to different light wavelengths. These inputs are supplied to a rendering tool (the *OpenGL* package), for visualization.

For our current modeling purposes, we use mass distribution values from a typical comet dust environment (Lisse *et al.*, 2004) for the ejecta plume, although we expect that the actual observed plume will have a coarser particle distribution (and be correspondingly less opaque). When the actual observations are made, parameter searches using this forward model will be performed to better constrain these ejecta plume properties.

6.5. INSTRUMENT IMAGE SEQUENCE SIMULATIONS

We also used this final form of the ejecta plume model in the planning of the instrument image sequences for the comet flyby spacecraft. This is done by modeling

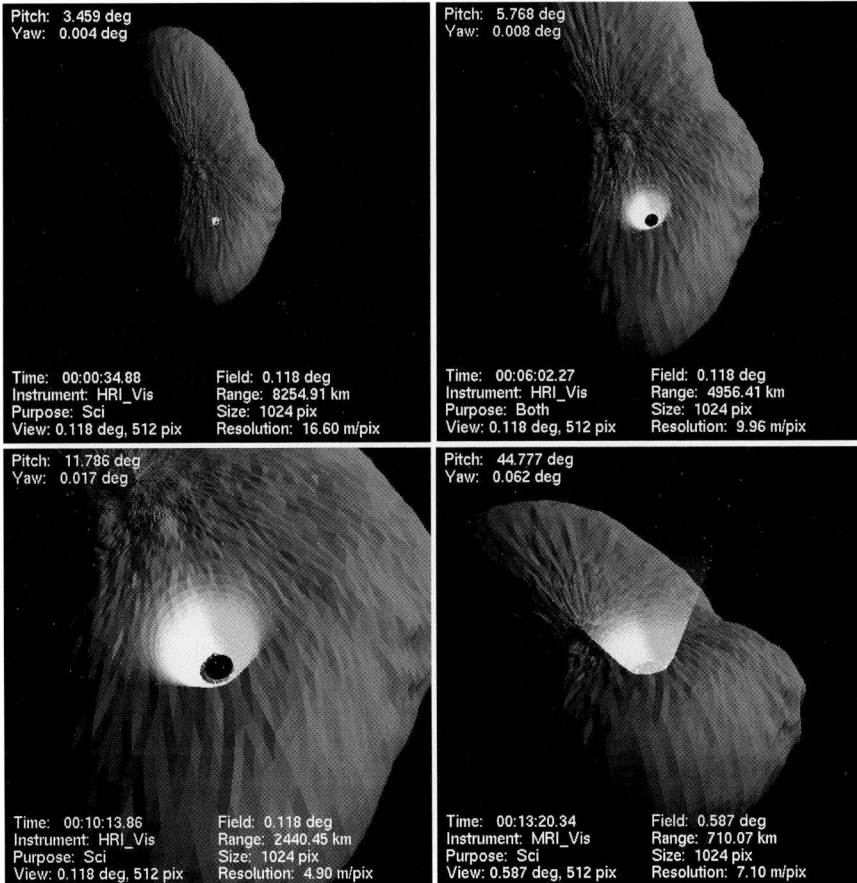

Figure 11. A few sample images from an instrument sequence simulation for the Deep Impact mission High-Resolution Instrument (HRI), showing false color renderings of a modified-Borelly comet shape-model *grey*, transient crater surface area *black*, ejecta plume *white*, and several hundred random ejecta fragments *white*. The first image (*upper left*) shows the view about 30 s after impact, showing the forming crater and ejecta plume. The second image (*upper right*) shows the view shortly after transient crater formation is complete. The third image (*lower left*) shows what some of our best views of the interior of the crater might look like as the point of closest approach is rapidly passed. The fourth image (*lower right*) shows the last possible image of the comet and ejecta plume as seen by the Medium-Resolution Instrument (MRI) just before putting the spacecraft in Safe Mode at closest approach. This sequence depicts one of our best possible scenarios: with a comet presenting a large face-on profile, an excellent hit by the impactor, a well-behaved gravity-dominated cratering event, very little image smear, and excellent impact site tracking for instrument pointing. What we actually see will most likely not be this ideal!

an impact on a shape-model target body and viewing it through a specifically designed display module that simulates the flight path of the comet and flyby spacecraft, comet orientation and sunlight, spacecraft orientation, and instrument field-of-view and mode information (Figure 11). This image sequence modeling can also be used after the actual encounter to forward model many of the observed features from the impact and flyby.

7. Conclusion

Impact cratering theory and modeling play an important role in the Deep Impact mission, from initial inception to final data analysis. Experimentally derived impact crater scaling laws provide us with our best estimates for the crater diameter, depth, and formation time. Cratering theory has strongly influenced the impactor design, producing a craft that should produce the largest possible crater on the surface of Tempel 1 under a wide range of scenarios. Numerical hydrocode modeling allows us to estimate the volume and thermodynamic characteristics of the material vaporized in the early stages of the impact (a mixture of impactor and comet material) with a view towards disentangling these two components when the actual impact occurs. Hydrocode modeling will also aid us in understanding the observed crater excavation process, especially in the area of impacts into porous materials. Finally, experimentally derived ejecta scaling laws and modeling provide us with a means to predict and analyze the observed behavior of the material launched from the comet during crater excavation, and may provide us with a unique means of estimating the magnitude of the comet's gravity field and by extension the mass and density of comet Tempel 1. Together with laboratory cratering experiments (Schultz and Ernst, this volume), impact cratering theory and computational modeling will provide us with important tools toward understanding the results of this unique impact experiment on comet Tempel 1.

References

Anderson, C. E.: 1987, *Int. J. Impact Eng.* **5**, 33.
Anderson, J. L. B., Schultz, P. H., and Heineck, J. T.: 2003, *J. Geophys. Res. (Planets)* **108**, 13.
Artemieva, N. A. and Ivanov, B. A.: 2001, *Lunar Planet. Inst. Conf. Abst.* **32**, 1431.
Asphaug, E. and Benz, W.: 1994, *Nature* **370**, 120.
Brandt, J. C. and Chapman, R. D.: 2004, *Introduction to Comets*, Cambridge University Press, Cambridge, USA.
Brownlee, D. E., Horz, F., Newburn, R. L., Zolensky, M., Duxbury, T. C., Sandford, S., *et al.*: 2004, *Science* **304**, 1764.
Chamberlain, J. W. and Hunten, D. M.: 1987, *Theory of Planetary Atmospheres*, Academic Press, San Diego.
Cintala, M. J., Berthoud, L., and Hörz, F.: 1999, *Meteor. Planet. Sci.* 34, 605.
Collins, G. C., Melosh, H. J., and Ivanov, B. A.: 2004, *Meteor. Planet. Sci.* **39**, 217.

Croft, S. K.: 1981, in Schultz, P. H. and Merrill, R. B. (eds.), *Multi-Ring Basins*, Pergamon Press, New York, p. 207.

Geissler, P., Petit, J. M., Durda, D. D., Greenberg, R., Bottke, W. F., and Nolan, M. C.: 1996, *Icarus* **120**, 140.

Grady, D. E. and Kipp, M. E.: 1987, in Atkinson, B. K. (ed.), *Fracture Mechanics of Rock*, Academic Press, San Diego, p. 429.

Heiken, G. H., Vaniman, D. T., and French, B. M.: 1991, *Lunar Sourcebook*, Cambridge University Press, Cambridge, USA.

Holsapple, K. A.: 1980, *Lunar and Planetary Science Conference*, Vol. 11, Pergamon Press, New York, p. 2379.

Holsapple, K. A. and Schmidt, R. M.: 1982, *J. Geophys. Res.* **87**, 1849.

Housen, K. R., Schmidt, R. M., and Holsapple, K.A.: 1983, *J. Geophys. Res.* **88**, 2485.

Jessberger, E. K.: 1999, *Space Sci. Rev.* **90**, 91.

Larson, D. B.: 1977, *The Relationship of Rock Properties to Explosive Energy Coupling*, UCRL-52204, University of California Research Labs.

Lisse, C. M., A'Hearn, M. F., Fernandez, Y. R., McLaughlin, S. A., Meech, K. J., and Walker, R. J.: 2004, *Icarus*, in press.

Love, S. G., Hörz, F., and Brownlee, D. E.: 1993, *Icarus* **105**, 216.

McGlaun, J. M., Thompson, S. I., and Elrick, M. G.: 1990, *Int. J. Impact Eng.* **10**, 351.

Melosh, H. J.: 1989, *Impact Cratering: A Geologic Process*, Oxford University Press, New York.

Melosh, H. J.: 2000, *Lunar Planet. Inst. Conf. Abst.* **31**, 1903.

Melosh, H. J. and Ivanov, B. A.: 1999, *Ann. Rev. Earth Planet. Sci.* **27**, 385.

Nolan, M. C., Asphaug, E., Melosh, H. J., and Greenberg, R.: 1996, *Icarus* **124**, 359.

Nordyke, M. D.: 1962, *J. Geophy. Res.* **67**, 1965.

Peale, S. J.: 1989, *Icarus* **82**, 36.

Pierazzo, E., Artemieva, N. A., and Spitale, J. N.: 2001, *ESF-IMPACT 5: Catastrophic Events and Mass Extinctions: Impacts and Beyond*, Granada, Spain.

Poorman, K. L. and Piekutowski, A. J.: 1995, *Int. J. Impact Eng.* **17**, 639.

Rubin, M. B., Vorobiev, O. Y., and Glenn, L. A.: 2000, *Int. J. Solids Struc.* **37**, 1841.

Sagdeev, R. Z., Elyasberg, P. E., and Moroz, V. I.: 1988, *Nature* **331**, 240.

Scotti, J. and Melosh, H. J.: 1993, *Nature* **365**, 733.

Schmidt, R. M. and Housen, K. R.: 1987, *Int. J. Impact Eng.* **5**, 543.

Schultz, P. H. Anderson, J. L. B., and Heineck, J. T.: 2002, *Lunar Planet. Inst. Conf. Abst.* **33**, 1875.

Schultz, P. H. and Ernst, C.: 2005, *Space Sci. Rev.*, this volume.

Shuvalov, V. V.: 1999, *Shock Waves* **9**, 381.

Taylor, S. R.: 1982, *Planetary Science: A Lunar Perspective*, Lunar and Planetary Institute, Texas.

Thompson, S. L. and Lauson, H. S.: 1972, *Improvements in the Chart-D radiation hydrodynamic code III: Revised analytical equation of state*, Sandia National Laboratories Report SC-RR-710714, Albuquerque, New Mexico, 119 pp.

Werner, R. A.: 1994, *Celest. Mech. Dyn. Astron.* **59**, 253.

Wroth, C. P. and Bassett, R. H.: 1983, *Géotechnique* **33**, 32.

Wünnemann, K. and Ivanov, B. A.: 2003, *Planet. Space Sci.* **51**, 831.

EXPECTATIONS FOR INFRARED SPECTROSCOPY
OF 9P/TEMPEL 1 FROM DEEP IMPACT

JESSICA M. SUNSHINE[1,*], MICHAEL F. A'HEARN[2], OLIVIER GROUSSIN[2],
LUCY A. McFADDEN[2], KENNETH P. KLAASEN[3], PETER H. SCHULTZ[4]
and CAREY M. LISSE[2]

[1] *Science Applications International Corporation, Chantilly, VA, U.S.A.*
[2] *Department of Astronomy, University of Maryland, College Park, MD, U.S.A.*
[3] *Jet Propulsion Laboratory, California Institute of Technology, Pasadena, CA, U.S.A.*
[4] *Department of Geological Sciences, Brown University, Providence, RI, U.S.A.*
(*Author for correspondence; E-mail: sunshinej@saic.com)

(Received 23 September 2004; Accepted in final form 23 December 2004)

Abstract. The science payload on the Deep Impact mission includes a 1.05–4.8 μm infrared spectrometer with a spectral resolution ranging from $R \sim 200$–900. The Deep Impact IR spectrometer was designed to optimize, within engineering and cost constraints, observations of the dust, gas, and nucleus of 9P/Tempel 1. The wavelength range includes absorption and emission features from ices, silicates, organics, and many gases that are known to be, or anticipated to be, present on comets. The expected data will provide measurements at previously unseen spatial resolution before, during, and after our cratering experiment at the comet 9P/Tempel 1. This article explores the unique aspects of the Deep Impact IR spectrometer experiment, presents a range of expectations for spectral data of 9P/Tempel 1, and summarizes the specific science objectives at each phase of the mission.

Keywords: coma, comets, Deep Impact, infrared spectroscopy, nucleus, Tempel 1

1. Introduction: The Importance of 1–5 μm Infrared Spectroscopy

The science payload on the Deep Impact mission includes a 1.05–4.8 μm infrared spectrometer. As described in more detail by Hampton *et al.* (2005), this spectrometer (HRI-IR) is included in the high resolution imaging system and consists of a dual prism and a 1024×512 pixel HgCdTe focal plane array (which typically is re-binned to 512×256 pixels). The HRI-IR has a minimum spectral resolution ($\lambda/d\lambda$) of ~ 200 at 2.63 μm and a maximum spectral resolution of ~ 900 at 1.05 μm. The instantaneous field of view (IFOV) is 10 μrad.

Interplanetary missions, like Deep Impact, have frequently carried near-infrared spectrometers to a wide variety of targets. The use of near-infrared spectrometers is driven by the value of this spectral region for understanding cool-to-warm bodies (vs. hot bodies, *i.e.*, stars). For objects at distances from $\sim 1/2$ to several AU from the sun, this spectral region includes the transition from reflected sunlight to thermal emission. The 1–5 μm region also includes absorption features from most ices, silicates, and organics, as well as the vibrational transitions of many compositionally

Space Science Reviews (2005) 117: 269–295
DOI: 10.1007/s11214-005-3388-2

diagnostic gaseous molecules that are easily excited at the temperatures of objects at these solar distances.

The goal in designing the spectrometer is to address the key scientific objectives of the mission (A'Hearn, 2005). At the highest level, these include studying the compositional variation with depth in the nucleus and the physical process of cratering on a cometary nucleus. Since ices are the most likely component of the nucleus to show variations with depth, observations of volatiles are a necessity. In designing the Deep Impact HRI-IR spectrometer, we set a requirement for observing the fundamental vibrational transition of CO at 4.7 μm, which in turn sets the practical limit for the long-wave end of the spectral range. CO is one of the diagnostic, moderately abundant, volatiles in comets and is therefore particularly important for study. However, the product of CO abundance and the g-factor for this band is lower than for the strongest bands or blends of the other abundant volatiles. Furthermore, the fundamental vibrational band is at a wavelength at which thermal emission from the instrument itself is a problem. Taken together, this makes CO the most difficult of the abundant volatiles to observe. The IKS infrared channel of the TKS instrument on the *Vega* mission to comet Halley (Krasnopolsky *et al.*, 1986) went to a slightly longer, but quite similar long-wave limit. A sample spectrum from the IKS instrument is shown in Figure 1 (Combes *et al.*, 1988), highlighting the gaseous species producing identifiable emission features. The CO fundamental is the weakest of the clearly identified features (the OCS being a questionable detection).

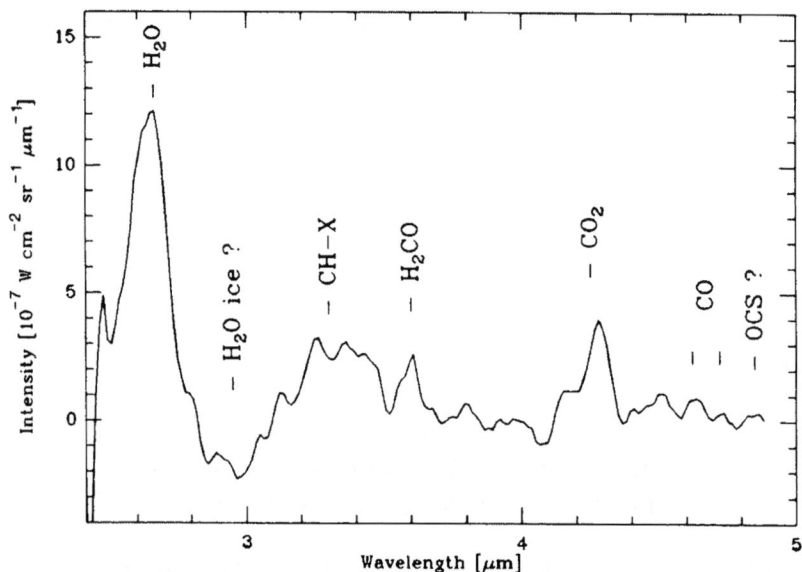

Figure 1. The *Vega* IKS spectrum of 1P/Halley 1986, taken at a distance from the nucleus of 10,000 km and at a spectral resolving power $R \sim 50$. Following Combes *et al.*, 1988, lines due to the C-H aliphatic stretch, formaldehyde, carbon monoxide and carbon dioxide are clearly present, with a possible detection of water ice absorption.

Spectroscopy of the coma *via* remote sensing is commonplace, but observations from Earth are nearly always of a very extended volume of the coma. Earth-based observations that are limited to the region near the nucleus are very rare, and are achievable only when comets approach very close to Earth. Spectra of comae include three components: (1) the solar spectrum reflected by the solid grains in comae (mostly refractory grains for remote sensing, but possibly including icy grains for observations from a close flyby), (2) thermal emission by the dust grains (which allows us to measure the superheat, or excess over the blackbody temperature, of the dust), and (3) emission-line spectra from gaseous species. If icy grains represent a significant proportion in the innermost component, as implied by models for their lifetime, then absorption features in the reflectance spectra of the dust may differ from the features in the spectra of nuclei. As with spectra of nuclei, the longer wavelengths of coma spectra will be dominated by thermal emission rather than by reflection. The exact wavelength of the break will be determined by a combination of the size distribution and the wavelength dependent indices of refraction of the individual particles. The observations from IKS (Figure 1) and observations made with a wide variety of remote sensing equipment, from ground-based telescopes through space-based telescopes like ISO, have revealed a wealth of emission features in the near-infrared.

Unlike cometary coma spectra, spectra of bare cometary nuclei are rare. An extremely limited number of telescopic spectra currently exist (e.g., Licandro *et al.*, 2002; Abell *et al.*, 2003, 2005), and only one space mission, *Deep Space-1 (DS-1)*, has ever obtained spectra of a nucleus (19P/Borrelly) without significant interference from the coma (Soderblom *et al.*, 2004b). *DS-1* dramatically documented the variation in temperature on the nucleus. The interpretation of those results will be very important in building up our understanding of the physical properties of the nucleus, particularly of the properties that influence the effective thermal inertia. However, as discussed in detail in subsequent sections, data from *DS-1* and the existing telescopic spectra indicate that on a global scale very low albedo components dominate the spectra of comet nuclei and suppress most of the plausible characteristic absorption features of common ices and silicates. While this result was important in showing that dark, presumably organic, material dominates the spectral reflectivity, it means that global spectral analysis of the nuclear reflectivity is unlikely to be diagnostic of nuclear composition. This is in contrast to objects such as icy satellites, which have significantly less dark material. Thus, for the nucleus, the ability to determine surface temperatures with good spatial resolution may be at least as important as measuring spectral absorptions.

Below, we examine in more detail previous 1–5 μm spectroscopy of cometary nuclei and comae. The Deep Impact HRI-IR spectrometer will provide measurements at previously unseen spatial resolution, before, during, and after a cratering event. We explore the unique aspects of this experiment and present a range of expectations for data of 9P/Tempel 1. Finally, we summarize the HRI-IR observation plan and highlight the specific science objectives at each phase of the mission.

2. Review of Comet Spectra from 1–5 μm

2.1. PREVIOUS RESULTS FROM OBSERVATIONS OF COMETARY NUCLEI

Our current understanding of the spectra of cometary nuclei is very limited. To date, we are aware of only five studies of the near-infrared spectral properties of bare cometary nuclei. These include four Jupiter Family comets and one Halley Family comet. Due to a new generation of instruments for telescopes and spacecraft, it is now possible to observe faint comet nuclei with reasonable signal-to-noise ratios (SNR) in the 1–2.5 μm region. However, as of yet, no spectral data on comet nuclei exist in the 2.5–5 μm region, but this region will be measured by Deep Impact's HRI-IR instrument, as described in subsequent sections.

   The first published near-infrared spectra of bare nuclei were obtained by Licandro *et al.* (2002, 2003) who measured the Jupiter Family comets, 124P/Mrkos and 28P/Neujmin-1 from 0.9 to 2.3 μm using the Near-Infrared Camera Spectrometer (NICS) at the 3.56 m Telescopio Nazionaleo Galielo (TNG). The spectrum of 124 P/Mrkos was found to be moderately sloped, and within the signal-to-noise and spectral range, was devoid of any specific spectral features (see Figure 2). In particular, there was no evidence for water ice absorptions. These results were interpreted as probable evidence for mantling of the ice that is widely assumed to be a large component of cometary nuclei. Licandro *et al.* (2003) also note a possible

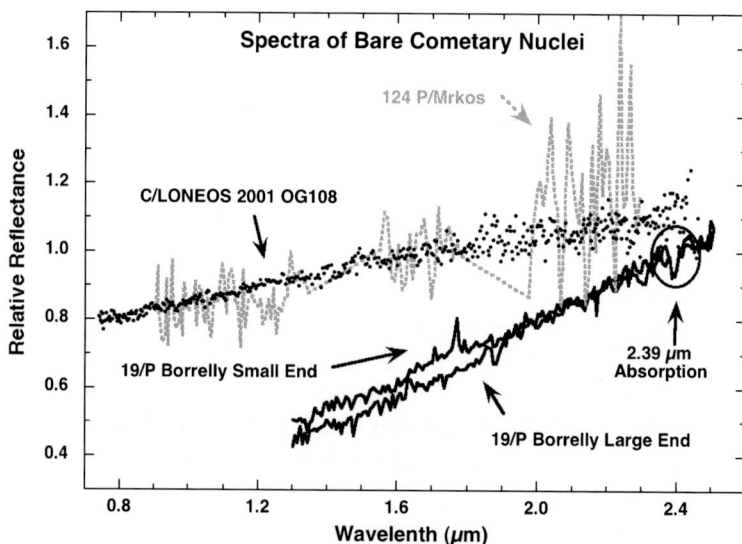

Figure 2. Existing near-infrared spectra of bare cometary nuclei. The data of 19/P Borrelly (black line) are from the *Deep Space*-1 mission (Soderblom *et al.*, 2004b). Telescopic spectra of C/LONEOS 2001 OG108 (black points) are from Abell *et al.* (2003, 2005) and 124 P/Mrkos (grey dashed line) are from Licandro *et al.* (2003).

small change in slope between two nights of their observations. The NICS spectrum of 28P/Neujmin-1 is similarly featureless. However, Campins *et al.* (2003) preliminarily report significant differences in slopes between their two observations in 2001 and 2002.

The first spectrum of a bare cometary nucleus observed at the NASA Infrared Telescope Facility on Mauna Kea was measured by accident (Abell *et al.*, 2003, 2005), as part of an on-going survey of asteroids. Using the SpeX instrument (Rainer *et al.*, 2003), Abell *et al.* (2003, 2005) observed the nucleus of Halley family comet C/LONEOS OG108 at a phase angle of 8° from 0.7 to 2.5 μm. At the time of observation, the object was assumed to be an asteroid, as cometary activity had yet to be observed. Even with relatively high signal-to-noise, the near-infrared spectrum of C/LONEOS OG108 was also found to have no discernible absorption features, but to be linearly increasing with wavelength, with a very similar slope to the spectrum of 124P/Mrkos (Figure 2). Based on these observations, Abell *et al.* (2003) conclude that this comet is spectrally similar to D-type asteroids.

Ground-based spectra of one other bare Jupiter family cometary nucleus have been reported. Delahodde *et al.* (2002) observed 90P/Gehrels from 1.4–2.5 μm using the Very Large Telescope. Their preliminary comparisons suggest that there may be differences in spectral slopes and/or unspecified spectral features among their data.

Deep Space-1 (DS-1), a technology demonstration mission, provided the only space-based near-infrared spectral observations of a bare nucleus. *DS-1* encountered the Jupiter Family comet 19P/Borrelly in September of 2001 (Nelson *et al.*, 2004). Although complicated by pointing, rapidly changing Sun angles, saturation of the spectrometer signal over high albedo regions, and uncertainties in the wavelength calibration of the spectrometer, the SWIR channel of the MICAS (Miniature Integrated Camera and Spectrometer) instrument collected valuable spectra of 19P/Borrelly from 1.3–2.5 μm (Soderblom *et al.*, 2004b). After removing instrument and thermal effects, Soderblom *et al.* (2004b) found an extremely sloped nucleus spectrum with one spectral feature at 2.39 μm (see Figure 2). In the absence of any other absorption features, specific identification of this feature has proven to be elusive. However, the absorption at 2.39 μm was present in all 46 spectra of 19 P/Borrelly and is generally consistent with organic materials, and possibly specifically with nitrogen-bearing organics. This possibility is intriguing as Kissel *et al.* (2004) report nitrogen enrichment in grains from another Jupiter family comet (81P/Wild 2) compared to grains at 1P/Halley. Although the *DS-1* spectra are much higher spatial resolution than disk averaged telescopic spectra (160 m by the width of the nucleus), there are only slight changes in spectral slope ($\sim$5%) from the small to the large end of the comet, as indicated on Figure 2.

Taken together these existing near-infrared observations show that on a global scale comet nuclei are spectrally featureless. Notably absent in these spectra are evidence of water ice absorptions. Instead, spectra with a variety of slopes are observed, as indicated on Figure 2. While differences in slope between comets (and perhaps between different observations of the same comet) are clear and consistent with the range of slopes seen in visible (0.4–0.9 μm) spectra (e.g., Luu, 1993), they are not yet understood. The only example of spatially resolved data (for 19P/Borrelly) includes one absorption feature, possibly due to organics. As discussed below, even if no absorption features are present, the data from the HRI-IR spectrometer on Deep Impact, with increased spatial resolution, improved SNR, and extended spectral range, will provide new constraints on our understanding of the surface composition of comet nuclei.

2.2. PREVIOUS 1–5 μM DATA OF COMETARY COMAE

The spectra of cometary dust between 1 and 5 um are mostly dominated by the continuum produced from the scattering of sunlight and the re-emission of absorbed sunlight as thermal radiation. The 1–5 μm continuum provides useful constraints on models of dust grain size and composition. The observed radiation depends both on the ability of the dust to scatter and emit light, and on the number of such scatterers and emitters. For dust particle mass distributions and compositions currently thought to be representative of comets (Gruen *et al.*, 1985; Lamy *et al.*, 1987; Jessberger *et al.*, 1988; McDonnell *et al.*, 1991; Lisse *et al.*, 1998), the scattered radiation (typically shortward of 3 μm) is most sensitive to the population of particles with radii <1 μm, while the thermal emission (typically longward of 3 μm) is most sensitive to particles with radii > 1 μm. A generally decreasing slope for all comets from 0.5–3 μm has been described as being consistent with a geometrical effect in scattering from ~1 μm particles (Jewitt, 1991), but Lamy *et al.* (1989) has attributed this result to compositional variations in larger grains. Variations in grain size from comet to comet suggest real differences in cometary origin and evolution. From 3–5 μm, deviations from a blackbody are very sensitive to the abundance of small, optically dark grains (Harker *et al.*, 2002; Lisse *et al.*, 2002), which can attain temperatures upwards of 600 K at 1 AU in the Sun's radiation field. Highly emissive comets like C/Hale-Bopp 1995 O1, C/Hyakutake 1996 B2, and 1P/Halley 1986 with large amounts of small particle emission have dust spectra in excess of a blackbody spectrum ("superheat"), while low activity, highly evolved comets, like 2P/Encke and 9P/Tempel 1, have dust spectra well-described by a blackbody curve.

In addition to the continuum, two specific absorption features in the 1–5 μm region have been attributed to cometary dust. A broad water ice absorption feature in the very active comet C/Hale-Bopp 1995 O1 was found by Lellouch *et al.* (1998). In addition, various authors including Combes *et al.* (1988) have discussed the possibility that the 3.2–3.5 μm emission in the VEGA-IKS spectrum of 1P/Halley (see Figure 1) is partially due to hydrocarbons in the solid phase. However, at the resolutions of the observations to date, neither of these measurements by themselves is uniquely diagnostic of the composition of the dust. The ice absorption lies on top of the liquid and gaseous water absorption and the purported 3.29 μm feature lies among the emission lines and their associated wings from the aliphatic C-H stretch exhibited by methanol, methane, and formaldehyde.

In contrast, gas emission features in the 1–5 μm region are highly diagnostic, and most of the molecules that have thus far been detected as emissions from comets have strong fundamental vibrational bands in this region. The first breakthrough in this wavelength region was in 1986 when the *Giotto* IKS-VEGA experiment observed the spectrum of the inner coma of comet 1P/Halley between 2.4 μm and 4.9 μm, with a spectral resolution $R \sim 50$ (Combes *et al.*, 1988; see Figure 1). As predicted by Crovisier and Encrenaz (1983) using synthetic spectra, *Giotto* results clearly indicate emission lines due to H_2O at 2.7 μm, CO_2 at 4.3 μm, CO at 4.7 μm, and some unidentified CH-X molecules at 3.2–3.4 μm. H_2O was also weakly detected at 1.38 μm using the IR channel of the TKS experiment onboard the *Vega 2* spacecraft (Krasnopolsky *et al.*, 1986). The C-H stretch features, as discussed above, may be due to solid grains. Additional potential features due to H_2CO at 3.6 μm, OCS detection at 4.85 μm, and the primary absorption band due to H_2O ice at 2.9 μm are more tenuously present in the data. The CO_2 observations are particularly noteworthy, because they cannot be observed from the ground (or even high in the atmosphere from SOFIA), due to strong telluric absorptions in this region.

The second leap forward was the Infrared Space Observatory (ISO), which observed comets C/1995 O1 (Hale-Bopp) (Crovisier *et al.*, 1999a) and 103P/Hartley 2 (Colangeli, 1999; Crovisier *et al.*, 1999b) from 2.5–4.9 μm, using the spectrographs ISO/SWS with $R \sim 1500$ and PHT-S with $R \sim 90$. Emissions due to H_2O at 2.7 μm and CO_2 at 4.3 μm were detected on both comets. Emission due to CO at 4.6 μm was also detected, but only on C/1995 O1 (Hale-Bopp) (Crovisier *et al.*, 1999a). The resolution of the ISO was not sufficient to resolve the numerous emission lines possibly present in the 3.2–3.6 μm region. However, Bockelée-Morvan *et al.* (1995) analyzed spectra of seven comets observed with the IRTF and/or UKIRT with a high spectral resolution of 70–400 from 3.2–3.6 μm. They concluded that emission features in this region are mainly due to CH_3OH (methanol) fluorescence emission and probably unsaturated hydrocarbons such as PAH (Polycyclic Aromatic Hydrocarbons).

The third major advance was with the development of new instruments with very high spectral resolution $R \sim 25000$ like NIRSPEC at the Keck telescope. Using this instrument in the wavelength range 3.0–3.4 μm, Mumma *et al.* (2001) were able to identify several new emission lines. For example, on comet C/1999 H1 (Lee) they detected C_2H_6, OH, CH_3OH, CH_4, HCN, C_2H_2, and NH_2. Several emissions lines remain unidentified in the 3.3–3.5 μm region, but are most probably due to radicals or complex molecules. Finally, many molecules are known to have emission lines between 1.0 and 4.9 μm, but as of yet have only been detected at UV and radio wavelengths. One of the most interesting species is the emission of SO_2 at 4.0 μm. This feature is very weak; however, it is located in a region of the spectrum without any contamination from other emission lines. Thus while the short-lived SO_2 molecule has not yet been detected at 4.0 μm, as discussed below, the high sensitivity and spatial resolution of Deep Impact's HRI-IR spectrometer may allow for its detection at 9P/Tempel 1.

3. Expectations for Deep Impact

Here we present what we might observe with the Deep Impact's HRI-IR spectrometer (from 1–5 μm) for the nucleus, dust, and gas coma of comet 9P/Tempel 1. Expectations are based on our current knowledge of this comet, on past IR spectroscopic observations of other comets from ground, space, and *in situ* observations, and on synthetic spectral modeling. For the first time, we will have the opportunity to observe a comet before, during, and after an impact event and at spatial scales previously unattainable. In subsequent sections, we present expectations for the nucleus pre-impact (the exterior), during the impact event, and from the post-impact crater (the interior). Similarly, we examine pre-impact expectations for the coma, from a few weeks before impact when the distance from the nucleus is several million kilometers, until a few minutes before the impact when the distance is only a few hundred kilometers. We then discuss the expectations for dust and gas emissions at the time of impact and in the subsequent minutes. Finally, we explore the range of post-impact coma phenomena that we may observe from a few minutes after impact through "look-back" observations that extend for several days.

3.1. EXPECTATIONS FOR OBSERVATIONS OF THE NUCLEUS

3.1.1. *The Pre-Impact Exterior*
The Deep Impact mission offers a unique opportunity for near-infrared observations of the surface of the nucleus of 9P/Tempel 1. As we begin to resolve the nucleus, as in the *DS-1* mission, we will obtain spectra with increasingly smaller, and ultimately negligible, coma contributions. The HRI-IR spectra will increase in resolution until a maximum of 7 m per pixel at closest approach. This represents a significant increase in spatial resolution compared to *DS-1* SWIR data of 19P/Borrelly, which had a maximum per-pixel resolution of 160 m by the width of the comet.

At these higher spatial resolutions, we expect to resolve morphologic features due to jetting and natural craters, such as those seen in the clear filter high-resolution *DS-1* images of 19P/Borrelly (Soderblom *et al.*, 2004a) and *Stardust* images of 81P/Wild 2 (Brownlee *et al.*, 2004). Differences in albedo, overall slope, and/or composition may be correlated with specific morphologies, particularly when the HRI-IR spectra are combined with 7-color visible HRI images (see Thomas *et al.*, 2005). Even in the absence of spectral features, understanding whether spectral slope varies, and if so, if it is correlated to specific processes, is of keen interest as it may help explain the variation in slopes observed in disk-integrated spectra of cometary nuclei.

The search for specific spectral features will be facilitated by a large and spatially detailed dataset that extends from 1–5 μm. The combination of increased spatial resolution, spectral range, and number of spectra available for co-adding to increase our signal-to-noise, raises the likelihood of detecting absorptions should they exist. Given the known components of cometary nuclei: silicates, organics and ices, we might observe, for example, ice-rich bright areas exposed by recent jetting or slumping that have spectral features of water, CO_2 and/or CO ice. Alternatively, we might expect dark areas, presumably rich in refractory materials, to have spectral signatures in the 1–2 μm region (due to various silicates) or in the 3.2–3.5 μm region (due to organics). We will of course examine the 2.39 μm region for the absorption band that was observed on Borrelly's nucleus (Soderblom et al., 2004b) and search for additional bands at longer wavelengths that might aid to identifying its origin.

Finally, we will use the thermal signatures captured by the HRI-IR spectrometer to determine thermal variations across the nucleus. With knowledge of the rotation state (Belton et al., 2005), we can relate the observed surface temperature to both the local instantaneous insolation and past history to constrain thermal inertia.

3.1.2. The Impact Event

The Deep Impact collision may produce a significant 'flash'. Observations of hypervelocity impacts into a wide range of material types (Ernst and Schultz, 2003, 2004; Schultz and Ernst, 2005) reveal that the luminous efficiency (radiant energy scaled to initial impact kinetic energy) may range from 10^{-5}–10^{-3} and will depend on the nature of the upper surface of the comet. Laboratory experiments impacting silicate targets with velocities from 5–6 km/s resulted in peak temperatures exceeding 6000 K that rapidly decayed to temperatures of around 2800 K to 3200 K over more than 1 millisecond. This thermal evolution results from ejecta that includes heated silicate particles and melt droplets that line the crater floor. Integrating the total energy over time and wavelength yielded a cumulative radiant energy of about 0.7 joules, which represents about 0.035% of the initial kinetic energy. The radiation is a thermal component related to heated particulates from the target. As described in more detail by Schultz and Ernst (2005), the cumulative radiant energy can be scaled to the Deep Impact collision over the 1–5 μm range of the HRI-IR, which yields an integrated radiant energy of 5 mega-joules in about 100 milliseconds. However, the total duration of the impact flash may be as long as 500 milliseconds to 1 s, depending on composition. Thus, as detailed in Schultz and Ernst (2005), this observation, may provide an estimate of the silicate-to-volatile ratio of 9P/Tempel 1.

The spatial field-of-view of the HRI-IR spectrometer slit allows for tracking of the evolution of different ejecta components through time. Time-exposed spectra (seconds) of vapor plumes from hypervelocity impacts in the laboratory (lasting less than 50 microseconds) illustrate the successful application of this strategy, as long as there is sufficient radiant flux (Schultz, 1996). As the imaged plume passed the slit, evolving spectra were successfully recorded. For the Deep Impact HRI-IR spectrometer, the projected slit width at the time of impact will be 85 m, while the slit length will extend across ~5.4 km of the comet surface (Klaasen et al., 2005). The first several spectra after impact should capture the vapor phases and thermal components. Rapid expansion of the plume should then smear the source at certain locations in the slit during the exposure.

For Deep Impact, the horizontal velocity component of the initial plume (combined with its optical depth) will control the observed spectra. Even though the initial plume expansion velocities may exceed 10 km/s, most of the early stages will have significantly lower velocities. If the near surface of the comet is highly porous, vapor phases developed within the cavity initially will be directed upwards (toward the Deep Impact spacecraft), thereby reducing lateral growth (e.g., see Schultz and Ernst, 2005). In this case, much of the earliest evolution of the plume could be captured within the HRI-IR slit with an intensity limited by the optical depth of the ionized gases. If the upper surface layers (depth of ~1–5 m) have significant strength, then it is likely that the HRI-IR will capture radiating components in streaks. Although not time resolved, the spatially restricted radiating sources will be captured.

It is expected that the vapor plume will then evolve with time (next several seconds) due to changing temperatures and chemical reactions. However, subsequent spectra will record cooler ejecta from depth, including primary volatile species. This component will emerge closest to the observation point as the higher temperature constituents move toward either end of the slit. Solar fluorescence may then reveal primary atomic and molecular species of interest. Even though the radiating phases may have left the region of impact, their contribution to the spectra may persist depending on the horizontal component of the velocity. If the surface is composed of silicates, these spectra also will exhibit a strong thermal background and provide a direct measure of the temperature, which will be useful to characterize the nature of the surface (see Schultz and Ernst, 2005).

During the subsequent ~300–800 s after crater formation, ejecta from the nucleus will pass over the HRI-IR spectrometer slit. Over time, these ejecta will originate from greater and greater depths, and will be spatially separated along the slit. As such, observing ejecta as a function of time will provide data on composition as a function of depth in the nucleus. Our expectation is to first see refractory materials coming from the surface (organics and silicates), followed by an increase in volatile species (e.g., H_2O, CO, CO_2), which will likely include some ices. Over this time period, the temperature of ejecta will vary due to competing effects of the energy released by the impact and the intrinsic temperature profile of the nucleus, which arise from changes in solar insolation and thermal inertia.

3.1.3. The Post-Impact Crater Floor

After the crater formation event, and with some luck in our pointing, the HRI-IR spectrometer will be able to observe the interior of our newly created crater and its surrounding ejecta deposits. This unique observation, will for the first time, measure the exposed interior of a comet nucleus. If the surfaces of comet nuclei are covered by a mantling process (ionic space weathering or residual buildup of refractory materials) as is commonly thought (e.g., Brin and Mendis, 1979; Johnson *et al.*, 1987), then the Deep Impact crater should reveal more pristine materials. The candidate materials for near-surface comet interiors include ices (water, CO_2, methane, etc.), silicates, and various organic materials.

The HRI-IR spectrometer is well suited to detect these materials should they exist. The primary silicates, olivine and pyroxene, if they include even minor amounts of iron, have diagnostic absorptions in the 1 and 2 μm regions due to electronic transitions of Fe^{2+} ions in crystallographic sites within the silicate structures (Burns, 1993). Spectra of related hydrated silicates include overtones and combinations of overtones of vibrational, bending, and stretching bands of bound H_2O and OH- ions in the 1.4–2.4 μm region (Hunt, 1979; Hunt and Ashley, 1979). Planetary ices have several vibrational and overtone features in the near-infrared, which can vary with temperature and grain size (e.g., Clark, 1981). Major absorptions include: water ice at 1.25, 1.6, 2.0, 2.9, and 3.1 μm; methane ice at 1.7, 2.3, 2.6, and 3.3 μm; methanol ice at 2.27 μm; CO ice at 1.6, 2.3, and 4.7 μm; CO_2 ice at 2.0, 2.7, 3.3, and 4.3 μm, and ammonia ice at 1.5, 1.6, 2.0, 2.3, and 2.9 μm (e.g., Coradini *et al.*, 1998; Quirico *et al.*, 1999). Spectra of natural organics contain C-, H-, and O-vibrational modes that also lead to a variety of features in the 1.5–3.5 μm region (e.g., Cloutis, 1990; Moroz *et al.*, 1998).

While few comet exteriors and no comet interiors have been observed in the near-infrared, a number (although still relatively few) observations of Trans-Neptunian objects (TNOs), (*i.e.*, Kuiper Belt Objects or Edgeworth-Kuiper Belt Objects) have been made (e.g., Barucci *et al.*, 2005). The TNO's and Centaurs, a population derived from TNO's by perturbations with Neptune (Barucci *et al.*, 2002), are now widely accepted as the source region for short period comets (e.g., Levison and Duncan, 2001). As such, the composition of TNO's is highly relevant to the study of comets. The spectra of a number of TNO's include absorptions from a variety of the possible cometary nucleus materials, suggesting that if these materials are present and not masked by darkening agents, they should be detectable at 9P/Tempel 1 with the HRI-IR spectrometer.

Due to their faintness, very few near-infrared spectra of TNO objects exist. As recently summarized by Barucci *et al.* (2005), the signal-to-noise of these spectra are low, but those observed are diverse and suggest non-uniformities on the surfaces of some TNO's. Some are generally flat, yet featureless (e.g., 1996 TL66; Luu and Jewitt, 1998) while others show strong water ice absorptions (e.g., 19308/1996 TO66; Brown *et al.*, 1999, 2000; Licandro *et al.*, 2001). The spectra of one object 26375/1999 DE9 includes absorptions near 1.4, 1.6, 2.0, and 2.25 μm attributed to hydrous silicates and possibly a 1 μm olivine feature (Jewitt and Luu, 2001).

The spectra of Centaurs, as summarized by (Barucci *et al.*, 2002, 2005), are dominated by water ice absorptions. While featureless in the visible, the near-infrared spectrum of 2060 Chiron (Luu *et al.*, 2000) includes water ice features in the 1.5 and 2 μm regions, as does 10199 Chariklo (Brown and Koresko, 1998; Dotto *et al.*, 2003). Recent modeling by Groussin *et al.* (2004a) indicates that both Chiron and Chariklo contain 70–80% non-ice materials (silicates, carbon, or kerogen). 8405 Asbolus, while globally featureless in Earth-based spectra (Brown, 2000; Barucci *et al.*, 2000), showed strong water absorption in Hubble Space Telescope observations (Kern *et al.*, 2000).

Among all solar system objects, the Centaur 5145 Pholus stands out based on its extremely sloped near-infrared spectrum with absorptions at 2.04 and 2.27 μm, assigned to water ice and methanol ice, respectively (Cruikshank *et al.*, 1998). Cruikshank *et al.*'s radiative transfer modeling suggests a significant component of olivine, amorphous carbon (a darkening agent), and moderate amounts of organics (tholins). Taken together the existing spectra of TNO's and the derivative Centaurs suggest the range of spectral features that might be observable in the interior of 9P/Tempel 1. Deep Impact's HRI-IR, with extended coverage out to 5 μm, may reveal more features that might aid in identifying specific materials.

Remote observations of TNO's are, of course, of their exterior surfaces. The surfaces of TNO's are thought to be subject to a variety of processes that affect their spectral properties and in particular may explain their color diversity (Barucci *et al.*, 2005). Exposure to solar radiation, solar, galactic, and cosmic-ray irradiation, or space weathering, will produce slopes that increase toward longer wavelengths. In contrast, the collision-resurfacing hypothesis (Luu and Jewitt, 1996), suggests that more neutrally sloped spectra could be restored by regular resurfacing from mutual collisions. Recent work by Moroz *et al.* (2004) suggests that the spectra of organics may also become neutral when exposed to space weathering. Finally, sporadic cometary activity, for those objects closer to the sun (the Plutinos), may result in resurfacing by re-condensed ice (Hainaut *et al.*, 2000).

These same processes, in different relative proportions and acting over much different time scales, are likely to be present on comet nuclei. As we compare the interior of 9P/Tempel 1 to observations of its exterior, the Deep Impact experiment will help us understand these processes, at least for short period comets; the smallest, warmest, and youngest members of the TNO family. In addition, our inferences about the interior of 9P/Tempel 1 are likely to reveal a higher volatile content than inferred from observations of the nucleus exterior. These data may therefore provide a more useful comparison to the surfaces of TNO's, which have yet to be significantly devolatilized by heating in the inner solar system.

3.2. Expectations for Observations of the Coma

3.2.1. *Pre-Impact*

Comet 9P/Tempel 1 is a typical Jupiter family comet with a Tisserand parameter of 2.97 and an orbital period of 5.5 years. Combining the Spitzer infrared and HST visible observations from March 2004, HST visible observations from May 2004, and several H_2O production rates measured during the 1994 passage (A'Hearn et al., 1995; Fink et al., private communication), Groussin et al. (2004b) derived a nucleus mean radius of 3.3 ± 0.2 km, a visible geometric albedo of 0.04 ± 0.01, and an active fraction of $8.3 \pm 2.2\%$ for 9P/Tempel 1 (see Belton et al., 2005). With an 8% active fraction, 9P/Tempel 1 is a low activity comet compared to other Jupiter family comets like 22P/Kopff (35% active fraction; Lamy et al., 2002), 103P/Hartley2 (100% active fraction; Groussin et al., 2004c), or 46P/Wirtanen (85% active fraction; Groussin and Lamy 2003). Visible observations using narrow band filters (A'Hearn et al., 1995) and UV spectroscopic observations (Fink et al., private communication) allowed the detection of several coma species with relative abundances typical of Jupiter family comets. These include: OH (1.1×10^{28} molecules/s at perihelion), CN (log CN/OH = -2.8), C_2 (log C_2/OH = -2.9), C_3 (log C_3/OH = -4.1), NH (log NH/OH = -2.6), and NH_2 (log NH_2/OH = -2.8). A more detailed discussion of the dust and gas coma of 9P/Tempel 1, is presented in a companion paper by Lisse et al. (2005).

The only spectrum of a Jupiter family comet coma currently available at high-enough resolution to identify the numerous gaseous species in the 1–5 μm range is of 103P/Hartley 2 obtained with the ISOPHOT-S instrument (Colangeli, 1999). This comet is smaller than 9P/Tempel 1 with a radius of 0.71 km (Groussin et al., 2004c), but has a similar H_2O production rate of 1.2×10^{28} molecules/s at perihelion (Crovisier et al., 1999b), which allows for useful comparisons. Based on their presence in the spectrum of 103P/Hartley 2, we should be able to detect the emission lines of H_2O at 2.7 μm and CO_2 at 4.3 μm in the coma of 9P/Tempel 1.

The unique Deep Impact experiment will allow us to collect data at a higher spectral resolution than ISO ($R \sim 200$–900 vs. $R \sim 90$), a larger wavelength range (1–5 μm vs. 2.5–5.0 μm), and the highest spatial resolution ever obtained (only 7 m per pixel at closest approach). Taking advantage of the larger wavelength range, we should detect the weak emission lines of H_2O at 1.4 μm and 1.9 μm. The high spectral resolution will allow us to resolve the two peaks of CO emission at 4.7 μm (if it is present) and some emission lines in the 3.1–3.6 μm region. In particular, we expect to see broad CH_3OH emission lines at 3.4 μm and H_2CO at 3.6 μm.

The very high spatial resolution will also allow us, for the first time, to differentiate between jets and non-jet regions, providing an opportunity for direct spectral comparisons. Jets have already been observed on 9P/Tempel 1 (Lisse et al., 2005; Biver, private communication) and we expect to see differences in the absolute and relative composition between jets and non-jet regions. In particular, we expect more highly volatile species in the jets (e.g., CO and CO_2). As Deep Impact approaches the nucleus, the spatial resolution will increase and we will be able to explore the near nucleus coma, which is not resolvable from the ground. If they exist in enough quantity, this may allow for the detection of species with relatively short lifetime such as SO_2 at 4.0 μm or NH_3 at 2.3 μm. Both SO_2 and NH_3 have very weak emission lines but are located in a very clear region of the spectrum, which facilitates their detection.

In order to improve our predictions and illustrate the capacities of the Deep Impact HRI-IR spectrometer, we developed a preliminary model to calculate synthetic spectra of 9P/Tempel 1 as they may appear during the mission. A similar analysis was performed on the VIRTIS instrument on *Rosetta* (Coradini *et al.*, 1998). Our model is adapted from the formalism of Crovisier and Encrenaz (1983). Taking into account resonance fluorescence for the gas emission and a Haser's model for the gas density, we selected the most abundant species in comets for which molecular band position, intensity, and shape are known. These species are given in bold in Table I. For each species we selected the fundamental and combination bands located in the 1–5 μm region and their line intensities from the HITRAN database, except for CH_3OH which is adapted from Bockelée-Morvan *et al.* (1995). For the combination bands, we assume an emission rate equal to the excitation rate, *i.e.* we do not take into account the de-excitation of combination bands through intermediate levels (hot bands). This assumption leads to an over-estimate of the flux from combination bands like H_2O at 1.38 μm and 1.88 μm, and OCS at 2.44 μm and 4.0 μm. The dust continuum is modeled with a density proportional to $1/r^2$ (where r is the distance from the nucleus) and a size distribution function for the grains with a relative slope of -3.7, a minimum grain size of 0.1 μm, and a maximum grain size of 1 cm (Lisse *et al.*, 1998). For the nucleus continuum, the visible light is reflected on the nucleus assuming a Lambertian surface with a Bond albedo of 0.03 and the thermal emission is calculated assuming a constant surface temperature of 320 K (upper limit). Initial compositions are set to those of C/1995 O1 Hale-Bopp, for which we have the largest dataset of detected species (see Table I). The model includes many parameters that can then be modified including composition, distance to the nucleus, gas velocity, nucleus temperature, grain escape velocity, etc. . . . We chose a reasonable set of parameters, given in Table II, to support our goal of investigating likely scenarios for the Deep Impact spectrometer.

Figures 3 and 4 illustrate the results for two modeled spectra with a slit on the nucleus calculated 10 days and 1 day before impact, respectively. In both cases, the signal is strongly dominated by the nucleus continuum and very few emission lines are visible. At 10 days out (Figure 3) some coma emissions of H_2O at 1.4, 1.9 and 2.7 μm and CO_2 at 4.3 μm are detectable (but CO_2 is very weak). However, 1 day before impact (Figure 4) the nucleus so dominates that only H_2O at 2.7 μm is observed. These results indicate that for maximum scientific return on the coma a few days before impact, we must observe the coma away from the nucleus in order to increase the S/N ratio.

Such an observation is illustrated by Figure 5, which shows a modeled spectrum 1 day before impact with a slit ~90 km off the nucleus. In the absence of the nucleus, we can identify several species. The main emission lines are due to H_2O at 1.4, 1.9, and 2.7 μm, CH_3OH at 3.4 μm, H_2CO at 3.6 μm, CO_2 at 4.3 μm, and CO at 4.7 μm (double peaks). We also see weaker emission lines of NH_3 at 2.3 μm and

TABLE I

A non-exhaustive list of emission lines of molecules detected in comets (many of which have yet to be detected in the 1–5 μm region).

Molecule[a]	Wave number (cm^{-1})	Wavelength (μm)	X/H$_2$O for Hale-Bopp
HCO	9297	1.08	–
CN	9117	1.10	–
N$_2$+	9016	1.11	–
C$_2$	8268	1.21	–
H$_2$O	**7250**	**1.38**	**1**
C$_2$	6928	1.44	–
C$_3$	6482	1.54	–
C$_2$	5633	1.78	–
H$_2$O	**5331**	**1.88**	**1**
NH$_3$	**5070**	**1.97**	**0.007**
NH$_3$	**4400**	**2.27**	**0.007**
CH$_4$	**4340**	**2.30**	**0.006**
C$_2$H	4106	2.44	–
OCS	**4101**	**2.44**	**0.004**
C$_2$H	4011	2.49	–
H$_2$S	**3846**	**2.60**	**0.015**
C$_2$H	3785	2.64	–
C$_2$H	3773	2.65	–
H$_2$O	**3756**	**2.66**	**1**
CO$_2$	**3715**	**2.69**	**0.06**
HDO	3707	2.70	–
C$_2$H	3692	2.71	–
CH$_3$OH	**3681**	**2.71**	**0.024**
HNC	3653	2.74	–
CO$_2$	**3612**	**2.77**	**0.06**
OH	**3568**	**2.80**	**0**
HNCO	3538	2.83	–
H$_3$O+	3530	2.83	–
H$_3$O+	3514	2.85	–
NH$_3$	**3333**	**3.00**	**0.007**
HCN	**3311**	**3.02**	**0.0025**
NH$_2$	3301	3.03	–
C$_2$H$_2$	**3295**	**3.03**	**0.001**
C$_2$H$_2$	**3282**	**3.05**	**0.001**
H$_2$O+	3259	3.07	–
NH$_2$	3219	3.11	–

(*Continued on next page*)

TABLE I

(*Continued*)

Molecule[a]	Wave number (cm^{-1})	Wavelength (μm)	X/H_2O for Hale-Bopp
H_2O+	3213	3.11	–
NH	3127	3.20	–
$HCO+$	3089	3.24	–
$HCOOCH_3$	3045	3.28	–
H_2CS	3025	3.31	–
CH_4	**3019**	**3.31**	**0.006**
CH_3CN	3009	3.32	–
CH_3OH	**2999**	**3.33**	**0.024**
C_2H_6	**2985**	**3.35**	**0.003**
H_2CS	2971	3.37	–
CH_3OH	**2970**	**3.37**	**0.024**
$HCOOCH_3$	2969	3.37	–
$OH+$	2956	3.38	–
CH_3CN	2954	3.39	–
$HCOOCH_3$	2943	3.40	–
OCS	2918	3.43	–
CH_3OH	**2844**	**3.52**	**0.024**
H_2CO	**2843**	**3.52**	**0.011**
HC_3N	3327	3.55	–
H_2CO	**2782**	**3.59**	**0.011**
$CH+$	2754	3.63	–
CH	2733	3.66	–
HDO	2727	3.67	–
H_2CO	**2719**	**3.68**	**0.011**
SO_2	**2500**	**4.00**	**0.0023**
CO_2	**2349**	**4.26**	**0.06**
CO_2	**2337**	**4.28**	**0.06**
HC_3N	2272	4.40	–
HNCO	2269	4.41	–
CH_3CN	2268	4.41	–
$CO+$	2184	4.58	–
$HCO+$	2184	4.58	–
CO	**2143**	**4.67**	**0.23**
OCS	**2062**	**4.85**	**0.004**

[a]For most of these molecules, the shape and/or the intensity of the emission lines are unknown and we cannot include them in our model (e.g., Figure 6). Only the most abundant species (indicated in bold) are currently included in the model. Data are taken from HITRAN (Rothman *et al.*, 2003) and Crovisier's databases (http://wwwusr.obspm.fr/~crovisie/basemole/), and from Bockelée-Morvan *et al.* (1995) for CH_3OH. The X/H_2O production rate ratios for C/1995 O1 Hale-Bopp come from Bockelée-Morvan *et al.* (2000).

TABLE II

A list of the model parameters used to generate the synthetic HRI-IR spectra.

Spacecraft /comet distance	8.8×10^9 m 10 days before impact
	8.8×10^8 m 1 day before/after impact
Comet/Sun distance (at time of impact)	2.25×10^{11} m
Phase angle (at time of impact)	$63°$
Nucleus radius	3.4 km
Nucleus Bond albedo	0.03
Nucleus IR emissivity	0.95
Nucleus sub-solar temperature	320 K
Minimum size of dust grains in the coma	1.0×10^{-7} m
Maximum size of dust grains in the coma	0.01 m
Slope of the dust grains differential size distribution	-3.7
Dust grain Bond albedo	0.08
Dust grain IR emissivity	0.92
Dust grain temperature	245 K
Dust grain expansion velocity	300 m/s
Dust grain density	2500 kg/m^3
Gas expansion velocity	1000 m/s
Gas temperature	296 K (HITRAN database)
Dust to gas ratio	1 before impact, 3 after impact
Water production rate (at time of impact)	1.0×10^{28} molecules/s
Composition relatively to water	Similar to Hale-Bopp[a]

[a]See Table I.

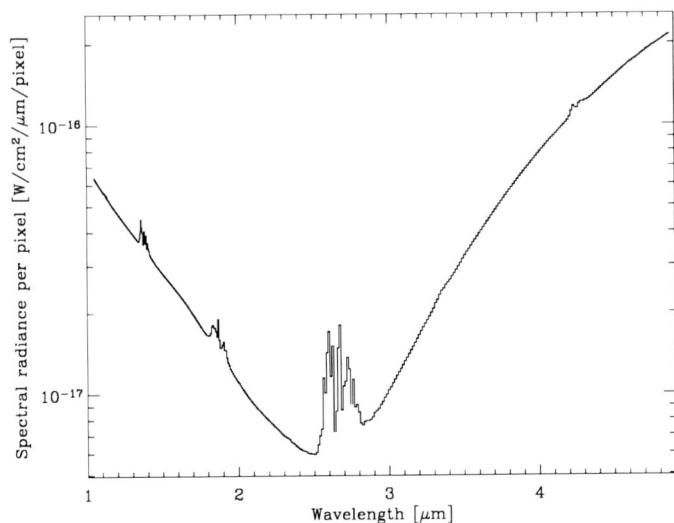

Figure 3. Modeled spectrum of comet 9P/Tempel 1 as seen with the Deep Impact HRI-IR spectrometer 10 days before impact with the slit on the nucleus. Emission lines from H_2O at 1.4, 1.9 and 2.7 μm are clearly visible. Emission lines from CO_2 at 4.3 μm are extremely weak.

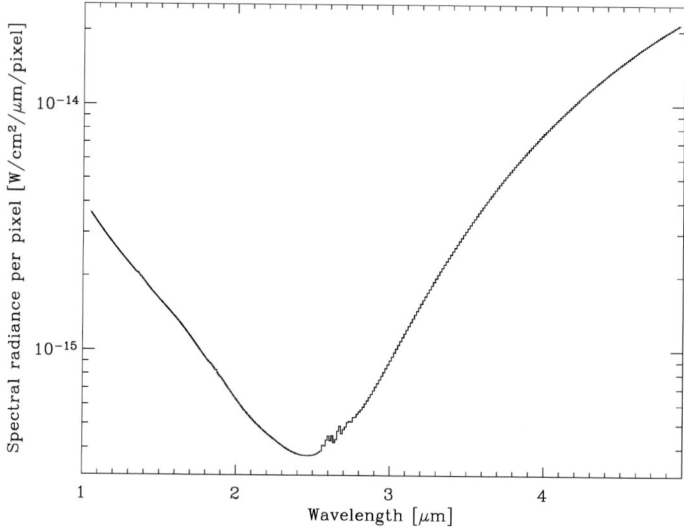

Figure 4. Synthetic spectrum of comet 9P/Tempel 1 as seen with the Deep Impact HRI-IR spectrometer, 1 day before impact with the slit on the nucleus. Only emission lines from H_2O at 2.7 μm are visible.

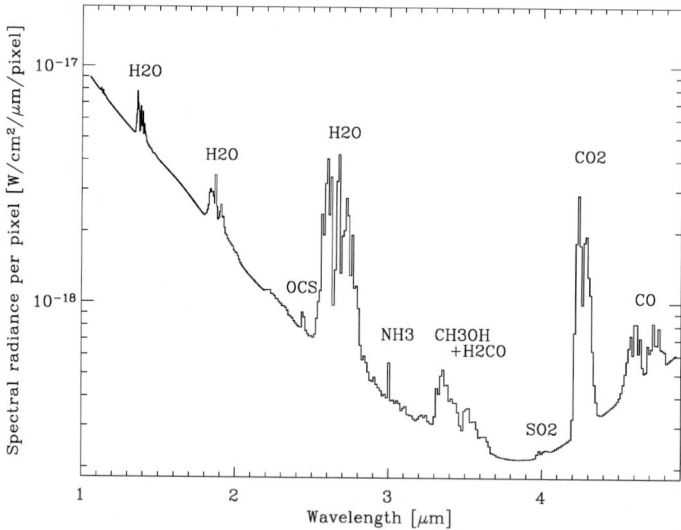

Figure 5. Synthetic spectrum of comet 9P/Tempel 1 as seen with the Deep Impact HRI-IR spectrometer 1 day before impact with a slit $\sim$90 km off the nucleus. Emission lines from H_2O at 1.4, 1.9 and 2.7 μm, CH_3OH at 3.4 μm, H_2CO at 3.6 μm, CO_2 at 4.3 μm, the two peaks of CO at 4.7 μm, NH_3 at 3.0 μm, and OCS at 2.44 μm are clearly visible. SO_2 at 4.0 μm is weakly visible. Note that for illustrative purposes, the ratio NH_3/H_2O, OCS/H_2O and SO_2/H_2O were artificially increased by a factor of 20 (compared to C/1995 O1 Hale-Bopp). As described in the text, we do not think their actual detection at 9P/Tempel 1 is very probable.

3.0 μm, OCS at 2.44 μm, and broad emission lines of SO$_2$ at 4.0 μm. However, it is important to note that for illustrative purposes we have increased the ratio of NH$_3$/H$_2$O, OCS/H$_2$O and SO$_2$/H$_2$O by a factor of 20, compared to C/1995 O1 Hale-Bopp (see Table I). For 9P/Tempel 1, we expect ratios similar to those of C/1995 O1 Hale-Bopp or lower, and thus detection of NH$_3$, OCH, and SO$_2$, will in fact be very challenging and improbable.

As discussed above, most information on dust will come from the analysis of the dust continuum. The best regions to estimate this continuum are in the spectral regions that are free from emission lines. As indicated on Figure 3, such regions include 1.2–1.3 μm, 1.5–1.8 μm, 2.0–2.2 μm, and 3.7–3.9 μm. Modeling reflected light from dust grains and their thermal emission allows us to fit the observed continuum and derive some physical properties of the grains, such as their temperature and size distribution function (Lisse et al., 1998). However, the solution is often non-unique and the derived properties are usually model dependent. Prior to impact, we will carry out many observations at different distances from the nucleus and at different positions in the coma, from which we expect to solve the model ambiguities and build a globally consistent overview of the dust properties (mainly temperature and size distribution). As with gas emissions, one of Deep Impact's most unique dust measurements will be to take advantage of our very high spatial resolution to compare the dust in jets and non-jet regions. More details on dust observations with Deep Impact can be found in Lisse et al. (2005).

3.2.2. Impact Event

The impact event should lift off a significant amount of dust particles from the surface and interior of the nucleus. Large grains that usually cannot lift off from the nucleus by gas sublimation and drag forces alone will escape the nucleus because of the impact energy. We expect to see significant changes in the dust continuum at the time of impact. In particular, the impact will produce ejecta at high temperature that will strongly increase the thermal component of dust continuum in the 3–5 μm region. Subsequent cooling rates will depend on grain size, thermal conductivity, density, and composition, which may be determined from studying the temporal variations of the dust continuum after impact. In addition, from our observations we should be able to estimate the maximum size of a grain that can be ejected from the nucleus, a very important constraint on activity models (e.g., Groussin and Lamy, 2003). At the same time, if the reflected component of the dust continuum increases strongly, it will reveal the presence of numerous micron and sub-micron size particles on the surface and interior of the nucleus, indicating the presence of very friable refractory materials, a low cohesion force between the grains, and/or very low grain density.

The chemistry of gases at the time of impact will be difficult to sort out. Over the short-term, the physics is quite complex. With excess heat and the presence of exotic materials, a wide range of species may be produced. During the late stages of excavation, however, relatively unaltered materials from the nucleus should be added to the coma. Over a longer time scale (see below), we expect to monitor the increased thermal signature and potentially detect changes in emission ratios and even new emission features.

3.2.3. *Post-Impact*

The major objective for post-impact spectroscopic observations of the coma is to monitor activity and look for any changes in the gas and dust coma after impact. As discussed above, the determination of the physical properties and composition of the coma is model dependant, and many observations before, during, and after the impact are necessary to better constrain the models and reduce their uncertainties.

For the gas coma, we expect the impact to reveal fresh volatile materials, which will cause increases in the production rates and will likely significantly change the relative abundances of different molecules. The highly volatile species (CO and CO_2) are expected to be located under the surface where they are protected from direct insolation. As such, the impact should produce an increase in CO and CO_2 relative to water. While it is impossible to quantitatively predict this enrichment, a factor of 10–100 relative to water is possible. This enrichment will be stronger in the jets resulting from impact (if they exist) than in the other region of the coma. Figure 6 illustrates what a spectrum would look like 1 day after impact with a slit ~90 km off the nucleus, assuming an enrichment of all species (except CH_3OH, see below) relative to water by a factor 20 compared to C/1995 O1 Hale-Bopp. We have

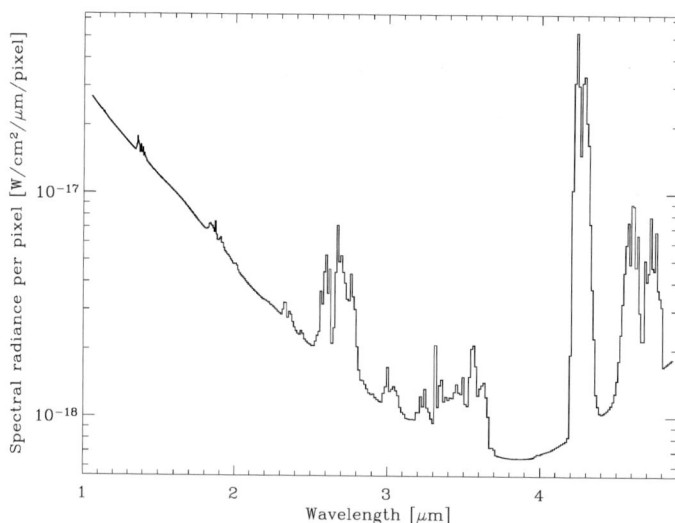

Figure 6. Synthetic spectrum of comet 9P/Tempel 1 as seen with the Deep Impact HRI-IR spectrometer, 1 day after impact with a slit ~90 km off the nucleus. Emission lines from H_2O at 1.4, 1.9 and 2.7 μm, CH_3OH at 3.4 μm, H_2CO at 3.6 μm, CO_2 at 4.3 μm, CO at 4.7 μm (double peak), NH_3 at 2.2–2.3 μm, OCS at 2.44 μm, HCN at 3.0 μm, and CH_4 at 3.31 μm are identified. As discussed in the text, the ratio of all molecules (except CH_3OH) were increased by a factor of 20 relative to water, while the dust to gas ratio was increased by a factor of 3 relative to pre-impact values.

also increased the dust-to-gas ratio by a factor of 3 relative to pre-impact values, to represent the larger amount of dust in the coma after impact. A comparison with Figure 5, which represents the same observational conditions before impact, indicates that SO_2 at 4.0 μm is no longer visible, due to the increase of a factor of 3 of the dust continuum. However, we are now able to detect new species such as HCN at 3.00–3.05 μm and CH_4 at 3.31 μm. The detection of CH_4 will only be possible if CH_3OH does not increase during impact, otherwise the CH_3OH emission dominates the 3.2–3.6 μm region and masks the detection of other species with weaker emission lines. With a spectral resolution $R \sim 200$, the interpretation of features from 3.0–3.6 μm, a region very rich in emission lines, is expected to be difficult.

After impact, molecules never before identified in the 1–5 μm region may appear in the coma. Table I presents a non-exhaustive list of molecules identified in comets that have emission lines in the infrared region, many of which have yet to be detected in the infrared. For most of these molecules, the shape and the intensity of the emission lines are unknown, and we therefore cannot include them in our model (e.g., Figure 6). Only the most abundant species are currently included in the model (in bold in Table I). The large number of molecules, more than 40, illustrates the wealth of potential in the 1–5 μm region. However, this also presents an interpretive challenge with several emission lines likely to overlap, suggesting the complexity awaiting our analysis, as the model solutions are unlikely to be unique. It is also possible that Deep Impact will detect emission lines not listed in Table I, leading to the identification of a new molecule in comas. Finally, if no changes in the gas coma are observed after impact, it would call into question our current model of cometary nuclei.

For the dust, we expect to see significant changes in the reflected and thermal continuum, which will allow us to better constrain the physical properties of the grains. As the properties of the dust are currently poorly known, it is very difficult to make predictions on what we will actually observe. However, the impact should increase the dust production rate, rapidly increase the temperature (which will then decrease post-impact), and modify the size distribution of the grains. All these phenomena should be observable by studying the dust continuum post-impact at different times and positions in the coma. Determining if the impact has a global effect on the coma or just a localized effect near the impact region, is of particular interest. As with the gas coma, observing no changes in the dust after impact would be very surprising, yet very interesting, and would also oblige us to completely review our current view of comets.

4. Summary

A comprehensive set of nucleus and coma observations using the HRI-IR spectrometer is planned for Deep Impact, before, during, and after the impact event to support the science described above. The measurements to be made will allow a number of important experiments to be conducted at 9P/Tempel 1. Most of the planned observations will provide information for multiple spectroscopy objectives, since the spectrometer slit often includes both the nucleus and the nearby coma at the same time. The spectroscopy science goals can be broadly categorized as:

- Variations in coma and the disk-integrated nucleus spectrum versus rotation phase
- Pre-impact coma composition, structure, and evolution at high spatial resolution
- Disk-resolved nucleus composition and temperature over one hemisphere
- Thermal profile of the impact event
- Impact ejecta composition (both gas and dust) versus time from impact
- Composition of the interior of the nucleus as exposed by the crater and its ejecta deposits
- Post-impact coma composition, structure, and evolution

Table III summarizes the planned HRI-IR spectrometer observations and the science goals that are expected to be addressed with each observation. A more detailed discussion (with additional measurement specifics) of the anticipated HRI-IR spectrometer data set is given in the companion paper by Klaasen *et al.* (2005).

A typical exposure time of 2.88s, yields the predicted mean signal levels in instrument data number (DN) units and the SNRs listed in Table IV for spatially resolved nucleus and coma scenes. For this calculation, the nucleus is assumed to have an albedo 5% and phase coefficient at 63° of .047. As described in Hampton *et al.* (2005), the central third of the HRI-IR spectrometer includes an anti-saturation filter, designed to block thermal contributions above 3.5 μm when observing the nucleus. SNRs at the longer wavelengths are given both for a 200 K surface outside the anti-saturation filter (which covers the central 1/3 of the spectrometer slit) and for a 300 K surface behind the anti-saturation filter. A 300 K surface outside the filter will be saturated. The predicted coma SNRs are highly uncertain; the numbers in Table IV assume that the coma is 20 times dimmer than the nucleus and has the same spectrum. In these calculations, the coma is assumed to be no warmer than 200 K. Coma SNRs can be increased at the expense of spatial resolution by averaging over spatial blocks of pixels during post-processing. Nucleus SNRs can similarly be boosted, by spectral summing, particularly over the 1.0–2.5 μm range, which covers the first half of the focal plane array. Predicted mean DN levels scale linearly with exposure time. SNR scales approximately linearly with exposure at low signal levels (<50 DN) and approximately as the square root of exposure at high signal levels (>200 DN).

The Deep Impact IR spectrometer was designed to optimize, within engineering and cost constraints, observations of the dust, gas, and nucleus of 9P/Tempel 1. The wavelength range of 1–5 μm includes absorption and emission features from ices, silicates, organics, and many gases that are known to, or expected to be, present on comets. Deep Impact is a unique experiment in that it will allow us to observe the

TABLE III

A summary of the Deep Impact HRI-IR spectrometer observation plan at each mission phase.

Time[a]	Pixel Scale (m)	Scientific goals	Data description	Comments
Far Approach				
I−60 d to I−25 h	>9,000	Rotation Pre-impact coma	Spatial scans with longer exposures every 4 h	Includes a full resolution scan of extended coma
I−25 h to I−11 h	>4,000	Rotation Pre-impact coma nucleus	Spatial scans every 2 h	
I−10 h		Calibration	Dark levels all instrument modes	Last pre-impact calibration
Approach				
I−9 h to I−1 h	>400	Pre-impact coma nucleus	Spatial scans	Includes full resolution scan of inner coma at I-1 h
I−36 m to I−19 m	>200	Pre-impact coma nucleus	Full nucleus scans	
I−14 m to I−10 m	>145	Pre-impact coma nucleus	Partial coma scan	3 full resolution at edges and nucleus center
I−6 m to 1−66 s	>95	Pre-impact coma	Full nucleus scans	Last pre-impact scan
I−35 s to I−9 s	87	Pre-impact coma nucleus	Exposures every 2.88 s	
Impact				
I−6 s to I+15 s	85	Impact thermal signature ejecta monitoring	Very short exposures every 0.72 s	
I+15 s to I+32 s	83	Ejecta monitoring	Short exposures every 1.44 s	
I+32 s to I+50 s	81	Ejecta monitoring	Exposures every 2.88 s	
I+50 s to I+67 s	79	Ejecta monitoring nucleus interior	Crater scan	
I+70 s to I+78 s	78	Ejecta monitoring post-impact coma	3 exposures	
I+87 s to I+107 s	76	Ejecta monitoring nucleus interior	Crater scan	
I+110 s to I+67 s	70	Ejecta monitoring post-impact coma	Exposures every 10% of time from impact	

(*Continued on next page*)

TABLE III

(*Continued*)

Time[a]	Pixel Scale (m)	Scientific goals	Data description	Comments
I+180 s to I+490 s	68−37	Nucleus interior post-impact coma	Full nucleus scan including off edges for inner coma	3 full resolution at edges and nucleus center
I+497 s to SM−142 s	37−21	Nucleus interior	Scan of central 1/4 nucleus	
SM-130 s to SM-110 s	18	Post-impact coma	Three coma observations	
SM-98 s to SM−26 s	16−9	Nucleus interior	Exposures every 2.88 s plus scan 3× crater diameter	Last partial nucleus scan
Shield mode				
SM−23 s to SM	8–7	Nucleus interior	Exposures every 2.88 s	Highest resolution before shield mode
SM+3 s to SM+1 m	7	Post-impact coma	Long (7-14 s) exposures for faint molecules	Significant smear at any exposure
Look back				
SM+33 m to SM+62 m	200–400	Nucleus interior post-impact coma	Scan across nucleus plus 3 different coma positions	Repeated 3 times in this time interval
SM+2 h to SM+6 h	>700	Post-impact coma	3 different coma positions taken every 2 h	Full resolution
SM+12 h to SM+60 h	>4300	Post-impact coma	3 different coma positions taken every 6 h	Full resolution

[a]I: Impact; SM: Shield mode.

comet from 1–5 μm prior to impact at higher spatial resolution and over longer time scales than ever before. In addition, we will use the HRI-IR spectrometer to monitor our impact experiment obtaining data on ejecta over time, and changes in dust and gas emission. Finally, by observing the crater and its surrounding ejecta deposits, we hope to directly observe the interior of the comet. In the above discussions, we largely speculate on what we might observe. The only certain outcome, should the Deep Impact successfully complete its mission, is that our current understanding of comets will undoubtedly change.

TABLE IV

The predicted HRI-IR spectrometer mean signal levels and SNRs[a].

Wavelength (μm)	Mean nucleus DN	Mean nucleus SNR	Mean coma DN	Mean coma SNR
1.2	49	27	2	1
2	199	55	5	2
2.8 (300 K, filt)	700	102		
3.6 (300 K, filt)	5500	287		
4.6 (300 K, filt)	20	17		
2.8 (200 K, no filt)	115	42	2	1
3.6 (200 K, no filt)	92	37	2	1
4.6 (200 K, no filt)	199	55	3	1

[a]Filt: behind the anti-saturation filter; no filter: outside the filter, see text.

Acknowledgments

This research is supported by the Deep Impact Discovery mission, Science Mission Directorate, NASA Headquarters. The Deep Impact science team is managed by Dr. Michael A'Hearn through the University of Maryland. The authors are particularly indebted to the hard work and dedication of the HRI-IR spectrometer engineering team led by Don Hampton (Ball Aerospace and Technology Corporation) and Dennis Wellnitz (University of Maryland). Reviews by J. Crovisier and an anonymous reviewer are greatly appreciated and helped clarify this manuscript.

References

Abell, P. A., *et al.*: 2003, *LunarPlanet. Sci. Conf.* **34**, abstract 1253.

Abell, P. A., *et al.*: 2005, *Icarus*, submitted.

A'Hearn, M. F., Millis, R. L., Schleicher, D. G., Osip, D. J., and Birch, P. V.: 1995, *Icarus* **118**, 223.

Barucci, M. A., de Bergh, C., Cuby, J.-G., Le Bras, A., Schmitt, B., and Romon, J.: 2000, *Astron. Astrophys.* **357**, L53.

Barucci, M. A., Cruikshank, D. P., Mottola, S., and Lazzarin, M.: 2002, in Bottke, *et al.* (eds.), *Asteroids III*, University of Arizona Press, Tucson, p. 273.

Barucci, M. A., Doressoundiran, A., and Cruikshank, D. P.: 2005, in Festou, M., Keller, H. U., and Weaver, H. A. (eds.), *Comets II*, University of Arizona Press, Tucson, in press.

Belton, M. J., *et al.*: 2005, *Space Sci. Rev.*, this volume.

Bockelée-Morvan, D., *et al.*: 2000, *Astron. Astrophys.* **353**, 1101.

Bockelée-Morvan, D., Brooke, T. Y., and Crovisier, J.: 1995, *Icarus* **116**, 18.

Brin, G. D., and Mendis, D. A.: 1979, *Astrophys. J.* **229**, 402.

Brown, M. E. and Koresko, C. C.: 1998, *Astrophys. J. Lett.* **505**, L65.

Brown, R. H., Cruikshank, D. P., and Pendleton, Y.: 1999, *Astrophys. J.* **519**, 101.

Brown, M. E.: 2000, *Astron. J.* **119**, 977.

Brownlee, D. E., *et al.*: 2004, *Science* **304**, 1764.

Burns, R. G.: 1993, *Mineralogical Applications of Crystal Field Theory*, Cambridge University Press, New York, p 551.

Campins, H., Licandro, J., Guerra, J., Chamberlain, M., and Pantin, E.: 2003, *AAS/DPS* **35**, abstract 47.02.

Clark, R. N.: 1981, *J. Geophys. Res.* **86**, 3087.

Cloutis, E. A.: 1990, *Science* **245**, 165.

Colangeli, L.: 1999, *Astron. Astrophys.* **343**, L87.

Coradini, L., *et al.*: 1998, *Plant. Space Sci.* **46**, 1291.

Combes, M., *et al.*: 1988, *Icarus* **76**, 404.

Crovisier, J. and Encrenaz, T.: 1983, *Astron. Astrophys.* **126**, 170.

Crovisier, J., *et al.*: 1997, *Science* **275**, 1904.

Crovisier, J., *et al.*: 1999a, *ESA-SP* **427**, 137.

Crovisier, J., *et al.*: 1999b, *ESA-SP* **427**, 161.

Cruikshank, D. P., *et al.*: 1998, *Icarus* **135**, 389.

Delahodde, C., Hainaut, O. R., Romin-Martin, J., and Lamy, P. L.: 2002, *Astr. Comets Met. Confer.*, abstract 20.08.

Dotto, E., Barucci, M. A., Leyrat, C., Romon, J., de Bergh, C., and Licandro, J.: 2003, *Icarus* **164**, 122.

Ernst, C. and Schultz, P.: 2003, *Lunar Planet. Sci. Conf.* **34**, abstract 2020.

Ernst, C. and Schultz, P.: 2004, *Lunar Planet. Sci. Conf.* **35**, abstract 1721.

Groussin, O. and Lamy, P.: 2003, *Astron. Astrophys.* **412**, 879.

Groussin, O., Lamy, P., and Jorda, L.: 2004a, *Astron. Astrophys.* **413**, 1163.

Groussin, O., *et al.*: 2004b, *COSPAR* **35**, abstract B1.1 002204.

Groussin, O., Lamy, P., Jorda, L., and Toth, I.: 2004c, *Astron. Astrophys.* **419**, 375.

Gruen, E., Schwehm, G., Massonne, L., Fertig, J., and Graser, U.: 1985, *Lunar Planet. Sci. Conf.* **16.**, 28.

Hainaut, O. R., *et al.*: 2000, *Astron. Astrophys.* **356**, 1076.

Harker, D. E., Wooden, D. H., Woodward, C. E., and Lisse, C. M.: 2002, *Astophys. J.* **580**, 579.

Hampton, D., *et al.*: 2005, *Space Sci. Rev.*, this volume.

Hunt, G. R.: 1979, *Geophyiscs* **44**, 1974.

Hunt, G. R. and Ashley, R. P.: 1979, *Econ Geol* **74**, 1613.

Jessberger, E. K., Christoforidis, A., and Kissel, J.: 1988, *Nature* **332**, 691.

Jewitt, D. C.: 1991, in Newburn, R. L., Neugebauer, M., and Rahe J. H. (eds.), *Comets in the Post-Halley Era*, Kluwer Academic Publishers, Dordecht, Holland, p. 1043.

Jewitt, D. C. and Luu, J. X.: 2001, *Astrophys. J.* **122**, 2099.

Johnson, R. E., Cooper, J. F., Lanzerotti, L. J., and Strazzulla, G.: 1987, *Astron. Astrophys.* **187**, 889.

Kern, S. D., McCarthy, D. W., Buie, M. W., Brown, R. H., Campins, H., and Rieke, M.: 2000, *Astrophys. J.* **542**, L155.

Kissel, J., Krueger, F. R., Silen, J., and Clark, B. C.: 2004, *Science* **304**, 1774.

Klaasen, K. P., Carcich, B. Carcich, G., and Grayzeck, E. J.: 2005, *Space Sci. Rev.*, this volume.

Krasnopolsky, V. A., *et al.*: 1986, *Nature* **321**, 269.

Lamy, P. L., Gruen, E., and Perrin, J. M.: 1987, *Astron. Astrophys.* **187**, 767.

Lamy, P. L., Malburet, P., Llebaria, A., and Koutchmy, S.: 1989, *Astron. Astrophys.* **222**, 316.

Lamy, P., Toth, I., Jorda, L., Groussin, O., A'Hearn, M. F., and Weaver, H. A.: 2002, *Icarus* **156**, 442.

Lellouch, E., *et al.*: 1998, *Astron. Astrophys.* **339**, L9.

Levison, H. F. and Duncan, M. J.: 2001, *Astrophys. J.* **121**, 2253.

Licandro, J., Oliva, E., and Di Martino, M.: 2001, *Astron. Astrophys.* **373**, L29.

Licandro, J., *et al.*: 2002, *Earth Moon Planets* **90**, 495.

Licandro, J., Campins, H., Hergenrother, C., and Lara, L. M.: 2003, *Astron. Astrophys.* **398**, L45.

Lisse, C. M., *et al.*: 1998, *Astophys. J.* **496**, 971.

Lisse, C. M., Fernandez, Y. R., and Biesecker, D. A.: 2002, *DPS* **34**, abstract 12.06.

Lisse, C. M., A'Hearn, M. F., Farnham, T., Groussin, O., Meech, K. J., and Fink, U.: 2005, *Space Sci. Rev.*, this volume.

Luu, J. X.: 1993, *Icarus* **104**, 138.

Luu, J. X. and Jewitt, D. C.: 1996, *Astrophys. J.* **112**, 2310.

Luu, J. X. and Jewitt, D. C.: 1998, *Astrophys. J. Lett.* **494**, L117.

Luu, J. X., Jewitt, D. C., and Trujillo, C.: 2000, *Astrophys. J.* **531**, L151.

McDonnell, J. A. M., Lamy, P. L., and Pankiewicz, G. S.: 1991, in Newburn, R. L., Neugebauer, M., and Rahe J. H. (eds.), *Comets in the Post-Halley Era*, Kluwer Academic Publishers, Dordecht, pp. 1043.

Moroz, L. V., Arnold, G., Korochantsev, A. V., and Wasch, R.: 1998, *Icarus* **134**, 253.

Moroz, L. V., *et al.*: 2004, *Icarus* **170**, 214.

Mumma, M. J., *et al.*: 2001, *Astrophys. J.* **546**, 1183.

Nelson, R. M., Rayman, M. D., and Weaver, H. A.: 2004, *Icarus* **167**, 1.

Nemtchinov, I. V., Shuvalov, V. V., Artemieva, N. A., Kosarev, I. B., and Trubetskaya, I. A.: 1999, *Int. J. Impact Eng.* **23**, 651.

Quirico, E., *et al.*: 1999, *Icarus* **139**, 159.

Rayner, J. T., *et al.*: 2003, *Astron. Soc. Pac.* **805**, 362.

Rothman, L. S., *et al.*: 2003, *J. Quant. Spect. Rad. Trans.* **82**, 5–44 (http://www.hitran.com).

Soderblom, L. A., *et al.*: 2004a, *Icarus* **167**, 4.

Soderblom, L. A., *et al.*: 2004b, *Icarus* **167**, 100.

Schultz, P.: 1996, *J. Geophys. Res.* **101**(E9), 21117.

Schultz, P. and Ernst, C.: 2005, *Space Sci. Rev.*, this volume.

Sugita, S., Schultz, P. H., and Hasegawa, S.: 2003, *J. Geophys. Res.* **108**(E6), 5140.

Thomas, P., *et al.*: 2005, *Space Sci. Rev.*, this volume.

THE DEEP IMPACT EARTH-BASED CAMPAIGN

K. J. MEECH[1,*], M. F. A'HEARN[2], Y. R. FERNÁNDEZ[1], C. M. LISSE[3],
H. A. WEAVER[3], N. BIVER[4] and L. M. WOODNEY[5]

[1]University Hawaii, Institute for Astronomy, 2680 Woodlawn Drive, Honolulu, HI 96822, U.S.A.
[2]Department of Astronomy, University Maryland, College Park, MD 20742-2421, U.S.A.
[3]Space Department, Johns Hopkins University, APL, 11100 Johns Hopkins Road,
Laurel, MD 20723-6099, U.S.A.
[4]Observatoire de Paris-Meudon, 5 Place Jules Jansses, 92190 Meudon, France
[5]Department of Physics, University of Central Florida, P.O. Box 162385, Orlando,
FL 32816-2385, U.S.A.
(*Author for correspondence, E-mail: meech@ifa.hawaii.edu)

(Received 15 September 2004; Accepted in final form 3 February 2005)

Abstract. Prior to the selection of the comet 9P/Tempel 1 as the *Deep Impact* mission target, the comet was not well observed. From 1999 through the present there has been an intensive world-wide observing campaign designed to obtain mission critical information about the target nucleus, including the nucleus size, albedo, rotation rate, rotation state, phase function, and the development of the dust and gas coma. The specific observing schemes used to obtain this information and the resources needed are presented here. The *Deep Impact* mission is unique in that part of the mission observations will rely on an Earth-based (ground and orbital) suite of complementary observations of the comet just prior to impact and in the weeks following. While the impact should result in new cometary activity, the actual physical outcome is uncertain, and the Earth-based observations must allow for a wide range of post-impact phenomena. A world-wide coordinated effort for these observations is described.

Keywords: deep impact, earth observing

1. Introduction

The selection of the *Deep Impact* mission target was driven primarily by launch and orbital dynamics considerations (A'Hearn *et al.*, 2005; Blume, 2005) and not by what was known about the potential target. Nevertheless, the success of the mission is greatly enhanced if much is known about the target *a priori*. This is of benefit not only for planning the encounter sequences, but is especially important in the case of the *Deep Impact* mission target, 9P/Tempel 1, because the goal is to look for post-impact changes in the outgassing which will be indicative of the properties of the pristine nucleus interior. To this extent, more so than for any other mission, it is very important to establish a good baseline of observations which characterize the nucleus prior to the encounter. Furthermore the pre-impact ground observations will verify the capability of ground-based techniques for basic cometary nucleus reconnaissance.

Space Science Reviews (2005) 117: 297–334
DOI: 10.1007/s11214-005-3382-8

The *Deep Impact* mission has been designed to provide good observing conditions from Earth. Although the flyby spacecraft of the *Deep Impact* mission will
make unique *in situ* measurements, the constraints of space missions limit us to
imaging and near-infrared spectroscopy in an 800-s interval from time of impact
until the flyby spacecraft has flown past the point of observability of the impact site.
A unique aspect of this mission is the observing program planned from Earth and
Earth-orbit at the time of impact. These observations are designed to complement
the spacecraft data during the period surrounding encounter, and will continue long
after the event, since long-lived changes in the behavior are a plausible outcome
of the experiment. However it is important to note that the specific outcome of
the impact is unknown; indeed, that is one significant motivation for the mission
in the first place. We and our collaborators have set up a ground-based observing
campaign that will allow us to observe whatever phenomena are created by the
impact.

1.1. SELECTION OF DI TARGET – WHAT WE KNEW

At the time of target selection, 9P/Tempel 1 was known to be a typical Jupiter-
family low-activity comet (A'Hearn *et al.*, 1995) rarely bright as seen from Earth,
and was well placed for observing only every other apparition ($\sim$11-year intervals).
Although the comet was discovered in 1867 (Yeomans *et al.*, 2005), there have thus
been relatively few physical observations of the comet prior to its selection for the
Deep Impact mission. From the inclusion of the non-gravitational parameters in
the orbit solution for the comet, it was suggested that the pattern of outgassing
was relatively unchanged during the last seven apparitions. Most of the existing
physical observations were of the gas and dust coma around the time of the two
perihelion passages of 1983 and 1994 (Lisse *et al.*, 2005), along with some limited
information about the nucleus rotation (Belton *et al.*, 2005).

In order to prepare for the encounter, the *Deep Impact* team has undertaken a large
observing campaign of nucleus characterization. The sections below discuss the
rationale and opportunities for the observations; many of the results are presented
in Belton *et al.* (2005).

2. The Pre-Encounter Period

2.1. SIZE AND ALBEDO

Knowledge of the size and shape of the nucleus and its albedo is important for the
autonomous targeting software and in order to calculate instrument exposure times.
To ensure targeting that is not too close to the limb such that the impactor might
miss, it is *crucial* to know the size and shape of the nucleus. If the albedo is higher

than assumed, the size will be correspondingly smaller and we would have to adopt
a strategy that ensures that we hit the target without optimizing the observability.
If the albedo is lower than assumed, the size will be correspondingly larger and we
can optimize the targeting for observability of the crater at closest approach.

Simultaneous optical and thermal infrared observations of the bare nucleus can
give an estimate of both the instantaneous nucleus size and geometric albedo. The
technique relies on the fact that the flux in the optical is proportional to the nucleus cross section, albedo and phase function, whereas the thermal flux is related
to the nucleus size, thermal phase function and nucleus emissivity (Lebofsky and
Spencer, 1989). As seen in Figure 1, which is a light curve combining CCD data
since 1999 with data from the *International Comet Quarterly*, the onset of activity
typically occurs for this comet between 600–400 days pre-perihelion, at a heliocentric distance r around 3–4 AU. However, signal-to-noise calculations for detecting
a nucleus of the size of 9P/Tempel 1 in the thermal IR at the Keck 10 m telescope
indicated that this would only be feasible inside $r < 2.5$ AU, at a time when the
comet would be active.

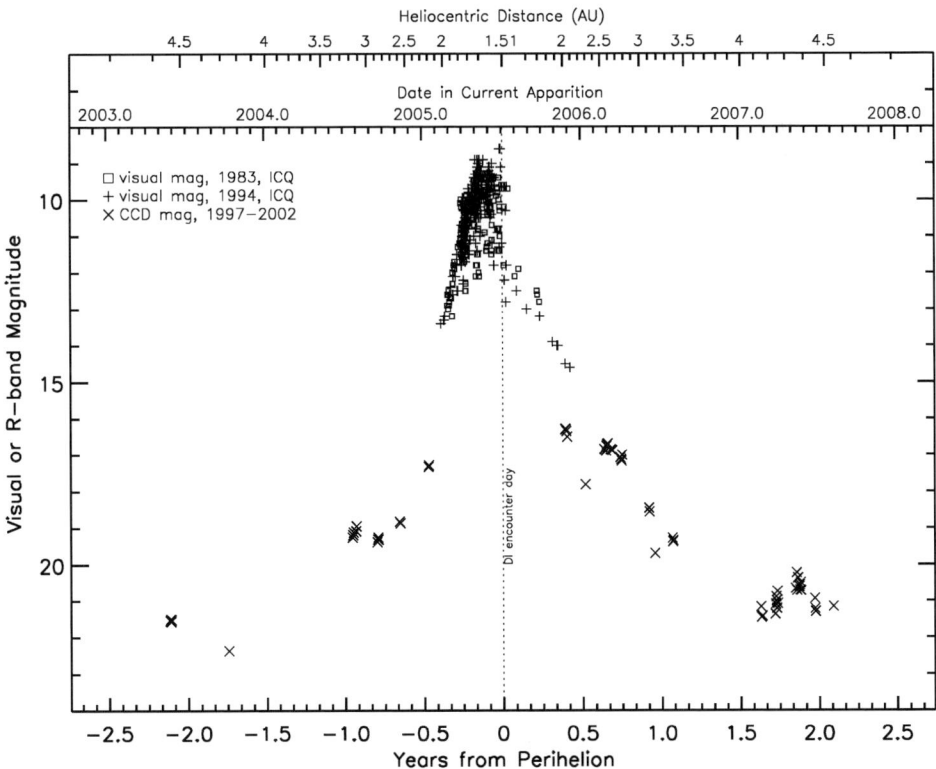

Figure 1. Light curve of comet 9P/Tempel 1 combining data from the ground-based program since
1999 and data from the *ICQ*. Note that the brightness peaks about 2 months pre-perihelion.

TABLE I

Spitzer Space Telescope observing opportunities.

UT dates	r (AU)[a]	Δ (AU)[a]	α $(°)$[b]	Coma?[c]	Observing time
10/03/03–11/13/03	4.272–4.157	4.287–3.542	13.4–11.6	No	None
02/27/04–04/06/04	3.809–3.657	3.281–3.717	13.6–15.6	No	Awarded
12/24/04–02/23/05	2.321–1.967	2.088–1.137	25.6–20.7	Yes	Awarded
07/02/05–10/14/05	1.506–1.798	0.882–1.818	41.6–32.0	Yes	Awarded

Note. This includes only the windows that satisfy the solar elongation constraints for Spitzer observing.
[a]Heliocentric and geocentric distance.
[b]Phase angle.
[c]Existence of observable coma.

Thus, when completely devoid of coma, the nucleus would be too faint to detect in the thermal IR even with the 10-m Keck telescope. Therefore a compromise was made, and data were obtained post-perihelion in August 2000 when the comet was at $r = 2.54$ AU and still very active, necessitating significant modeling to remove the coma contribution and to determine the rotational phase at the time of observations (Fernández *et al.*, 2003; Belton *et al.*, 2005).

The only other opportunity to obtain data on the bare nucleus meant using the Spitzer Space Telescope. Director's discretionary time was awarded for this project so that we could obtain rotationally resolved IR fluxes during early 2004. Table I presents a summary of the opportunities for observing the comet that are compatible with the Spitzer Space Telescope observing windows. We were also awarded General Observer time in Cycle 1 to observe the comet in 2005 after activity had started.

2.2. ROTATION STATE

Observed brightness variations of a comet may be caused by: the activity, which changes the effective scattering area; the changing geocentric and heliocentric distances; the rotation of the nucleus (contributing a brightness modulation due to changing shape or surface albedos); and the changing solar illumination and scattering angles. Knowledge of the rotation period is important both for thermal modeling, as well as mission planning. In addition, the rotation state is a fundamental property of the nucleus and an understanding of the rotation can be used to derive constraints on the nucleus bulk strength and density. We need to be able to ensure that the impact occurs on the largest face of the nucleus (giving a higher probability for success), and the last time to totally control the time of impact was at launch. Now after launch, we are able to make only small adjustments because of fuel limitations.

TABLE II

Optimal observing times for rotation campaigns.

Dates	r (AU)[a]	Mag$_{nuc}$	Solar elongation ($^\circ$)	α ($^\circ$)[b]	Coma?[c]
02/99	3.0	19.3	120	9	Probably
10/00	2.7	18.4	-9	6	Yes
11/01	4.3	20.8	43	2	Probably not
12/02	4.7	21.1	70	3	No
12/03	4.1	22.0	100	1	No
04/05	1.7	15.9	-169	14	Yes

[a]Heliocentric distance.
[b]Phase angle.
[c]Existence of observable coma; observations were obtained for all dates prior to 4/05 as of January 2005.

Inverting a photometric light curve for an asteroid or an inactive comet is non-trivial. In increasing order of difficulty, one can obtain the (i) sidereal rotation period and spin axis, (ii) the shape, and (iii) the light scattering properties (Magnusson et al., 1989). Ideally, the nucleus should be inactive, but as bright as possible and near opposition in order to obtain the maximum observing time per night. In order to place constraints on the rotation pole, we want at least four good light curves without coma contamination; a unique pole solution can be obtained in the ideal case only with at least three light curves. These should be at low phase angle ($\alpha < 20^\circ$, although it is better if $\alpha < 10^\circ$) so that it is possible to associate light curve features with the shape of the nucleus rather than the specifics of the light scattering from the particulate surface. In addition, we should obtain the observations over a range of ecliptic longitudes in order to sample different geometries. Unfortunately, most comets begin to become active out near $r \sim 5$–6 AU (Meech and Svoreň, 2005), and as the nuclei are quite small with low albedos (Meech et al., 2004), this means that the inactive nucleus is quite faint, requiring large telescopes to achieve observations of adequate S/N.

The best periods for obtaining data for the rotational light curve are shown in Table II. Because the rotation period was known to be long and that aliasing effects from the daily sampling could be a problem, we tried to coordinate observations between observatories separated by longitude. Intensive international observing campaigns were conducted during the 2000, 2001 and 2002 opportunities. The results of these efforts are discussed in (Belton et al., 2005).

2.3. PHASE FUNCTION

Most bodies in the solar system with particulate light-reflecting surfaces have been observed to exhibit a nonlinear surge in brightness at low phase angles ($\alpha < 7^\circ$).

This is caused by a combination of particles in the surface covering their shadows and a contribution from coherent backscattering. In principle, observations over a wide range of α can be used with detailed photometric models of the multiply scattered light in rough particulate surfaces to get information about the particle single scattering albedos, the porosity of the surface, the particle sizes, compaction of the surface and the macroscopic roughness of the surface. Detailed modeling requires high-quality data over a wide range of phase angles. The parameters which describe the small-scale surface scattering properties of the optically active regolith (porosity, particle size distribution, and compaction) can be well constrained from observations at phase angles $\alpha < 12°$, whereas macroscopic roughness can only be ascertained by large phase angle observations. Less compacted, fluffy regoliths have more pronounced surges. Macroscopically rough features (clumps to mountains, craters and ridges), can alter the local incidence and emergence angles which has an effect on brightness. This term can provide information about the impact and outgassing history of the nucleus and its effect on topology. For comets, knowledge of the particle sizes in the surface helps in the understanding of the degree to which the surface has been subjected to microphysical processes, micrometeorite impacts and annealing, and will be important for understanding and modeling the heat transfer into the nucleus.

For comets these observations are particularly difficult because not only is knowledge of the rotation important (to remove its signature), but the comet must not have any activity to compete with the small changes in brightness with phase. The phase-dependent photometric behavior of comets is currently poorly understood. There have been attempts to determine the phase functions for the nuclei of 2P/Encke and 28P/Neujmin 1 (Fernández et al., 2000; Delahodde et al., 2001) from the ground and 19P/Borrelly (Buratti et al., 2004) in situ. Both for the practical issues related to understanding the likely brightness of the nucleus at the high phase angles at encounter as seen from the spacecraft (63°) and to understand the nucleus surface properties prior to impact, there were two observing seasons pre-encounter for which it was possible to undertake a campaign to acquire the necessary phase function data. Observations require that the comet is without coma, and that there is good coverage of the phase curve (at least 3–4 data points between about 15° to 2–3°, and then dense coverage for $\alpha < 2°$). Furthermore, the comet should be at opposition so that long time-series observations can be made to remove the rotational signature from the phase function data.

The epochs and geometric circumstances are shown in Table III. A major international campaign was undertaken for the Fall 2000 opportunity, producing an excellent set of data. However, because of weather, insufficient information was obtained to remove the rotation signature using this dataset alone. Although also plagued again by poor weather, data from the Fall 2003 campaign have been calibrated and should yield good phase function data as the rotational state is solved. This phase function analysis is ongoing (Hsieh et al., personal communication).

TABLE III

Phase function observing windows.

Dates	r (AU)	α (°)	Mag$_{nuc}$
08/13/01–02/02/02	4.13–4.53	14.2–1.4	23.5–21.9
10/01/03–01/03/04	4.27–4.00	13.4–1.8	23.0–21.9

Note. Minimum α occurred on 11/07/01 at $\alpha = 1.27°$, and 01/02/04 at $\alpha = 1.84°$.

2.4. GAS PRODUCTION

2.4.1. *Optical, Ultraviolet and Infrared Wavelengths*

The surfaces of comets are mostly inactive, and because of this, the impactor will probably hit an inactive region. The likely outcome of the impact is that the crater will become a new active area, which will lead to new outgassing, lasting days to months. During the campaign to observe the Shoemaker-Levy 9 impact into Jupiter, there were many pre-impact predictions. Many spectroscopic observations showed emissions from unexpected species, and there were a number of expected emission features that were absent (Noll *et al.*, 1995). The goal of observing the gas production is to determine the changes in natural activity due to the impact. At optical wavelengths, this means monitoring the onset and development of the usual daughter species CN (3883 Å, 7873 Å), C_2 (5165 Å, 7715 Å), C_3 (4040 Å), NH_2 (5700 Å), and CH (4315 Å). These species have been monitorable from large telescopes since October/November 2004, when the comet left solar conjunction. The comet had a gas production rate of approximately 3×10^{27} molecules per second (A'Hearn *et al.*, 1995).

From spectroscopic surveys of comets in the optical, comet 9P/Tempel 1 has been found to have typical composition ratios of OH, CN and C_2 (A'Hearn *et al.*, 1995). We therefore expect the ultraviolet spectrum of the comet to be similar to other comets that have been observed in the UV by both HST and IUE. These observations will be possible only near the time of encounter, when the comet is bright (see Section 4.3.10). In the near-IR spectral region, spectral observations will only be possible for this comet at encounter.

2.4.2. *Radio and Sub-mm Wavelengths*

The molecules of primary interest in the radio and sub-mm wavelength regimes are H_2O (OH), CO, CH_3OH and HCN. H_2O is the main driver of cometary activity, and one rotational line at 557 GHz is currently observable from space satellites, while its main photo-dissociation product, OH, is observable from the ground at 18 cm (1.7 GHz) or with narrowband UV filters on optical telescopes. The brightness of the OH lines is related to the total production of water in the coma and the spectral line shape contains information on the velocity of the outflowing gas, and the distribution of gas production from the nucleus. CO abundance relative to water is very variable from comet to comet (1–25%) and can be depleted in the outer layers of cometary nuclei due to its high volatility. Methanol (CH_3OH) is the fourth most abundant molecule in comets (after H_2O, CO and CO_2), sometimes even more abundant than CO, and has been detected in the sub-mm in over 15 comets (Biver et al., 2002). Methanol lines are numerous and some can be observed simultaneously to probe the cometary gas temperature.

HCN is of lesser abundance relative to water (0.1% on average) but it is generally the easiest species detectable in comets from ground based sub-mm observatories. HCN can be observed via vibrational transitions in the infrared or via rotational transitions in the millimeter ($J = 1$-0 at 88.6 GHz/3.4 mm) and submillimeter ($J = 3$-2 at 266 GHz/1.1 mm or $J = 4$-3 at 355 GHz/0.85 mm). It has been detected in over 24 comets. Its abundance relative to water also displays a smaller variation than for CO and CH_3OH and thus it constitutes a good reference. 9P/Tempel 1 is not intrinsically bright, so these observations can begin only as the comet approaches perihelion.

OH measurements will just be possible between March and May 2005. The OH 18-cm maser line arises from the ground state Λ doublet. The excitation of the two energy levels is dominated by pumping by solar UV radiation (a UV photon is absorbed followed by rapid cascading to the ground state) and depends on the relative velocity between the comet and the sun. This process leads to either strong inversion ($i = (n_{up} - n_{low})/(n_{up} + n_{low}) > 0$) or anti-inversion ($i < 0$) of the population of the Λ doublet and the cometary OH coma then either amplifies or absorbs the background radiation. The pumping process strongly depends on the solar spectrum seen by the comet. Due to the complex structure of solar UV spectrum, the cometary velocity relative to the Sun will determine the inversion value. OH line intensities are proportional to i in the first approximation, which will be between $-0.3 < i < -0.2$ in March to May 2005 but near $i = 0.02$ at perihelion. Under normal activity, OH lines in comet 9P/Tempel 1 will only be detectable before perihelion. A 10-fold increase in production rate from the impact (or an unusual excitation process) will be necessary to detect OH. Daily observations of OH at Nançay (France) can be done, 1 h per day when the comet is close to meridian. Other facilities like the VLA (USA) and Parkes (Australia) should be able to observe the OH lines in comet 9P/Tempel 1 a few hours per day.

The HCN(3-2) or HCN(4-3) lines are expected to be detectable most of the time between April and July 2005 (S/N > 5 per observing run for any sub-mm telescope) for "normal" cometary activity. HCN observations will thus be useful to monitor cometary activity and enable precise measurement of the gas outflow velocity thanks to a good S/N ratio and the very high spectral resolution of the heterodyne technique. Because of intrinsically weak lines, CO may not be detectable before impact if its abundance relative to water is below 5%. Methanol should be detectable with long integration times to retrieve the gas temperature, while other molecular species like

TABLE IV

Example pre-encounter emission line strengths.

Species	Abundance (%)[a]	Line	IRAM $\int T \, dv$[b]	S/N[c]	JCMT $\int T \, dv$[b]	S/N[c]	CSO $\int T \, dv$[b]	S/N[c]
HCN	0.1	$J = 1 \to 0$	27	3	–	–	–	–
		$J = 3 \to 2$	210	5	84	6	49	4
		$J = 4 \to 3$	–	–	86	5	47	2
CO	10	$J = 2 \to 1$	19	1	9	1	6	0.5
		$J = 3 \to 2$	–	–	22	1.5	15	1
CH$_3$OH	3	145 GHz	31, 20	3, 2	–	–	–	–
		157 GHz	22–10	~1	–	–	–	–
		241 GHz	71, 54	3, 2	41, 30	3, 2	30	2.5
		304/307 GHz	–	–	–	–	45, 54	4, 5
CS	0.1	$J = 3 \to 2$	21	2	–	–	–	–
		$J = 5 \to 4$	51	2	25	2	16	1
		$J = 7 \to 6$	–	–	22	1.5	13	1

Note. Predicted line strengths are based on expected gas production rates occurring immediately before impact.

[a] Abundance is expressed as a fraction of the H_2O abundance.

[b] Units of line strength are mK km/s, with T: temperature in the spectrum across the line as a function of relative speed v.

[c] Signal-to-noise ratio expected for about 2.4 h of actual integration time (which corresponds to about 4 h of clock time).

CS, H_2S, H_2CO, will likely be marginally detectable in the assumption of classical abundances and normal activity of comet 9P/Tempel 1.

A list of expected line strengths for some radio emission lines in commonly observed species is given in Table IV. Note that these calculations apply to the comet *before* impact. The baseline prediction is for the comet to become several times brighter immediately after impact in which case the line strengths will likewise increase. The signal-to-noise ratios S/N given in the table assume that the line is observed for about 2.4 h.

2.5. DUST DEVELOPMENT

We want to closely monitor the dust environment around the nucleus (i) to interpret other observations with respect to the amount of coma contamination; (ii) to predict the brightness and dust production pre-impact for comparison post impact; (iii) to model (using Finson–Probstein techniques) the near-nucleus dust environment and assess the impact hazard to the spacecraft.

Comet 46P/Wirtanen began its activity between $r = 3.5$ and 4.5 AU pre-perihelion and 81P/Wild 2 began to develop a coma between $r = 4.0$ and 4.5 AU, based on dust production models for the Stardust mission. Assuming similar behavior for 9P/Tempel 1, observations were planned regularly from aphelion at $r = 4.7$ AU (10/2002) to 3.5 AU (05/2004), every 1–2 months to get deep images to assess when activity first began.

2.5.1. *Finson–Probstein Dust Modeling*

From past apparitions, this comet typically exhibits a sharp rise in brightness caused by outgassing and development of an extensive dust coma ~200 days pre-perihelion (although lower levels of activity begin much sooner; see Figure 1). This occurred in December 2004 ($r = 2.5$–2.3 AU). Images taken on December 1 show a dust coma extending out beyond 2×10^4 km from the nucleus (Figure 2). By observing the extent and morphology of the dust coma and using dust-dynamical models (Finson and Probstein, 1968) we can determine the relative velocity distribution, size distribution and production rates of the dust as a function of distance. These models evaluate the motion of a suite of particles after leaving the nucleus under the influence of solar radiation pressure and gravity. The scattered light from the dust is added together and fit to the surface brightness of the observed coma. The larger the extent of the detected coma (i.e. observations must go to low surface brightness levels above sky, dictating dark time), the more constraining the models.

9P/Tempel 1 is a good comet to model because beginning in February 2005 through encounter there are large orbital geometry changes from month to month. Although the geometry is not as good in the fall 2004 (we are looking edge on to the tail and cannot separate the motions of the different grains), we needed to

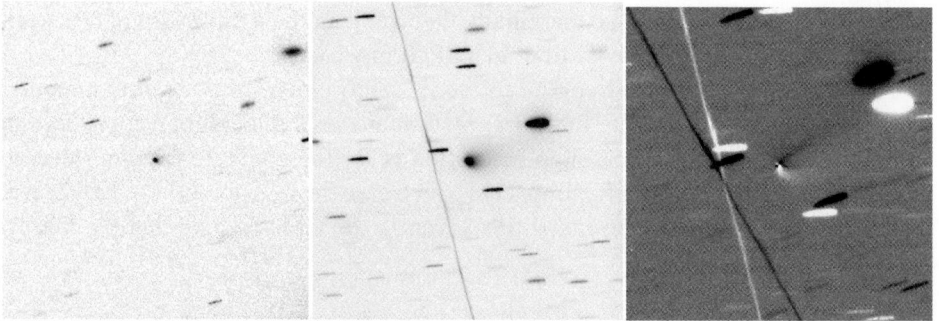

Figure 2. Left: 400-s image of comet 9P/Tempel 1 obtained using the University of Hawaii 2.2-m telescope on Mauna Kea with a CCD and R filter on December 1, 2004. The comet was at $r = 2.45$ AU and $\Delta = 2.49$ AU, $\alpha = 23.0°$ at the time. The dust coma extends over 2.0×10^4 km to the west. Center: 600-s image of comet 9P/Tempel 1 obtained using the UH 2.2-m telescope through an R filter on January 18, 2005 when the comet was at $r = 2.17$ AU and $\Delta = 1.65$ AU, $\alpha = 25.5°$. The dust coma extends over 1.1×10^5 km to the west. The field of view is 2.5×10^5 km; north is at top and east is at left for both images. Right: Enhanced image from January 18, 2005, showing the dust jets within the coma. The field of view is 1.5×10^5 km.

establish our first data point as a baseline for the onset of dust production as soon as the comet came out of solar conjunction in late October 2004. Some of the specific criteria for the optimization of wide field R-band images of the dust coma include the following.

- Small particles move rapidly away from the field of view, so we need many images closely spaced in time to follow them.
- Large grains, which move slowly along the orbit, need observations equally spaced over time periods of months for proper modeling. The large dust is of particular interest because of the potential hazard to the spacecraft as it passes through the orbit plane approximately 15 min after impact.
- We would like to have the predicted dust trajectories (syn-curves) widely spaced in time to get better constraints on the particle sizes, dust ejection velocities, and dust production start and end.
- Small particles tend to lie in anti-solar direction whereas large particles lag behind in orbit.
- Very circular orbits tend to have widely spaced syn-curves, but it is then harder to see changes. Comets on very elliptical orbits have problems inbound when the large particles also fall along the sun-to-comet radius vector (and overlap the syn-curves for small ones). 9P/Tempel 1 is a good comet in that its orbit has an intermediate eccentricity.

It is anticipated that as the comet brightens in the spring of 2005, the Small Telescope Science Program observers (McFadden *et al.*, 2005) will contribute greatly to the dataset of wide field dust images needed for the modeling.

2.5.2. *Features in the Dust Coma*

The Earth passes through 9P/Tempel 1's orbit plane twice a year, on roughly May 31 and December 1 (± 1 day). At these times we have the best chance for imaging the dust trail of large particles from previous apparitions, as has been done optically for comets 22P/Kopff and 81P/Wild 2 (Ishiguro *et al.*, 2002, 2003). Finson–Probstein dust-dynamical analysis of the grains in both dust trails suggested particles between millimeter-to-centimeter sizes. A peak intensity at a mean anomaly 0.02° from the comet for 22P/Kopff suggested that most of the large dust was emitted near perihelion.

The IRAS satellite detected a dust trail for 9P/Tempel 1 (Sykes and Walker, 1992), however, no optical trail has been found so far for 9P/Tempel 1. Knowledge of the dust distribution in the trail for 9P/Tempel 1, i.e. if most of the hazard is from material emitted at perihelion, will be important for the *Deep Impact* mission. We crossed the orbit plane of the comet on 12/1/04, and we observed the comet at the UH 2.2 m telescope (Figure 2) to do deep imaging to search for the trail. We will also use the Spitzer telescope (see Section 4.3.10 below) dust observations in 2005 to look for the dust trail.

Deep Impact Ground Based Observing Opportunities
P/Tempel 1

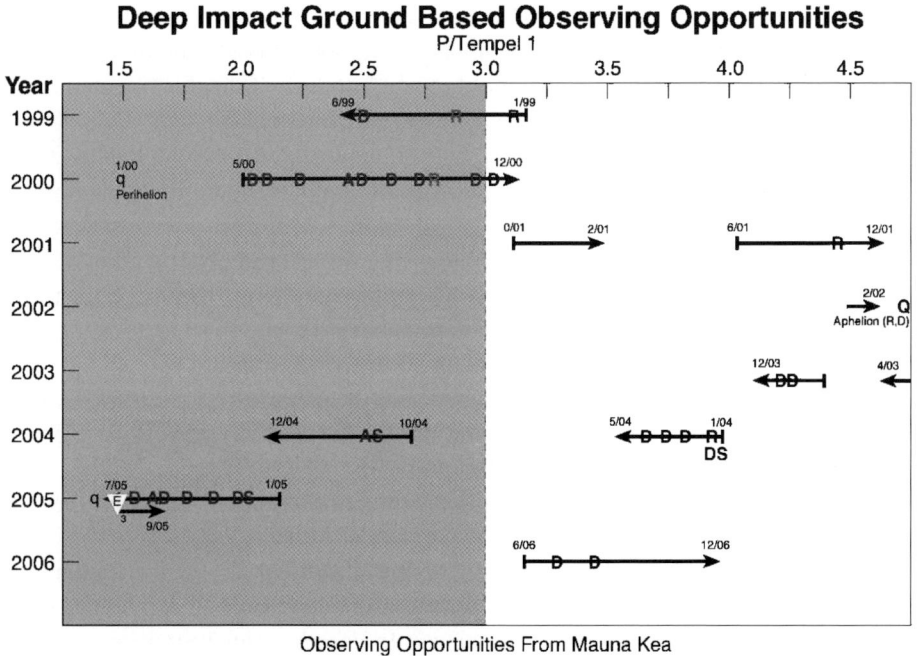

Observing Opportunities From Mauna Kea

Figure 3. Timeline for critical 9P/Tempel 1 observations. Heliocentric distance in AU is shown at top. R, rotation determination; A, albedo (simultaneous optical/IR); S, spectroscopic observations; D, imaging the dust coma for particle dynamical models. Shading indicates the distances at which significant coma is expected. With the exception of spectroscopy in January 2004, all of the noted observing opportunities have been (or will be) taken advantage of.

Figure 3 summarizes the visibility of the comet since the time of the mission selection, through the year of impact. The periods of visibility are shown as a function of the changing heliocentric distance of the comet by year (as arrows). The shaded part of the figure roughly shows the time period when the comet is expected to possess significant coma. Letters on the figure indicate when there were or will be targeted campaigns to measure the comets rotation (R), dust coma development or disappearance (D), onset of gas production (S), and albedo/size measurements (A). It can be seen that the 2000 perihelion passage was not well suited for observations of the active comet, but that we had excellent access to the inactive phases of the comet near aphelion.

3. Status of the Observations

3.1. SUMMARY OF THE OBSERVING RUNS

To date, we and our collaborators have obtained optical imaging data on 250 nights using 13 telescopes at 9 ground-based observatories world-wide in order to characterize the nucleus rotation state, size, albedo and dust development, in addition to data from the Keck 10 m and Spitzer telescope for albedo and nucleus size measurement and the Hubble Space Telescope for rotation determination. The level of cooperation among planetary colleagues as well as from colleagues outside the field has been excellent. In additional to our formal collaborators, there have been many observers who have given up some large telescope time in order to help out the mission, and they have made a tremendous contribution to our understanding of the nucleus properties. A summary of these runs through 2004 is included in Table V.

3.2. DEVELOPMENT OF THE DI DATABASE

In order to facilitate access to the growing amount of *Deep Impact* ground-based imaging, we designed and developed a relational database using IBM's DB2 database management system, which can be queried using a web interface developed in ColdFusion MX.

In its current form, the database contains reduced data products (photometry) from the ground-based CCD imaging program. Routines developed in perl are used to parse the text files containing reduced data and insert the data into the database. Full information about the observations obtained on each night, including information about the observers and observing conditions, seeing, the instrument and telescope used (along with filter transmission curves and CCD quantum efficiencies) are stored in the database. Each observation is also tagged with the instantaneous values of the heliocentric and geocentric distances (r and Δ) and phase angle (α), as well as the sun-centered and Earth-centered state vectors. These values are obtained automatically by communicating with the Jet Propulsion Laboratory Solar System Dynamics ephemeris routines when new data is uploaded to the database.

This *Deep Impact* database can be queried using a password protected web interface (see Figure 4). The returned data can be sorted on several key parameters, such as date, time, filter, photometry aperture, by telescope, observer or instrument, and returned as either an HTML table, tab-delimited dataset, an excel spreadsheet or CSV file format.

4. Coordination During the Encounter

The goal of having a focal point of coordination for the mission, rather than allowing a completely open competitive process drive the Earth-based science, is to ensure that mission-critical observations are made which are necessary for the interpretation of the spacecraft data, and to avoid duplication of effort.

At the time of encounter, the impact will be observable in dark skies from longitudes as far west as New Zealand, and as far east as Arizona. Latitudes farther north

TABLE V

Major participating observatories – pre-encounter datasets.

Year	Telescope	# Nts	PIs	Observers	Program
1999	UH 2.2 m	15	KM	KM, GB	Dust, Rotation
2000	UH 2.2 m	30	KM, JP, GB	KM, JP, BB, RW, CT, GB, AE, YF, PC	Rotation, Dust, Calib
	Keck 10 m	3	MA	YF, MB2	Size, Albedo
	Lowell 1.8 m	12	MB	MB	Rotation
	McDonald 2.7 m	8	TF	TF	Rotation
	TNG 3.6 m	4	GT	GT	Rotation
	Danish 1.5 m	5	TS	TS	Rotation
	ESO 2.2 m	2	OH	OH, KM	Rotation
	VLT 8.0 m	1	LB	LB, KM, HB	Rotation
2001	UH 2.2 m	42	KM, JP, YF, GB, SS	KM, JP, MK, YF, HH, DT, SM, GB, SS	Rotation
	CTIO 4 m, 1.5 m	12	RM, MM, KK	RM, MM, KK	Rotation
	KPNO 4 m, 2.1 m	11	PM, NS, MB	PM, NS, TL, MB	Rotation
	Lowell 1.8 m	3	MB	MB	Rotation
	TNG 3.6 m	6	GT	GT, JL	Rotation
	McDonald 2.7 m	1	TF	TF	Rotation
	Bohyunsan 1.8 m	1	YC	YC	Rotation
	Wise 1.1 m	8	YC	YC	Rotation
2002	UH 2.2 m	20	KM, JP, YF	KM, HH, JP, YF	Rotation, Phase Fn
	KPNO 2.1 m	3	BM, NS	BM, NS	Rotation, Phase Fn
	TNG 3.6 m	4	GT	GT	Rotation, Phase Fn
2003	UH 2.2 m	18	KM, JP, DK, HH, YF	KM, DH, JP, YF, DK, HH	Phase Fn
	KPNO 4 m	3	RM, LW	RM, LW	Phase Fn
	CTIO 4 m	2	NS2	PC2, CA	Phase Fn
	NTT 3.6 m	1	KM	CF	Phase Fn
2004	UH 2.2 m	10	KM, JP, YF	JP, KM, YF, NM	Rotation, Phase Fn
	VLT 8 m	4	KM	KM	Rotation, Phase Fn

Note. Observer affiliations are appropriate to the time of the observations. AE, A. Evans; BB, B. Barris; CT, C. Trujillo; DC, D. Kocevski; DT, D. Tholen; HH, H. Hsieh; GB, J. Bauer; JP, J. Pittichová; KM, K. Meech; MK, M. Kadooka; NM, N. Moskowitz; PC, P. Capak; PH, P. Henry; RW, R. Wainscoat; YF, Y. Fernández (IfA); BM, B. Mueller; CA, C. Aguilera; MB2, M. Belton; NS, N. Samarasinha; NS2, N. Suntzeff; PC2, P. Candia (NOAO); CF, C. Foellmi; HB, H. Boehnhardt; OH, O. Hainaut; TS, T. Sekiguchi (ESO); LW, L. Wasserman; MB, M. Buie; RM, R. Millis (Lowell); DH, D. Hampton (Ball Aerospace); GT, G.-P. Tozzi (Oss. Astrofisico di Arcetri); JL, J. Licandro (Inst. de Astrofísica de Canarias); LB, L. Barrera (Univ. N. Chile); MM, M. Mateo (U. Mich); MA, M. A'Hearn; SM, S. McLaughlin (UMD); TF, T. Farnham (Univ. TX); YC, Y.-J. Choi (Univ. Tel Aviv).

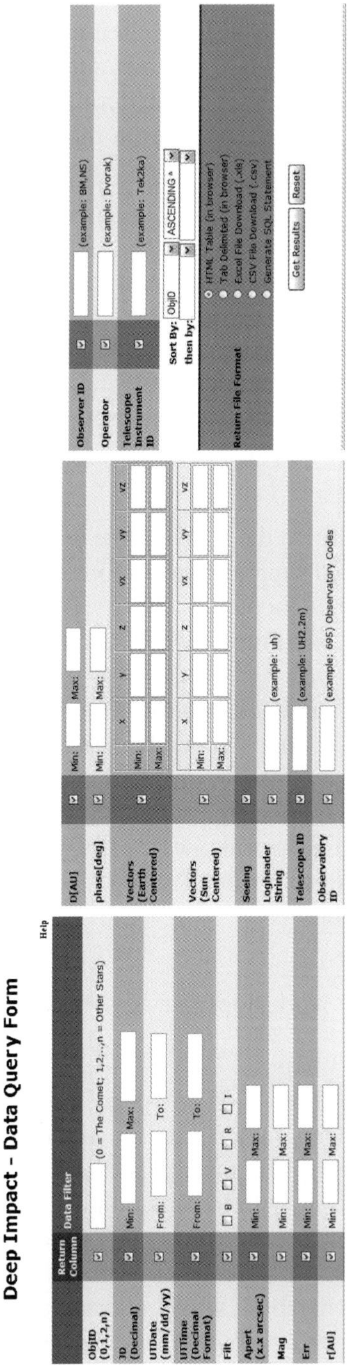

Figure 4. Screen shots of the *Deep Impact* database web browser query form.

than $\phi > 50°$ will not be able to observe the comet at all, as it will have a southern declination (RA $= 13^h37^m52^s$, $\delta = -09°4'07''$), and because it will not get completely dark at these latitudes. The observing circumstances for major world-wide facilities, arranged in order of longitude are shown in Table VI. New moon occurs on 6 July 2005, so the encounter will occur during dark skies (lunar illumination 4.9%). The comet will be $129°$ away from the moon, at a galactic latitude of $52°$.

4.1. THE OBSERVING WORKSHOPS

In order to educate the astronomical community about both the science goals of the *Deep Impact* mission, as well as the technical challenges, the spacecraft observing constraints, and the desired ground-based observations, a series of community workshops were organized. The goals of these workshops were as following.

- To update the community on the status of the mission and the science goals;
- To discuss the key ground support observations that will be needed at various wavelengths at the time of encounter;
- To discuss the unique capabilities of world-wide facilities to participate in the *Deep Impact* science;
- To form collaborations and begin the work for writing key project proposals for both the pre-impact science (November 2004 through impact) and the encounter science. Most major observatory proposal deadlines for the encounter period were late September/early October 2004; and
- To make the different observatories aware of the public interest in the mission, and share some of the Education and Public Outreach materials that have been developed for the mission.

Table VII summarizes the workshops that have been held.

4.1.1. *The DI Collaborator Webpage*
As a result of the workshops, a series of web pages for *Deep Impact* collaborators have been developed: http://deepscience.astro.umd.edu/collab/. The purpose of the web site is to support and coordinate the Earth-based observing program, by providing the observing community with information necessary for proposal writing; giving the community access to meeting presentations, as well as information about the instruments and facilities available at different observatories. In addition, the website provides access to a database where planned and scheduled observations are listed for all participating observers, to enable real-time coordination while observing. This website is password protected. To gain access to the website, please email Stef McLaughlin stefmcl@astro.umd.edu.

TABLE VI

World observability.

Observatory	Code[a]	λ^b	ϕ^c	χ^d	ATwiBeg[e]	ATwiEnd[e]	$\chi < 2.5^f$	Note
Nançay	–	000:12:00	+47:22:00	Set	21:35	22:50	16:15–21:25	1
Meudon	005	002:13:53	+48:48:18	Set	21:40	23:20	16:20–21:00	1
Calar Alto	493	002:32:45	+37:13:25	Set	20:44	21:26	15:38–22:23	1
Klet	046	014:17:17	+48:51:48	Set	21:00	22:30	15:30–20:10	1
Boyden	074	026:24:21	−29:02:17	Set	16:30	16:50	12:30–21:30	1
Kiev	085	030:30:08	+50:27:10	Set	20:10	00:00	14:40–18:50	1
Wise	097	034:45:45	+30:35:44	Set	17:50	18:20	12:50–20:00	1
Majdanak	188	066:52:48	+38:43:19	Set	16:20	16:50	11:10–17:30	1
Kavalur	220	078:49:35	+12:34:34	Set	14:10	14:30	09:30–17:40	1
Yunnan	286	102:47:24	+25:01:32	Set	13:10	13:30	08:10–15:40	1
Perth	323	116:08:06	−32:00:31	3.0	10:30	10:50	06:40–13:50	2
Purple Mtn	330	118:49:15	+32:04:00	5.2	12:20	12:50	07:20–14:20	2
Lulin	D35	120:52:25	+23:28:07	3.8	11:50	12:10	06:50–14:40	2
Bohyunsan	344	128:58:36	+36:09:53	3.4	12:00	12:30	06:50–13:40	2
Siding Spring	260	149:03:58	−31:16:37	1.4	08:10	08:30	04:20–13:20	2
Mt. John	474	170:27:54	−43:59:15	1.3	06:30	06:50	03:10–11:50	3
Mauna Kea	568	204:31:40	+19:49:34	1.2	06:00	06:20	01:10–09:10	4
Haleakala	608	203:44:31	+20:42:30	1.2	06:00	06:20	01:10–09:10	4
San Pedro	679	224:32:11	+31:02:39	2.5	03:55	04:25	22:50–06:15	5
Palomar	675	243:08:28	+33:21:26	2.5	04:10	04:40	23:10–06:10	5
Goldstone	–	243:12:18	+35:14:48	2.6	04:20	04:50	23:10–06:10	6
KPNO	695	248:24:19	+31:57:32	2.9	03:40	04:10	22:40–05:50	6
Lowell	688	248:27:52	+35:05:46	3.1	04:00	04:30	22:50–05:40	6
McDonald	711	255:58:43	+30:40:17	3.8	03:10	03:30	22:10–05:20	6
CTIO	807	289:11:39	−30:10:09	Set	23:00	23:20	19:00–04:00	1
Pachon	–	289:15:48	−30:14:26	Set	23:00	23:20	19:00–04:00	1
ESO	809	289:16:13	−29:15:26	Set	23:00	23:20	19:00–04:00	1
Paranal	309	289:35:48	−24:37:32	Set	23:00	23:20	19:00–04:00	1
Arecibo	251	293:14:49	+18:20:36	Set	14:00	14:20	09:50–18:50	1
La Palma	950	342:07:03	+28:45:37	Set	21:20	21:40	16:20–23:40	1
Armagh	981	353:21:08	+54:21:10	Set	23:40	01:40	17:50–20:30	1
IRAM-30 m	J86	356:36:26	+37:03:58	Set	20:45	21:30	15:40–22:25	1

Note. (1) Not above horizon at impact time. (2) Above horizon at impact time, but in daytime. (3) Above horizon at impact time, but in civil or nautical twilight. (4) Above horizon at impact time, but in astronomical twilight. (5) Above horizon at impact time, and in darkness, and at $\chi < 2.5$. (6) Above horizon at impact time, and in darkness, but at $\chi \geq 2.5$.
[a]Minor Planet Center observatory code.
[b]East longitude.
[c]Geocentric latitude.
[d]Airmass at time of encounter, nominally 06:10 UT on 4 July 2005.
[e]UT at beginning and end of astronomical twilight closest to observability window.
[f]UT time interval that the comet is observable at airmasses $\chi < 2.5$.

TABLE VII

Observing workshops and meetings for 9P/Tempel 1 coordination.

Workshop	Dates	#People	Location
Division for Planetary Sciences	09/01/03	50	Monterey, CA
European Southern Observatory	02/14–15/04	17	ESO HQ, Garching
NOAO/KPNO	02/20/04	20	NOAO HQ, Tucson, AZ
Institute for Astronomy	04/30/05	38	IfA Hilo, Hawaii
National Central University	05/28/04	30	NCU, Jhong-Li City, Taiwan
American Astron. Society Mtg.	05/30/04	4	Denver, CO
NOAO/CTIO	06/24/04	110	NOAO HQ, La Serena, Chile
Anglo Australian Observatory	09/06/04	22	Macquarie Univ., Sydney Australia

4.2. ENCOUNTER SCIENCE FROM EARTH

Key Earth-based observations at the time of encounter include both wavelength and timescale regimes which are inaccessible to the *Deep Impact* spacecraft, and in particular to observe the development of any long-term activity changes which may occur as a result of the impact. Table VIII summarizes the key science goals from Earth at encounter, and the wavelength regimes for observations.

TABLE VIII

Earth-based encounter science.

Science goal	Wavelength	Details	Comments
Parent volatiles	UV	CO, CO_2, H_2O	CO_2 requires modeling. CO Cameron band observations
	IR 2–5 μm	H_2O, many organics	High resolution needed for many molecules
	IR 5–9 μm	organics, PAH	Cannot be done from ground; new regime
	Sub-mm, mm	H_2O, HCN, OH, CS	
Isotopic abundances	Optical 0.3–0.9μm	$^{14}N/^{15}N$, $^{13}C/^{14}C$	Requires high resolution spectroscopy
	UV & Radio	D/H	
Ortho–para ratios	Optical 0.3–0.9μm	for NH_2	Requires high resolution $\rightarrow$ spin temperature
	IR	H_2O	
Dust evolution	Optical 0.3–0.9μm	Particle sizes silicate evolution	Wide field imaging and dust dynamical models
	IR 9–45μm		

4.3. DISCUSSION OF THE WORLDWIDE PLANS

Many of the national and international observatories had an open competition for observations at the time of impact, rather than a dedicated observing campaign. However, a few observatories have declared the night of impact as director's discretionary time (e.g. Keck, Subaru, Gemini), and the NASA IRTF has a long dedicated campaign. Below we discuss the various strategies and approaches that will be utilized for National and International observing facilities as well as the time that has been already allocated. Figure 5 summarizes these world wide plans for both the facilities which will directly view the encounter, and for those that will not. The primary Earth-support science goals are listed, along with the observing technique and wavelength regime required to achieve these goals, and the telescope facilities and instrumentation best suited to achieving these goals are listed.

While not all facilities will be able to observe the actual moment of impact, the possible new activity could take hours or even days to fully develop, so observatories at all longitudes will play an extremely valuable role in understanding the scientific implications of the impact. Therefore for our purposes here we refer to the "encounter" as including the impact as well as many days afterward. Table IX summarizes the allocation of telescope time at observatories world wide as known in January 2005.

At the time of the encounter, the observations will be coordinated with a control center at the summit of Mauna Kea (probably located in the NASA IRTF Facility). Telescopes will be connected via polycom connections to the IRTF control center, which will be connected via polycom to the mission operations at the Jet Propulsion Laboratory in Pasadena. This will ensure rapid and accurate dissemination of the events at encounter so that observers can revise observing strategies in real time.

4.3.1. *Mauna Kea and Haleakala Observatories*

The summit of Mauna Kea, Hawaii, will be the premier site for the real-time observations at encounter. There are 12 telescopes at the summit and in the sub-mm valley (counting the Smithsonian Array as a single telescope). Most of these facilities have multiple instruments which will be of use during the encounter. At most facilities, telescope time was awarded on the basis of open competition within the community. The Keck and IRTF TACs augmented their process with input from the *Deep Impact* science team. Gemini and Subaru are considering the three days of UT July 3, 4, and 5 to be director's discretionary (DD) time. Keck is considering UT July 3 alone to be DD time.

The current slate of instrumentation scheduled for Mauna Kea telescopes (at time of writing, mid-January 2005) is as follows.

- The Keck I and II telescopes will use HIRES and NIRSPEC, respectively, on UT July 4, 5, and 6. HIRES will allow high-spectral resolution capability in the optical and near-UV (particularly crucial for isotopic work). NIRSPEC

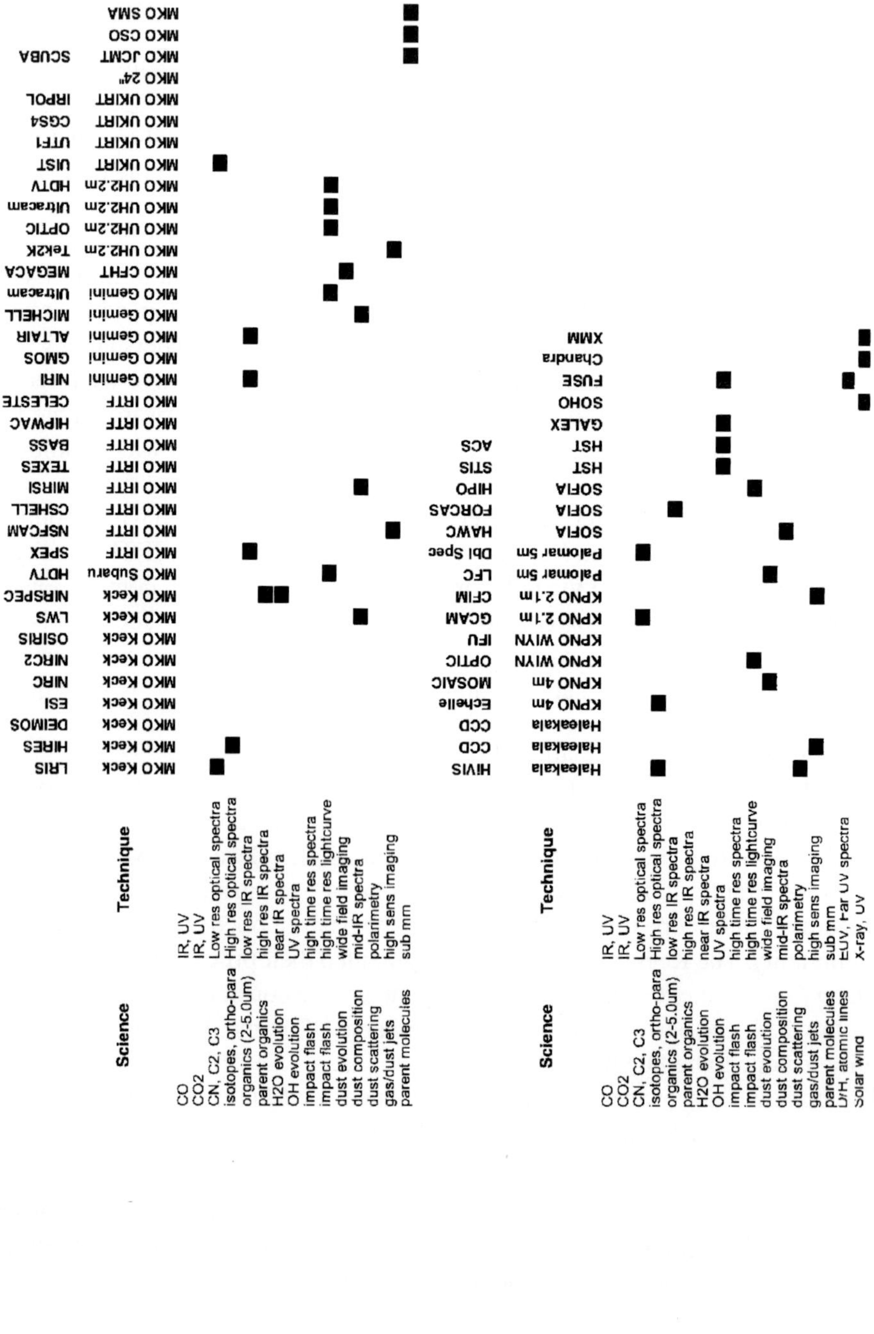

Figure 5. Breakdown of wavelengths and science goals that could be potentially covered by specific observatories.

TABLE IX
Observing allocations as of January 2005.

Telescope	D^a (m)	Local date[b]	t^c	Instrument	PI and PO[d]	Science goals
Mauna Kea Observatory						
Keck-I	10	Jun 7	0.5	LRIS	KM	Gas production baseline
,,	,,	Jul 3	1	HIRES	FC (DD)	Isotopes
,,	,,	Jul 4/5	2×0.5	,,	KM, AC, WJ	IR organics
Keck-II	10	Jun 2	1	NIRSPEC	MM2	
,,	,,	Jul 3	1	,,	FC (DD)	IR organics
,,	,,	Jul 4/5	2×0.5	,,	MM2	Coma structure evolution
CFHT	3.6	Queue	15 h	MegaPrime	KM, JP	TBD
Gemini-N	8.1	Jul 2–4	TBD	TBD	MM1 (DD)	TBD
Subaru	8.2	Jul 3	TBD	TBD	HK (DD)	Hot bands H_2O
UKIRT	3.8	Jul 2–4	3	CGS4	SM	Dust development
UH 2.2 m	2.2	Mar 9, 11	2×0.5	Tek2K CCD	KM	,,
,,	,,	Mar 31	0.5	,,	,,	,,
,,	,,	Apr 3	0.5	,,	,,	,,
,,	,,	May 11, 14	2×0.5	,,	,,	,,
,,	,,	May 31	0.5	,,	,,	Spatial behavior of gas
,,	,,	Jun 1	0.5	,,	,,	Dust development
,,	,,	Jul 1–8	8×0.5	SNFS	KM, KH	High-speed imaging
,,	,,	Jul 27, 29	2×0.5	Tek2K CCD	KM	Parent gas species
UH 0.6 m	0.6	Jul 2–5	4×0.5	OPTIC	KM, RW	,,
JCMT	15	Jun 8–9	2×0.5	Heterodyne	HB2, KM, LM	Isotopes and parent gas species
,,	,,	Jun 24–25	2×0.5	,,	,,	
,,	,,	Jul 3–9	7×0.5	,,	,,	

(Continued on next page)

TABLE IX
(*Continued*)

Telescope	D^a (m)	Local date[b]	t^c	Instrument	PI and PO[d]	Science goals
"	"	Jul 17–18	2×0.5	"	"	Parent gas species
"	"	Jul 23–24	2×0.25	"	"	"
"	"	Jul 28–30	3×0.25	"	"	"
CSO	10.4	Jul 3–4	2×0.5		DL	"
SMA	8 of 6 ea.	Jul 3–4	6		CQ	"
IRTF	3.0	Jun 23–24	2×0.5	BASS	MS	Dust properties baseline
"	"	Jun 25	0.5	SpeX	ND	Gas species baseline
"	"	Jun 26–27	2×0.5	NSFCam	MK	Polarization
"	"	Jun 28–29	2×0.5	HIFOGS, NSFCam	DW	Dust properties
"	"	Jun 30–Jul 2	3×0.5	MIRSI, SpeX	DW, CL	Dust properties
"	"	Jul 3	0.5	SpeX	CL, ND	High-speed spectroscopy of flash
"	"	Jul 4–5	2×0.5	HIFOGS, NSFCam	DW	Dust properties development
"	"	Jul 6–7	2×0.5	MIRSI, SpeX	DW, CL, ND	Dust development and parent species
"	"	Jul 8	0.5	MIRSI, CShell	DW, CL, ND	Dust development and parent species
"	"	Jul 9–10	2×0.5	NSFCam	MK, DW	Polarization
"	"	Jul 11	0.5	SpeX, CShell	ND, CL	Gas development
"	"	Jul 12, 16	0.5	MIRSI, CShell	DW, CL, ND	Dust and gas development
"	"	Jul 17	0.5	SpeX, CShell	ND, CL	Gas development
"	"	Jul 18–19	2×0.5	BASS	MS	Dust properties evolution
"	"	Jul 20–21	2×0.5	MIRSI, NSFCam	DW, CL	Dust development
"	"	Jul 27	0.5	MIRSI, NSFCam	DW, CL	Dust development
"	"	Aug 3	0.5	MIRSI, NSFCam	DW, CL	Dust development

(*Continued on next page*)

TABLE IX
(*Continued*)

Telescope	D^a (m)	Local date[b]	t^c	Instrument	PI and PO[d]	Science goals
Space-based Observatories						
Spitzer	0.85	Feb 3	1.7 h	MIPS	CL	Large dust grain environment
"	"	Feb 6	5.3 h	IRS	CL	"
"	"	Feb 10	0.4 h	IRS	CL	High-speed spectroscopy test
"	"	Jul 2–4	10 h	IRS	CL	High-speed spectroscopy, dust imaging
Hubble	2.4	Jul 3–4	17 orbits	ACS	PF	High-resolution imaging
Chandra	0.4	Jul 2–5	200 ks	ACIS-S	CL	Spectroscopy
XMM	0.4	Jul 3–4	63 ks	EPIC, RGS, OM	RS	Simultaneous X-ray and optical data
GALEX	0.5	Jul 3–5	150 ks	Spectrometer	PF	CO, CO_2
National Optical Astronomy Observatories						
Blanco	4	Jul 3–6	4×0.5	ISPI	KM, DH	Dust color and evolution
CTIO 1.5m	1.5	Jul 3–6	4×0.5	CPAPIR	KM	Dust color simultaneity
CTIO 0.9m	0.9	Jul 3–6	4×0.5	CFIM CCD	KM	Dust color simultaneity
CTIO Schmidt	0.6/0.9	Jul 3–5	3×0.5		PS	Wide field dust
Mayall	4	Jul 1–8	8	MOSAIC	MA	Narrowband imaging
KPNO 2.1m	2.1	Feb 1–4	4	CFIM CCD	MA	Dust development baseline
"	"	Mar 10–13	4	"	"	"
"	"	Apr 8–13	4	"	"	"
"	"	Jun 3–6	4	"	"	"
"	"	Jul 1–8	8	SQIID	"	Near-infrared dust imaging
WIYN	3.5	Jul 2–6	5	DSPKB	BM	Integral-field spectra
SOAR	4.1			Phoenix		

(*Continued on next page*)

TABLE IX
(Continued)

Telescope	D^a (m)	Local date[b]	t^c	Instrument	PI and PO[d]	Science goals
Gemini-S	8.1	Jul 2–4	TBD	TBD	MM1 (DD)	
"	"	Queue	5 h	T-ReCs	JD	Dust silicate properties
European Southern Observatory						
VLT Antu	8.2	Jul 2–10	8 × 0.5	FORS2	HR	Gas coma daughter species
VLT Kueyen	8.2	Jul 10–12	2 × 0.5	FORS1	HB1	Polarization
"	"	queue	10 h	UVES	HR	High-resolution spectroscopy, isotopes
"	"	Jul 2–10	8 × 0.5	"	"	"
VLT Melipal	8.2	Jul 2–6	4 × 0.5	VISIR	HB1	Dust silicate evolution
VLT Yepun	8.2	Jul 2–10	8 × 0.5	NACO	"	Near-IR coma structure
ESO 3.6 m	3.6	Jul 2–10	8 × 0.5	TIMMI2	"	Dust silicate evolution
NTT	3.6	Apr–Jun	3 × 0.5	EMMI	OH	Imaging (dust scattering), spectra (gas)
"	"	May 31–Jun 2	3 × 0.4	EMMI	"	Imaging (dust scattering), spectra (gas)
"	"	Jul 2–7	5 × 0.5	SOFI	"	Near-IR coma structure
"	"	Jul 7–10	3 × 0.5	EMMI	"	Imaging for dust scattering properties
ESO 2.2 m	2.2	Jul 2–10	8 × 0.5	WFI	"	Imaging for dust scattering properties
Observatorio del Roque de los Muchachos – La Palma						
WHT	4.2	Jul 3–7	5 × 0.5	OASIS/LCRIS	JL	Opt spectra, near IR imaging
TNG	3.6	Jul 1–8	8 × 0.5	DOLORES/SARG	JL	Opt/near-IR spectra
NOT	2.5	Jul 1–8	8 × 0.5	ALFOSC	JL	Low res spectra-gas
INT	2.5	Jul 1–7	7 × 0.5	WFC	C	Imaging for dust scattering properties

(Continued on next page)

TABLE IX
(Continued)

Telescope	D^a (m)	Local date[b]	t^c	Instrument	PI and PO[d]	Science goals
Calar Alto Observatory						
	2.2	Apr	5	CAFOS	LL	Optical spectroscopy
	2.2	Jan–Jul 1		CAFOS	LL	RI images every 2–3 days
	2.2	Jan–Jul 1		CAFOS	LL	Monthly optical spectroscopy
Las Campanas Observatory						
Baade	6.5	Jun 20–21	2×0.5	PANIC	DO	Near-IR imaging for dust
Baade	6.5	TBD	TBD	PANIC	DO	Near-IR imaging for dust
Clay	6.5	TBD	TBD	MAGIC	DO	Optical imaging for dust
duPont	2.5	Jun 28–Jul 10	13×0.5	WFCCD	DO, KVB	Optical imaging for dust
Swope	1.0	Jul 3–8	6×0.5	Retrocam	KVB	Spectroscopy
Lowell Observatory						
Perkins	1.8	TBD		MIMIR	MB	High-speed spectroscopy
Hall	1.1	TBD		SITe CCD		Narrowband imaging
31-Inch	0.8	TBD		Loral CCD		Dust development
Palomar Observatory						
Hale	5	Jul 2–6	5×0.5	PHARO	JB2, RD	High resolution imaging
Anglo-Australian Observatory						
AAT	3.9	Jul 1–8	8×0.5	UCLES	JB1	High-resolution visible spectroscopy
UKST	1.2	Jul 1–8	8×0.5	6dF	C	Multi-object spectroscopy
Lick Observatory						
Shane	3.0	Jul 1–5	5		RP	Spectroscopy of flash and coma
Nickel	1.0	TBD				

(Continued on next page)

TABLE IX
(*Continued*)

Telescope	D^a (m)	Local date[b]	t^c	Instrument	PI and PO[d]	Science goals
McDonald Observatory						
HET	9.2	TBD				
Smith	2.7	TBD				
Table Mountain Facility						
24-inch	0.4	TBD				
Mount Laguna Observatory						
40-inch	1.0	TBD				
Observatorio Astronómico Nacional de San Pedro Martir						
	2.1	TBD				
	1.5	TBD				

[a] D: diameter of primary mirror in meters.

[b] Observing run starts on this date in 2005, local time.

[c] t: time allocated, in units of nights unless otherwise stated; $n \times m$ indicates only a part of each night (m) has been allocated over n nights.

[d] PI: principal investigator; PO: primary observer; "DD" indicated Director's Discretionary time has been assigned. Initials correspond to: MA, Michael A'Hearn; JB1, Jeremy Bailey; JB2, James Bauer; MB, Marc Buie; HB1, Hermann Boehnhardt; HB2, Harold Butner; FC, Fred Chaffee; C, Coates; AC, Anita Cochran; JD, James De Buizer; RD, R. Dekany; ND, Neil Dello Russo; PF, Paul Feldman; OH, Olivier Hainaut; DH, Doug Hamilton; KH, Klaus Hodapp; WJ, William Jackson; HK, Hiroshi Karoji; MK, Michael Kelley; LL, Luisa-Maria Lara; JL, Javier Licandro; CL, Carey Lisse; DL, Dereu Lis; KM, Karen Meech; SM, Steven Miller; MM1, Matt Mountain; BM, Beatrice Mueller; MM2, Michael Mumma; DO, Dave Osip; JP, Jana Pittichova; RP, Richard Puetter; CQ, Chunhua Qi; HR, Heike Rauer; RS, Rita Schulz; MS, Michael Sitko; KV, Kaspar von Braun; RW, Richard Wainscoat; DW, Diane Wooden; LW, Laura Woodney.

allows high resolution near-IR spectroscopy to detect volatile organic parent molecules.

- With its closed loop adaptive optics system, the Gemini 8 m telescope is optimized for high spatial resolution in the near-IR. It is also equally well suited to performing diffraction limited mid-IR observations; one optimal use of this facility would be to use it for 10 and 20 μm spectroscopy using the Michelle instrument. This region is diagnostic of refractory silicate features which may trace the thermal history in the solar nebula.

- The UH 2.2 m telescope has a unique instrument that is permanently installed at the bent-cassegrain focus: the Supernova Integral Field Spectrograph (SNFS) which will allow us to obtain spatially resolved integral field spectroscopy over the 3000–10,000 Å range of the nucleus and inner coma of comet 9P/Tempel 1 before, during, and in the days after the impact. The spectral region covered by these data contains many molecular emission features that will be diagnostic of the temporal sequence of release of volatiles from the comet as a result of the impact. It may be possible to get adequate S/N at impact time to get spectra with time resolution of 1 min.

- In addition the UH 2.2 m will have a moderate-field ($\sim$7') CCD camera available for several nights before, during, and after the impact. The UH 0.6 m telescope will make use of an even wider field orthogonal-transfer CCD (Burke *et al.*, 1994). This camera can read out CCD images with tip-tilt adaptive optics at video rate.

- The Subaru 8 m telescope have several instruments that may reflect the optimal use of that facility. The Japanese community is strongly interested in utilizing two high speed high definition TV cameras, one to image the sky as visual observers will see the event.

- The UKIRT 3.8 m telescope will make use of its high-resolution infrared spectrometer, CGS4, on UT July 3, 4, and 5. Their priority is to observe the evolution of the "hot" bands of water before, during, and after the impact.

- The only instrument available at time of impact for the Canada–France–Hawaii 3.6 m telescope will be the wide field imager Megacam. Thus this facility will be well suited for watching the development of the dust coma and tail, especially post-impact.

- The robotically controlled Faulkes 2 m telescope atop Haleakala on Maui will be also observing before, during, and after the impact with a moderate-field ($\sim$ 4.6') CCD camera.

- The Air Force's 3.6 m AEOS telescope also atop Haleakala will be observing the impact with a high-resolution ($R \sim$ 17,000–49,000) visible spectrometer, covering the 6400–10,000 Å region.

4.3.2. *European Southern Observatory*

At the *Deep Impact* workshop held in Garching, Germany, during February 2004, a number of European planetary scientists met and discussed optimal observing strategies for ESO. Two European teams were created to lead in the development of coordinated large proposals to ESO for Period 74 (pre-impact) and 75 (encounter). Chile will not be able to see the comet until about 16 h after impact but the observatories will certainly be able to follow the short-term evolution of the comet's behavior.

The nucleus structures and dust team, lead by H. Boehnhardt and U. Käufl, has been awarded the equivalent of 19 nights of time spread out over six telescopes at La Silla and Cerro Paranal around the time of impact. They will: perform high resolution near-IR imaging to look for ejecta and jets (VLT UT4 with NACO; NTT with SOFI); obtain mid-IR images and spectra to investigate dust composition, in particular silicate evolution (VLT UT3 with VISIR, 3.6 m with TIMMI2); observe the comet in optical wavelengths to characterize polarization (VLT UT1 with FORS1) and to study dynamical effects and scattering properties of the dust (2.2 m with WFI, NTT with EMMI). Having complementary programs at both Cerro Paranal and La Silla will ensure some backup in case of poor weather.

The team studying the evolution of the gas coma, lead by H. Rauer, had been awarded the equivalent of eight nights of time at Paranal for optical and near-UV spectroscopy. High dispersion spectroscopy will be used to investigate isotopic ratios (VLT UT2 with UVES), and low dispersion spectroscopy will be used to get pre- and post- abundances of molecular species such as CN, C_2, C_3, etc. (VLT UT1 with FORS2).

4.3.3. *National Optical Astronomy Observatories*

Access to the primary facilities of the National Optical Astronomy Observatories (NOAO) on Kitt Peak and Cerro Tololo was via open competition through the normal proposal process. Time at other facilities where the US astronomical community has access through NOAO was generally likewise determined via open competition. However, the *Deep Impact* worked closely with the astronomical community to identify the areas where NOAO can make the greatest contribution.

The telescopes at Kitt Peak will be able to see the impact in real time, however, the comet will be at very high ($\chi > 2.5$) airmass. The moderate to large telescopes there – the Mayall 4 m, the WIYN 3.5 m, and the 2.1 m telescopes – have been allocated for 9P/Tempel 1 observations around the time of encounter. At the Mayall, the MOSAIC instrument will be used to obtain narrowband imaging and photometry. At the 2.1 m, the near-IR camera SQIID will give us a unique capability that is not available anywhere else: truly simultaneous photometric imaging in JHK (and possibly L, if weather cooperates). At the WIYN, we will be using an integral-field unit "Densepak" to obtain visible spectroscopy simultaneously at many different locations in the coma.

At Cerro Tololo, the Blanco 4 m telescope has been allocated to observe the comet using ISPI, a large-format near-IR camera. ISPI is the largest near-IR imager in South America, and would therefore make a unique contribution. Among other moderate and small telescopes at CTIO, we have been allocated time at the 0.9 m and the 1.5 m telescopes, which are run by the SMARTS consortium. At the 0.9 m, a classically scheduled run using CFCCD, a 2K-by-2K CCD camera, has been scheduled. At the 1.5 m, we will use CPAPIR, another wide-field large-format near-IR camera. With these three telescopes at CTIO, we have a capability not found elsewhere: the ability to obtain deep simultaneous images of the dust from U-band through K-band. Combined with dust dynamical modelling, thermal data planned from Spitzer and other telescopes, there is the potential to thoroughly characterize the dust component and to search for differences between the pre-impact dust (from the more evolved comet surface layers) and the post-impact dust excavated from the interior.

The SOAR 4 m and Gemini-South 8.2 m telescopes are located on Cerro Pachón. A possible optimal scenario that would complement the other observations being made from Mauna Kea at impact time is to use PHOENIX, a high resolution ($R = 50,000$) IR echelle spectrograph on SOAR to look for organics, and use the TReCS 10–20 μm spectrograph on Gemini to monitor the chemical and silicate evolution in the mid-IR.

4.3.4. *Lowell Observatory*

Lowell Observatory is a privately owned astronomical research institution located in Flagstaff Arizona. The observatory has three telescopes available at Anderson Mesa for research projects: the 1.8-m Perkins telescope, the 42-inch Hall telescope, and the 31-inch robotically operated telescope. The 1.8-m has optical imaging and spectroscopic capabilities as well as a newly commissioned near-IR imaging polarimetric spectrograph, *Mimir*. *Mimir* utilizes an Alladin III InSb 1024 × 1024 dectector which operates between 1 and 5 μm. It has two cameras providing a choice of plate scale of either 0.6 arcsec pix^{-1} or 0.18 arcsec pix^{-1}. There are four filter wheels that provide a variety of functions including standard bandpass filters (i.e., JHKLM), polarimetric elements (H-band only), and grisms for spectroscopy at resolving powers over a range from 20 to 2000. The wide field camera with a grism employs 5 arcminute slits from 1 to 10 pixels wide. This may make a unique contribution to the real-time encounter observations in that the system will be capable of obtaining high time resolution near-IR spectra which may be able to study the impact flash. The Hall telescope may be used with the KRON photometry or the SITE 2 K CCD for narrow band photometry or imaging of the coma to look at the production rates of various gaseous species. The 31-inch robotic telescope will play a critical role in monitoring the development of the dust coma on a nightly basis, weather permitting, starting early in the spring 2005, through encounter until after the impact when the comet moves into solar conjunction. This will be useful to watch for brightness fluctuations and outbursts in activity, and search for jets in the inner coma, and after the upgrade to a large format CCD, to obtain images for dust dynamical modelling.

4.3.5. *Palomar Observatory*

The Palomar Observatory located in California, is operated by the California Institute of Technology. Facilities include the Hale 200-inch telescope, the 60-inch, Oshin 48-inch, and the 18-inch schmidt telescopes. The 200-inch telescope will be utilized with the adaptive optics system and near-IR imaging to watch the evolution of the dust that is ejected during impact, and to look for evidence of water–ice in the near IR as the crater is excavated.

4.3.6. *Lick Observatory*

The University of California Observatories operates Lick Observatory on Mt. Hamilton, CA, for the University of CA community. The observatory has three major telescopes: the 120-inch Shane telescope, the Coude Aux Telescope and the Nickel 40-inch reflector, in addition to five smaller telescopes. The Shane telescope has been scheduled for five nights at the time of impact with a visitor instrument that can perform 0.5–2.5 micron low-resolution spectroscopy. The goal will be to monitor the evolution of the impact flash and development of activity in the gas and dust post-impact. The smaller telescopes are not yet scheduled for the encounter period, but plans are underway to utilize the 40-inch for imaging.

4.3.7. *Las Campanas Observatory*

The Carnegie Institution of Washington operates facilities at the Cerro Las Campanas observatory in Chile. The observatory facilities include the twin 6.5-m Magellan telescopes (Baade and Clay), the du Pont 2.5-m telescope and the Swope 1.0-m telescope. Near-IR and visible imaging is scheduled at Magellan (using PANIC the 1024 IR imager on Baade) two weeks prior to the encounter in order to establish baseline observations in collaboration with a thermal-IR project at Gemini south. Although the Magellan telescopes are currently only scheduled through June 27, 2005, the plan is to observe through encounter and the post-encounter period, monitoring the comet for eight rotation periods to monitor the freshly excavated active area and the evolution of the particles in the coma. Imaging and spectroscopy are scheduled at the 2.5-m duPont and 1.0-m Swope telescopes throughout the encounter period.

4.3.8. *Radio Facilities*

Ground based facilities where observations at sub-millimeter and/or millimeter wavelengths are planned and/or already scheduled include:

- the IRAM 30 m telescope in Spain, which can observe simultaneously four lines between 86 and 272 GHz;
- the IRAM Plateau de Bure interferometer in France, which has six 15 m telescopes and can observe at 86–115 GHz and at 205–250 GHz simultaneously (although autocorrelation mode will likely be more efficient to make detections than interferometric mode, thus reducing the interest of interferometers for such a modest activity comet);

- the 15 m JCMT and 10 m CSO atop Mauna Kea, which have receivers available at 210–275 GHz, 300–365 GHz (330–365 for JCMT), and 450–510 GHz;
- the APEX 12 m test antenna in Chajnantor (5000 m elev.), Chile (pending availability of heterodyne receivers);
- the Kitt Peak 12 m antenna (pending availability of 260 GHz receiver);
- the SMA, atop Mauna Kea, which consists of eight 6 m antennas that observe in the 230 and 345 GHz bands;
- the 100 m GBT in West Virginia, which can observe at a range of bands at millimeter, centimeter, and decimeter wavelengths;
- the Australian facilities, consisting of the Parkes 64 m (operating between 10 and 70 cm, up to 24 GHz) which can access lines of H, OH, CH_3OH, and CH_4, among others; the Australian Telescope Compact Array which operates between 85 and 105 GHz in the period from April through October; and the Mopra 22 m (operating at 3, 6, 13, and 20 cm, and 3 and 12 mm);
- the Very Large Array in Socorro, New Mexico, which consists of twenty-seven 25-meter dishes and can observe between 74 and 50,000 MHz (400–0.7 cm); and
- other radio-telescopes (e.g., Nobeyama 45 m, FCRAO 14 m) that can observe around 89 GHz, in which lies the HCN(1-0) line that is somewhat harder to detect than the HCN(3-2) line.

Note that CARMA (the BIMA+OVRO merged array) will not be ready in time, and that the comet's declination is unreachable from Arecibo at encounter.

The sub-mm fundamental line of H_2O at 557 GHz will be observable from space observatories such as ODIN. The ODIN satellite is a Swedish satellite with a 1m antenna, which has receivers at 119 GHz and four in the sub-mm band (480–580 GHz). The pointing constraints are such that the comet will be observable only between 60 and 120° solar elongation (June–September 2005). In addition, due to its low Earth orbit, the comet will be only visible for 60 min every 96 min orbit. Observing constraints for the SWAS satellite, also operating around 557 GHz, are similar: elongation > 75° (comet 9P/Tempel 1 will thus be visible most of the year 2005) but also about 1 h per 1.6 h orbit. While this would be within the capabilities of the SWAS satellite, it is not scheduled to be operational at this time (out of funding).

4.3.9. *Other Efforts*
Chinese astronomers at the National Central University in Taiwan are coordinating a large number of far eastern and eastern European observatories. Located at dark sites, these will fill in the gap in longitude coverage for this part of the world. The instrumentation at these facilities will be ideally suited to optical imaging and polarimetry of the dust. The observatories include Moletai (Lithuania), Gaumeigu (China), Yunnan (China), Lulin (Taiwan) and Majdanak (Uzbekistan).

A consortium of astronomers plans to utilize the facilities on La Palma both for extensive monitoring of the comet during the spring 2005 both for weekly monitoring of the development of the dust coma, and to monitor the onset of molecular gas production (low resolution optical spectra). At encounter, there are plans for doing near IR spectroscopy to look at organic species and high resolution JHK near-IR imaging.

In addition to participation through the Subaru telescope on Mauna Kea, the Japanese government has provided funds to build a new 1-m telescope on Ishigake Island (in the Okinawa Islands, just north of Taiwan). *Deep Impact* is the driver for completion and science readiness by July 2005. The first light instrument will be PICO, and optical polarimetric imager for dust observations.

Observers in Russia are planning to use the 6m SAO telescope, the 2.0 m telescope, and the 2.6 m telescopes for both imaging and medium and high resolution spectroscopy (optical) leading up to the encounter and at the time of encounter. Colleagues in Bulgaria will perform complementary narrow and wide band imaging of the dust coma of the comet.

At the Anglo-Australian Observatory, several nights around the time of impact have been allocated. The AAT will use UCLES, a visible-wavelength echelle spectrograph. The UK Schmidt Telescope will use 6DF, a multi-object visible-wavelength spectrograph with an enormous six-degree field of view.

At Calar Alto Observatory, the 2.2 m telescope will be used to monitor the comet with CAFOS, a visible-wavelength instrument capable of obtaining images and long-slit spectra. From early January 2005 through July 1, 2005, broad-band CCD images will be obtained of the comet every 2–3 days. Once a month spectra will be obtained to watch the development of the gas species, and an additional focussed run for both imaging and spectroscopy is scheduled for April 2005.

At Fort Davis, the optimal use of the 9.2 m Hobby–Eberly Telescope will be to obtain high-spectral resolution monitoring observations of the comet over the course of 2005 through June.

At the Observatorio del Roque de los Muchachos, on La Palma, the 4.2 m WHT has been scheduled to use OASIS, an integral-field visible-wavelength spectrograph combined with AO, and LIRIS, a moderate-resolution near-IR spectrograph. At the 3.6 m TNG, three instruments will be used: DOLORES, a moderate-field CCD imager; SARG, a high-resolution visible spectrograph; and NICS, a near-infrared camera. Finally, at the 2.5 m NOT, the plan is to use ALFOSC, a faint-object visible-wavelength spectrograph.

The PLANET consortium of telescopes is a global network for monitoring gravitational microlensing events to search for extrasolar planets. This consortium will be obtaining nightly R-band observations of the comet starting in early spring 2005, through the post-encounter period, primarily using the 1.5-m Boyden observatory in South Africa, and the 1-m Canopus telescope in Hobart Tasmania. Back up observatories in case of poor weather include the 2-m Pic du Midi telescope and the 1.5-m Danish telescope at La Silla. In addition, the consortium will be obtaining weekly BVRI color images.

4.3.10. *Unique Facilities and Special Arrangements*

Hubble Space Telescope. Cycle 13 time has been awarded to the *Deep Impact* team and collaborators to study the generation and evolution of the gaseous coma resulting from the impact. Time was awarded to obtain ultraviolet spectra with STIS before, during and after impact, with the goal of detecting highly volatile species such as CO, CO_2 and S_2 from newly exposed sub-surface layers. However, with the loss of the STIS instrument after the suspend state on 2004 August 3, a subset of these observations will be made with the ACS. In addition, we will be using the high resolution capabilities of the ACS/HRC in conjunction with the camera aboard the *Deep Impact* spacecraft two weeks prior to impact to use stereo imaging to determine the spatial orientation of active jets emanating from the comet's nucleus and to assess potential hazards of these jets to the spacecraft.

It will be possible to time the impact so that we guarantee that HST will be able to observe the comet at that exact moment, but given the uncertainties in the telescope's orbital dynamics, the exact moment of impact will not be decided until several weeks before UT July 4.

GALEX Ultraviolet Telescope. The Galaxy Evolution Explorer is an explorer class 0.5-m orbiting ultraviolet telescope launched 28 April 2003 for a 29 month mission. The instruments are designed for far- (1350–1750 Å) and near- (1750–2800 Å) UV imaging and spectroscopy (resolution: $R_{FUV} = 250$–300, $R_{NUV} = 80$–150). *Deep Impact* team members and collaborators have received GALEX telescope time to study the changes in the gas coma (in particular CO and CO_2) resulting from the excavation of subsurface material due to impact. GALEX cannot observe HI, OI or OH which are the principal dissociation products of H_2O, but it will be able to observe emissions of CI and the CO Fourth Positive system and the CO Cameron band system. The latter arises from prompt emission following photodissociation of CO_2, and can therefore provide a measure of CO_2 abundance. CO_2 cannot be measured from the ground in the infra-red. Changes in cometary activity from the quiescent state will be monitored using emission from the (0,0) band of CS at 2576 Å. The parent of CS (CS_2) photodissociates very rapidly to CS, so the CS abundance tracks the changes in outgassing with very little phase lag. Because of a strong g-factor compared to CO, this molecule is relatively easy to detect.

FUSE Observatory. The Far Ultraviolet Spectroscopic Explorer is a high-orbit extreme UV space observatory operating in the range 905–1195 Å with a spectral resolution of 15 km s^{-1} (i.e. about 0.05 Å). Several reaction wheels on the spacecraft are non-functional, so there are severe restrictions on the spacecraft pointing, but there are several important scientific objectives that can be addressed if some of the restrictions can be relaxed and the observatory pointed at 9P/Tempel 1. In particular the loss of the STIS instrument on HST leaves FUSE as the most capable facility for several items. A review of ultraviolet observations of cometary comae and the important questions is given by (Feldman *et al.*, 2005). FUSE can:

- search for argon emission and O VI emission after impact. An unconfirmed possible detection of argon has only been made in comet C/1995 O1 (Hale-Bopp). The argon abundance is dependent on the temperature environment at which the comet's ices condensed (Weaver et al., 2002; Notesco et al., 2003).
- provide a very sensitive probe of CO before and after the impact. While CO is commonly seen in long-period comets, an assessment of the remaining abundance within a highly evolved short-period comet (which has never been done) would give clues to the thermal and structural history of the nucleus.
- possibly detect N_2 for the first time in a comet (958.6 Å). Among nitrogen compounds this species should have been dominant in the solar nebula (Feldman et al., 2005) so a measurement of its abundance in comets would be fruitful.
- observe H_2 emission before and after the impact and watch its temporal variation. Detection of the emission (near 1600 Å) requires high spectral resolution. One important question is determining if there is any nuclear source of H_2 in addition to the extended source of dissociated H_2O.
- possibly observe the D/H ratio post-impact, which would be the first such measurement in a Jupiter-family comet, and hence could be a probe of D/H in a different region of the solar nebula.
- monitoring the abundances of atomic C, O, H, S and N in the coma before and after impact. In previous FUSE observations of comets, over 90 emission lines have been detected, but over half of these have yet to be identified. Because the FUSE bandpass covers the resonance transitions of nearly all atomic elements and many simple ions, there may be a rich set of post-impact observations.

Chandra X-ray Telescope and the XMM Newton Observatory. In the X-ray regime, the *Deep Impact* experiment allows for a controlled test of the charge exchange (CXE) emission mechanism that drives cometary X-ray emission (Lisse et al., 2001; Kharchenko and Dalgarno, 2001; Krasnopolsky et al., 2002). Previous ROSAT and Chandra observations studied cometary X-ray emission as the solar wind changed but the cometary emission remained constant. Here at a precise time, a fresh amount of neutral material will be injected into a finite volume of the extended cometary coma. This new material will directly increase the emission measure for the comet, passing from the collisionally thick to the collisionally thin regions of emission over the course of days. The DI experiment also allows for a direct search for prompt X-rays created by hyper-velocity impact processes.

The *Deep Impact* team has received 300 ks of Chandra time to observe 9P/Tempel 1 before, during and after the impact encounter. Pre-impact, we expect a modest X-ray signal from the comet from the interaction of its neutral gas coma with the 10^6 K solar wind plasma through the charge exchange mechanism. At encounter, the impactor vaporizes within 1 ms and is blown back out in a plume of vapor and plasma. Hypervelocity impact experiments at NASA Ames (Sugita et al., 2003) suggest the prompt production of abundant ionized and atomic species, strong magnetic fields and initial plasma temperatures of $\sim 10^6$ K. There may be a rapid non-equilibrium creation of an expanding plasma cloud with possible short-term X-ray generation. At Encounter, Chandra will spectroscopically observe the creation and evolution of the impact plume. Post-impact, the X-ray luminosity should increase 30-fold. The total integrated X-ray luminosity increase will be used as one of the best estimates of the total neutral gas mass ejected due to the impact. The differences in the optical and X-ray light curves will be used to probe the extent of the collisionally thick X-ray emitting region around the nucleus.

The X-ray Multi-Mirror (XMM) satellite is an ESA orbital X-ray observatory. XMM has three instruments which are effective over the energy range from 300 eV to 12 keV: EPIC which performs spatially resolved moderate resolution X-ray spectroscopy, RGS which performs high-spectral resolution soft X-ray spectroscopy and OM which can obtain optical and UV images below 5000 Å. While complementary, Chandra will not be able to get as good spectral quality data as XMM-Newton, and indeed only RGS can separate OVII and OVIII. However, time was not awarded on this facility.

SOHO Observatory. The Solar and Heliospheric Observatory (SOHO) is an international project to study the sun. Because it points at the Sun at all times, it will not be possible to use the LASCO imager or UVCS spectrometer to look at the comet, however, the observatory has an all-sky Lyman-α imager, the Solar Wind Anisotropy Monitor (SWAN) which could be used to track the comet's emission over time and orbit. The SOHO solar wind data will be important to understand the solar wind environment at the comet during the encounter. This will help us to understand the X-ray emission we see. This emission is dominated by lines of ionized and excited carbon, oxygen, sometimes neon, and possibly nitrogen.

Spitzer Space Telescope. Spitzer Space Telescope observations of comet 9P/Tempel 1 are possible at the end of Cycle 1, in a viewing window that starts 2 days before the encounter, ending 3 months later, and time has been awarded primarily for spectroscopy but also for imaging with the IRS instrument. Although there were many possibilities for the use of the Spitzer telescope, for observations at the exact time of impact the *Deep Impact* team made the decision to utilize the unique capabilities of the telescope in a wavelength regime which is largely inaccessible from the ground: the 5–8.7 μm window. The IRS data in combination with high-resolution spectroscopy from the spacecraft itself will let us measure the spectral energy distribution from 1 to 38 μm.

The IRS's wavelength region has heretofore been very poorly explored for comets because of the high terrestrial atmospheric opacity. There are a wealth of organic species that may be detected, including HCN, CS_2, CH_4, CH_3OH, CH_2O, CH_3OCH_3, H_2O, NH_3, CH_3NH_2, C_2H_6, C_2H_2 and C_2H_5OH. The neutral gas species will give us clues to the difference between surface and subsurface composition. For tens of seconds the plume could be hot enough for us to detect polycyclic aromatic hydrocarbons (PAHs) with features at 6.2, 7.7, and 8.6 μm, which have never been seen before in comets. We will search for the 6.85 μm solid state feature seen in dense molecular clouds, which may be due to a C–H deformation in hydrocarbon-containing ices or caused by carbonate materials. Lastly we will also monitor the 8–13 and 18 μm dust silicate features for changes in the degree of crystallinity. All these phenomena carry interesting potential diagnostics of cometary composition and formation.

We have requested continuous monitoring from the start of the encounter window through 2 days post-impact to observe the immediate effects of the impact. There will also be brief visits 5, 10, 20, 50 and 100 days post-impact to follow the long-term evolution of the comet.

In addition, we have been awarded time in the months leading up to the encounter to characterize the dust from the comet in the 8–38 μm window. Among other goals, this will let us determine the spacecraft hazard from large dust in the near-nucleus environment. Modeling of the dust spectral energy distribution will tell us the emissivity, grain size distribution, and mass of the large grains.

Our pre-encounter program also involves deep imaging at 24, 70, and 160 μm with the MIPS instrument to characterize the comet's dust trail.

NASA Infra-Red Telescope . NASA's IRTF will have an extended observing campaign of the comet in June–July 2005. Seven instruments – NSFCAM, MIRSI, SpeX, BASS, HIFOGS, CShell, and the new Apogee CCD camera – will be in use at various times from late June through early August, acquiring data for five PI programs. On the actual impact day, SpeX is expected to be in use, taking high frame-rate observations in the near-IR. Observations will usually be done by an support astronomer, with cometary scientist (or scientists) frequently on site. Because of the unpredictable nature of the event, and because of the need for coordination with the world-wide ground based observing campaign, the projects will be observed in queue mode, and the behavior of the comet after impact will dictate what is actually observed. The proprietary period for the data is 6 months, and the IRTF will provide PDS-compliant FITS headers, and will make the DI data archive available to the public.

Rosetta Spacecraft . The Rosetta spacecraft will be in a slightly better viewing condition than Earth-orbital satellites (comet geocentric distance 0.53 AU versus 0.89 AU in early July 2005) and the MIRO 0.3 m sub-mm telescope, although less sensitive than other Earth satellites, can be in permanent viewing of the comet at that time, to monitor the 557 GHz H_2O line. Rosetta can also offer a range of observing techniques for observing the impact, using OSIRIS for imaging, VIRTIS for near-IR spectra and ALICE for far UV spectra. The ALICE instrument on Rosetta does some of the same things as FUSE, albeit with lower sensitivity. The plan for *Rosetta* will be to begin observing the comet one week before the encounter and for two weeks after.

Sofia Telescope . The Sofia airborne observatory is an infrared-optimized telescope which is flown in the back of a Boeing 747 airplane. While the facility will not be flight ready in time for *Deep Impact*, there is the possibility that the telescope could be used on the ground, assuming the airplane can be moved from Waco TX, where the comet will be below the horizon at impact, and moved to NASA Ames. With the HIPO occultation instrument, wit will be possible to get simultaneous V and R observations with 20 Hz time resolution and $S/N \sim 100$.

5. Post-Encounter Observations

The comet will be observable into early September 2005. During the time from impact into September, there should be relatively frequent monitoring of the gas production and dust evolution in order to fully assess the long-term changes induced by the impact. National facilities where longer term queue scheduled observations are possible, and private observatories such at the University of Hawaii 2.2 m telescope on Mauna Kea, will be ideally suited for this sort of monitoring.

Acknowledgements

Support for this work was provided through University of Maryland and University of Hawaii subcontract Z667702, which was awarded under prime contract NASW-00004 from NASA.

References

A'Hearn, M. F., Millis, R. L., Schleicher, D. G., Osip, D. J., and Birch, P. V.: 1995, *Icarus* **118**, 223.

A'Hearn, M. F., Belton, M. J. S., Delamere, A., and Blume, W. H.: 2005, *Space Sci. Rev.* this volume.

Belton, M. J. S., Meech, K. J., A'Hearn, M. F., Groussin, O., McFadden, L., Lisse, C., *et al.*: 2005, *Space Sci. Rev.* this volume.

Biver, N., Bockelée-Morvan, D., Crovisier, J., Colom, P., Henry, F., Moreno, R., *et al.*: 2002, *Earth Moon Planets* **90**, 323.

Blume, W. H.: 2005, *Space Sci. Rev.* this volume.

Buratti, B. J., Hicks, D. M., Soderblom, L. A., Britt, D., Oberst, J., and Hillier, J. K.: 2004, *Icarus* **167**, 16.

Burke, B. E., Reich, R. K., Savoye, E. D., and Tonry, J. L.: 1994, *IEEE Trans. Electron. Devices* **41**, 2482.

Delahodde, C. E., Meech, K. J., Hainaut, O. R., and Dotto, E.: 2001, *Astron. Astrophys.* **376**, 672.

Feldman, P. D., Cochran, A. L., and Combi, M. R.: in press, in M. Festou *et al.* (eds.), *Comets II*, University of AZ.

Fernández, Y. R., Lisse, C. M., Kaufl, U., Peschke, H., Sibylle, B., Weaver, H. A., *et al.*: 2000, *Icarus* **147**, 145.

Fernández, Y. R., Meech, K. J., Lisse, C. M., A'Hearn, M. F., Pittichová, J., and Belton, M. J. S.: 2003, *Icarus* **164**, 481.

Finson, M. L., and Probstein, R. F.: 1968, *Astrophys. J.* **154**, 327.

Ishiguro, M., Watanabe, J., Usui, F., Tanigawa, T., Kinoshita, D., Suzuki, J., *et al.*: 2002, *Astrophys. J.* **572**, L117.

Ishiguro, M., Kwon, S. M., Sarugaku, Y., Hasegawa, S., Usui, F., Nishiura, S., *et al.*: 2003, *Astrophys. J.* **598**, L117.

Kharchenko, V., and Dalgarno, A.: 2001, *Astrophys. J.* **554**, L99.

Krasnopolsky, V. A., Christian, D. J., Kharchenko, V., Dalgarno, A., Wolk, S. J., Lisse, C. M., *et al.*: 2002, *Icarus* **160**, 437.

Lebofsky, L. A. and Spencer, J. R.: 1989, in R. P. Binzel, T. Gehrels, and M. S. Matthews (eds.), *Asteroids II*, University of AZ Press, Tucson, pp. 128–147.

Lisse, C. M., Christian, D. J., Dennerl, K., Meech, K. J., Petre, R., Weaver, H. A., *et al.*: 2001, *Science* **292**, 1343.

Lisse, C. M., A'Hearn, M. F., Farnham, T. L., Groussin, O., Meech, K. J., Fink, U., *et al.*: 2005, *Space Sci. Rev.* this volume.

Magnusson, P., Barucci, M. A., Drummon, J., Lumme, K., Ostro, S. J., Surdej, J., *et al.*: 1989, in R. P. Binzel, T. Gehrels, and M. S. Matthews (eds.), *Asteroids II*, University of AZ Press, pp. 66–98.

McFadden, L., Roundtree-Brown, M., Warner, E., McLaughlin, S., Behne, J., Ristvey, J., *et al.*: 2005, *Space Sci. Rev.* this volume.

Meech, K. J. and Svoreň, J.: in press, in M. Festou *et al.* (eds.), *Comets II*, University of AZ Press.

Meech, K. J., Hainaut, O. R., and Marsden, B. G.: 2004, *Icarus* **170**, 463.

Noll, K. S., McGrath, M. A., Trafton, L. M., Atreya, S. K., Caldwell, J. J., Weaver, H. A., *et al.*: 1995, *Science* **267**, 1307.

Notesco, G., Bar-Nun, A., and Owen, T.: 2003, *Icarus* **162**, 183.

Sugita, S., Schultz, P. H., and Hasegawa, S.: 2003, *JGR* **108**, 5140.

Sykes, M. V. and Walker, R. G.: 1992, *Icarus* **95**, 180.

Weaver, H. A., Feldman, P. D., Combi, M. R., Krasnopolsky, V., Lisse, C. M., and Shemansky, D. E.: 2002, *Astrophys. J.* **576**, L95.

Yeomans, D., Giorgini, J., and Chesley, S. R.: 2005, *Space Sci. Rev.* this volume.

DEEP IMPACT: THE ANTICIPATED FLIGHT DATA

KENNETH P. KLAASEN[1,*], BRIAN CARCICH[2], GEMMA CARCICH[2],
EDWIN J. GRAYZECK[3] and STEPHANIE MCLAUGHLIN[4]

[1] *Jet Propulsion Laboratory, California Institute of Technology, CA, U.S.A.*
[2] *Cornell University, Ithaca, NY, U.S.A.*
[3] *Goddard Space Flight Center, U.S.A.*
[4] *University of Maryland, College Park, MD, U.S.A.*
(*Author for correspondence; E-mail: kenneth.p.klaasen@jpl.nasa.gov)

(Received 18 August 2004; Accepted in final form 28 December 2004)

Abstract. A comprehensive observational sequence using the Deep Impact (DI) spacecraft instruments (consisting of cameras with two different focal lengths and an infrared spectrometer) will yield data that will permit characterization of the nucleus and coma of comet Tempel 1, both before and after impact by the DI Impactor. Within the constraints of the mission system, the planned data return has been optimized. A subset of the most valuable data is planned for return in near-real time to ensure that the DI mission success criteria will be met even if the spacecraft should not survive the comet's closest approach. The remaining prime science data will be played back during the first day after the closest approach. The flight data set will include approach observations spanning the 60 days prior to encounter, pre-impact data to characterize the comet at high resolution just prior to impact, photos from the Impactor as it plunges toward the nucleus surface (including resolutions exceeding 1 m), sub-second time sampling of the impact event itself from the Flyby spacecraft, monitoring of the crater formation process and ejecta outflow for over 10 min after impact, observations of the interior of the fully formed crater at spatial resolutions down to a few meters, and high-phase lookback observations of the nucleus and coma for 60 h after closest approach. An inflight calibration data set to accurately characterize the instruments' performance is also planned. A ground data processing pipeline is under development at Cornell University that will efficiently convert the raw flight data files into calibrated images and spectral maps as well as produce validated archival data sets for delivery to NASA's Planetary Data System within 6 months after the Earth receipt for use by researchers world-wide.

Keywords: Deep Impact, comet, Tempel 1, data set, calibration, data processing, data archive

1. Introduction

To achieve the Deep Impact (DI) mission goal of obtaining mankind's first look inside a comet, an instrument suite has been developed that will allow us to characterize the nucleus of Tempel 1 as it exists at the time we arrive, during the cratering event caused by our Impactor, and after the crater formation is complete. The compositional and dynamical nature of the crater ejecta will be determined, and the deep interior of the nucleus will be revealed in the walls and floor of the crater we create. The nature of Tempel 1's near-nucleus dust and gas coma will also be monitored before and after the impact event to help us understand its formation, geometry, composition, and evolution. The Deep Impact scientific data set will provide key measurements that will allow us to better determine what comets are made of and how they change throughout their lifetime.

To achieve Deep Impact's science objectives (see papers by A'Hearn *et al.*, Thomas and Veverka, Schultz and Ernst, Richardson *et al.*, Sunshine *et al.*, and Belton *et al.*, in this volume), measurements involving visible images and infrared (IR) spectral maps are required. Hampton *et al.* (this volume) describes the instruments being flown on Deep Impact and their measurement capabilities in a companion paper. The instruments on the main Flyby spacecraft comprise a High-Resolution Instrument (HRI) with a visible imager (HRIVIS) and an infrared spectrometer channel (HRI-IR) and a Medium-Resolution Instrument with a visible imager (MRI or MRIVIS). The Impactor carries a duplicate of the MRIVIS (except without a filter wheel) called the Impactor Targeting Sensor (ITS). The design of the Deep Impact mission and a summary of its spacecraft capabilities are contained in the paper by Blume *et al.* (this volume).

The measurements required for the DI instruments must span a wide range of time scales, spatial scales, and lighting conditions to fully characterize Temple 1 and the impact event. However, the extent of the data set is limited by the practical constraints of the spacecraft and instrument capabilities. Flying close to an active comet is also not without risk, and we have tried to ensure that we return all of the absolutely essential data prior to closest approach. Numerous tradeoffs have had to be made to define what we believe is the optimum data set that can be obtained within the existing constraints.

2. Constraints Limiting Data Acquisition

One constraint on the DI data set is imposed by the rate at which images and spectra can be read out of the instruments. Low-noise instrument performance can only be achieved if detector readout rates are limited. For the DI visible cameras, a full 1024×1024-pixel frame requires 1.42 s to read out. Overhead time to open and close the shutter, transfer the image from the active portion of the CCD array into its storage region, plus instrument internal timing margins increase the minimum frame time in this mode to 1.84 s even with zero-signal integration time. Adding filter wheel steps costs another 1.1 s per step. While frame times of a few seconds are normally perfectly acceptable, there are times, such as at the instant of impact, where the maximum time sampling rate is desired, requiring frame intervals of a fraction of a second.

Other camera modes are selectable that do provide shorter frame times at the expense of frame format size and the ability to use the shutter between frames to protect against any scene saturation causing charge to bleed into the storage register. Identical frames taken in immediate succession can be obtained slightly faster than successive frames with command parameters that are not identical. Minimum frame

TABLE I

DI VIS camera minimum frame times.

Frame format	Shuttered mode (s)	Unshuttered mode (s)
512×512	0.74	N/A
256×256	0.43	0.23
128×128	0.31	0.11
64×64	N/A	0.058

intervals for successive identical frames with zero integration time and no filter steps are given in Table I. The selected integration time must be added to these values to derive the actual frame intervals.

For the IR spectrometer, a full 512 spectral $\times$ 256 spatial frame takes 2.86 s with zero-added integration time (even with zero-added integration time, signal is integrated on the detector for the minimum frame time, if the detector is read out in an interleaved mode, or half the minimum frame time in the alternating readout modes). IR subframe modes with reduced spatial coverage are available that have shorter frame intervals: 512×128 takes a minimum of 1.43 s, and 512×64 takes 0.73 s. IR spectrometer data can be returned "unbinned" as well as in the normal 2×2 binned modes; see the companion paper by Hampton *et al.* (this issue) for more details on IR spectrometer data binning and its effects on spatial and spectral resolution.

The DI data set is also limited by the amount of data storage memory available onboard. When the data-taking rate is low and DSN coverage is ample, this buffer memory limit is not a real constraint. However, when we wish to take a large amount of data in a very short time, the buffer memory becomes a first-order constraint on what data can be captured. The Flyby spacecraft provides an allocation of 309 MB (megabytes) of science data storage for the Flyby instruments on each of two computer strings. Data can be directed to be stored on both strings (for redundancy in case a computer fails before the memory contents can be telemetered to the Earth) or to one or the other string. The Impactor spacecraft provides only about 100 MB of science data storage; however, data are not stored for any appreciable length of time on the Impactor during the time-critical period just prior to impact, so this limitation is of no concern. The Flyby computers have an additional allocation of 31 MB of storage for Impactor data; this allocation only becomes a constraint for times when we gather Impactor data while there is no ongoing downlink, e.g., during calibration activities.

The downlink data rates to the Earth determine how long it takes to empty the onboard data buffer. Data rates vary from a low of about 20 kbps (for example, when using the low-gain antenna shortly after launch or the high-gain antenna to a 34-m DSN station after encounter) to a maximum of 200 kbps (for example, over the high-gain antenna near encounter to a 70-m DSN station). Even at the maximum downlink rate, it takes about 4.5 h to empty one 309-MB memory. Since the high-priority impact-related observations are completed in less than 20 min, these data are limited to what can be stored in the onboard buffers. During the early portion of the approach to the comet when the mission plan provides only 34-m DSN coverage, mostly for only one 8-h pass per day, we are limited by the downlink rate to acquire no more than about 1 Gb/day. This limit goes up during the later portions of approach to about 3 Gb/day for continuous 34-m DSN coverage and to about 14 Gb/day for continuous 70-m DSN coverage.

After release of the Impactor, data are transmitted at 60 kbps from the Impactor to the Flyby spacecraft over an S-band radio link. Telemetry formatting overhead uses about 25% of this rate leaving no more than about 50 kbps for science data. During the time shortly before impact, the Impactor data are sent immediately to the ground by the Flyby as soon as they are received. This approach keeps the maximum amount of Flyby storage memory available for post-impact Flyby science data and also ensures that the Impactor images are returned to the Earth well before the most risky closest approach event. The Flyby imposes its own ~25% telemetry overhead on the Impactor data. Thus, to keep up with the flow of Impactor crosslink data, about 80 kbps of the available 200 kbps downlink rate must go to the Impactor up until the Impact event terminates the crosslink (this 80 kbps returns only 50 kbps of real Impactor science data).

The DSN tracking schedule in the current mission plan provides the coverage and data rates shown in Table II.

Spacecraft thermal constraints limit the allowed attitude of the vehicle with respect to the Sun direction. Normally, the spacecraft is oriented with its Y-axis pointed within 30° of the Sun. When the desired science targets require spacecraft

TABLE II

DI DSN tracking coverage plan.

Mission period	Allocated DSN (m)	Passes/day	S/C antenna	Data rate (kbps)
L to L + 7 d	34	3	Low gain	20
L+7 d to L+20 d	34	3	High gain	200
L+20 d to L+30 d	34	1	High gain	200
February 25	34	1	High gain	200
March 23	34	1	High gain	200
April 21	34	1	High gain	40
I-60 d to I-3 d	34	1	High gain	40
I-3 d to I-2 d	34	3	High gain	40
I-2 d to I+1 d	70	3	High gain	200
I+1 d to I+2 d	34	1	High gain	32
I+2 d to I+30 d	34	1/7 d	High gain	20

attitudes outside this range, the time permitted in those orientations is limited to 15 min out of any 4-h period to keep all subsystems within their allowable temperature range. This constraint comes into play during the early comet approach phase (prior to I-10 d) and for certain desirable calibration targets during cruise. The most significant impact of this constraint is that it imposes 4-h gaps in the rotation phase sampling of the nucleus during its early approach.

Finally, the pointing range of the high-gain antenna (HGA) is limited to the +X hemisphere. Therefore, real-time communication at high data rate is not possible for spacecraft orientations that place the Earth outside this hemisphere. This is the case during the post-flyby lookback phase. As a result, all lookback data must be stored in a buffer memory for later playback. This means that enough buffer memory must be either held open for lookback data (at the expense of impact crater data) or that some data stored in memory earlier must be played back and then written over by subsequent lookback data.

3. Data Return Strategy

The constraints on the DI data return strategy imposed by the available downlink data rates were discussed above. Our inability to return data as fast as we acquire it forces us to buffer most of the high-priority data associated with the impact event and its aftermath. The subset of data that we can return in near real time has been selected with care so that if the Flyby spacecraft should suffer damage due to a high-velocity dust impact near closest approach, we would have returned the data necessary to meet our mission success criteria. We call this our "live-for-the-moment" data return strategy. The data selected for immediate real-time playback include (a) the best-resolution color images and IR long-slit spectra of the entire nucleus pre-impact, (b) the best-resolution color images and IR long-slit spectra of the impact site pre-impact, (c) visible images and IR long-slit spectra of the impact site and the ejecta cone within ≤ 0.7 s of impact and at sampling intervals from 0.06 s to 30 s for the first 60 s thereafter plus two additional samples before I+660 s, (d) long-slit IR spectra of the coma before impact and after I+660 s, and (e) final highest-resolution images of the crater at resolutions of 2.1 and 1.4 m/pixel (HRI) and 10.3 and 7.2 m/pixel (MRI). All ITS data during the last hour prior to impact are also returned in real time.

The data selected for immediate real-time playback remain stored in buffer memory for a second playback after closest approach. The rest of the prime DI data that are not selected for real-time return are also stored in memory for later playback. The most critical data, including those right at impact, are stored redundantly on both computer strings. Some of the other data are stored on only a single string so as to increase the total amount of unique data returned in the most likely case that both computer strings remain alive and well. The data selected for non-redundant storage are all data that are functionally redundant. For example, we will chose to store every other sample of a time sampled series in alternate strings so that if one computer is lost, we still get reasonable time sampling, but if both computers survive, we get double the sampling frequency. Between I-1 h and closest approach, all the IR spectrometer data are stored redundantly, because a hardware feature disables the spectrometer data channel to the opposite computer string when data are stored only to one string. Resetting the data channel frequently was deemed too operationally complex to do during the critical encounter period.

Playback of the prime stored data (2×309 MB $= 618$ MB total) will only require about 9 h of downlink time to the scheduled 70-m DSN stations. The data should be successfully returned within about one day after closest approach interleaved between lookback imaging sessions.

4. Planned Data Set

The DI science data set that is planned is consistent with the mission and spacecraft capabilities and constraints and will, if successfully obtained, meet all of the science requirements of the mission. The data planned are summarized here by purpose and mission phase.

4.1. INSTRUMENT CALIBRATION

Instrument calibration data are planned to be obtained starting within a few days after launch, approximately monthly during the cruise to Tempel 1, and on either side of the impact event. The initial post-launch calibrations will take advantage of the only opportunity to observe targets that fill all, or a large fraction, of the instruments' fields of view (FOV), i.e., the Moon and the Earth. Radiometric, point-spread, modulation-transfer function, and in-field scattered light response calibrations will be performed with each instrument. Images of the Moon, the Earth, and at least one star will be acquired with the HRI and MRI cameras and the IR spectrometer nominally at L+3d (L = launch). The Moon and the Earth will overfill the HRI FOV, but will only partly fill the MRI and IR FOVs. These targets will be illuminated at about 90° solar phase angle. Small scans of the IR slit across portions of the targets are planned. A set of dark frames will be acquired with each instrument, and internal stimulus flat field images will also be acquired in each visible camera.

Scattered light tests involving long-exposure imaging with the Moon positioned at various angles close to, but outside, the instruments' FOVs are planned. This out-of-field scattered light calibration will be done with the MRI at L+9 d and by the HRI (both visible and IR) at L+30 d.

The Impactor spacecraft is not scheduled to be powered on until L+10 d, and the first ITS calibration occurs nominally on L+13 d. The ITS will also view the

Moon, the Earth, and a star; however, its FOV will be even less covered by the extended targets. Dark frames and internal stimulus images will be acquired by the ITS as well.

The focus of the HRI camera will be monitored using star images as the planned post-launch bakeout of moisture absorbed by its telescope composite structure progresses. Current plans include such images at L+14 d. In addition, calibrations using celestial bodies are planned by the optical navigation team at various times during the post-launch and cruise phases.

The cruise science calibrations will include dark frame sets, internal visible stimulus images, and imaging of various celestial sources. Star clusters will be included for calibrating any geometric distortion in the images. Radiometric standard sources will permit absolute radiometric response and linearity calibrations through each spectral filter as well as point-spread determinations of spatial resolution. A number of different standard sources of various spectral classes and brightnesses are used for cross correlation and to improve spectral response modeling across filters with widely varying response rates. Stars, nebulae, and galactic clusters with known spectral emission lines will be used for IR spectral calibration. The planned set of calibration sources is summarized in Table III.

TABLE III

Celestial sources planned for use in DI inflight instrument calibrations.

Target	Spectral class	V_{mag}	RA	Dec	Viewability	Notes
HD60753	B2	6.7	113.36	−50.58	Continuous	Hot standard, few absorption lines, VIS
i Car	B3	3.9	137.82	−62.32	Continuous	Hot standard, few absorption lines, VIS
bet Hyi	G2	2.8	6.44	−77.25	Continuous	Solar analog, VIS and IR
16 CygA	G1.5	5.96	295.45	50.53	Continuous	Solar analog, VIS
Vega	A0	0.03	279.23	38.78	Continuous	Bright standard, VIS and IR
Achernar	B3	0.5	24.43	−57.24	Continuous	Bright standard, VIS
Canopus	F0	−0.72	95.99	−52.70	Continuous	Bright standard, VIS
M11			282.77	−6.27	Feb–April 2005	Cluster for geometric calibration, VIS
Pleiades			56.5	24.2	Feb/Mar 2005	Bright cluster for geometric calibration, VIS and IR
NGC3114			150.68	−60.12	Continuous	Cluster for geometric calibration, VIS and IR
NGC7027			316.76	42.23	Apr 2005 and later	Best IR spectral calibrator
NGC6543			269.64	66.63	Continuous	IR spectral calibrator
Sirius	A1	−1.45	101.28	−16.72	Feb–May 2005	IR radiometric standard

The final pre-impact calibrations using celestial sources are planned for I-40 d and I-5 d. A final set of dark frames and internal visible stimulus flat fields is planned for each instrument at I-10 h. The final post-impact calibration will occur at about I+3 d and will again use celestial sources.

4.2. APPROACH DATA

Approach science data acquisition will begin at I-60 d. The scientific objectives of the approach phase are:

- Determine the shape of the nucleus
- Map albedo, color, and spectral variations over the entire surface for indications of heterogeneity
- Determine the rotational state of the nucleus
- Identify large-scale structures in the coma and trace them to their origin on the surface
- Monitor the nucleus to properly characterize its state of activity at the time of impact
- Search for satellites or escaping objects for a possible nucleus mass constraint
- Map the evolution of inner coma structure over a full rotation period.

Regular comet sampling with the Flyby instruments is planned covering the coma and the nucleus as it rotates. The solar phase angle of the nucleus increases by about 0.5°/day, from 28° at the start of the approach phase to 60° at I-7 d. The nominal 6-km diameter nucleus will first be spatially resolved (diameter equal to one pixel) by the HRI camera about 3 d before encounter. The nucleus is expected to remain unresolved by the MRI and IR spectrometer throughout this period. Table IV summarizes the data collection plan. All data are returned uncompressed within 24 h of acquisition.

TABLE IV

DI approach data collection plan.

Time period	Sampling frequency	HRIVIS	MRIVIS	IR	Data volume per sample (Mb)
I-60 d to I-3 d	1 every 4 h	256^2, 8-color	256^2, 8-color	64^2, 512 wavelengths	50
I-3 d to I-2 d	1 every 2 h	256^2, 8-color	256^2, 8-color	64^2, 512 wavelengths	50
I-2 d to I-1 d	1 every hour	512^2, 8-color	512^2, 8-color	512×50, 1024 wavelengths	487

4.3. IMPACTOR DATA

Shortly after its release from the Flyby at I-1 d, the Impactor ITS begins acquiring data and telemetering it over the S-band crosslink back to the Flyby. The initial data are for engineering and navigation purposes. Science imaging begins at I-22 h with a pair of full-frame images – one exposed for the nucleus and one exposed for the dimmer coma. Similar image pairs are obtained every 2 h up to I-12 h. Navigation imaging is obtained during the intervals between the science image pairs. The nucleus should equal one pixel in diameter at about I-16 h. At I-12 h, a demonstration of the final 2 min of science data acquisition before impact is conducted to verify that the Impactor will execute this critical portion of its observing sequence correctly and to allow for a fix attempt if for some unexpected reason it does not. At I-10 h, just prior to acquiring the standard image pair, an ITS dark-frame and internal stimulus calibration sequence is scheduled.

Table V summarizes the ITS observing sequence after I-20 h. Note that the ITS images serve to some extent as a backup to the MRI images from the Flyby except that the ITS has only a clear filter. The nucleus will fill the selected ITS FOV starting at about I-4 min. The ITS boresight is pointed at the expected impact

TABLE V

ITS data collection plan.

Time	Pixel scale (m)	Prime
I-20 h to I-12 h	7200–4320	1024^2 normal & long exp pair every 2 h
I-12 h	4320	Pre-impact demo; I-2 min to Impact
I-10 h		Calibration
I-10 h to I-8 h	3600–2880	1024^2 normal & long exp pair every 2 h
I-7 h to I-4 h	2520–1440	1024^2 normal & long exp pair every 1 h
I-3 h to I-1 h	1080–363	1024^2 normal & long exp pair every 0.5 h
I-1 h to I-34 m	363–210	1024^2 every 10% of time to impact (6 frames)
I-30 m	181	256^2 long exp
I-30 m to I-19 m	181–116	1024^2 every 10% of time to impact (6 frames)
I-17 m	105	512^2
I-16 m	95	512^2 normal and 256^2 long exp
I-14 m to I-9 m	87–54	512^2 every 10% of time to impact (6 frames)
I-8 m	49	512^2 normal and 256^2 long exp
I-7 m to I-5 m	45–30	512^2 every 10% of time to impact (5 frames)
I-278 s	28	512^2 normal and 256^2 long exp
I-250 s to I-70 s	25–7	256^2 every 10% of time to impact (17 frames)
I-70 s to I-17 s	7–1.8	128^2 every 10% of time to impact (16 frames)
I-16 s to I-13 s	1.6–1.3	64^2 about every 1 s (4 frames)
I-12 s to I+2 s	1.2–0.1	64^2 every 0.7 s (20 frames)

site until I-4 min, at which time it is repointed to align with the relative velocity
vector. During the last 30 s before impact, the ITS imaging rate is the maximum that
can be transmitted across the crosslink. The last image expected to be transmitted
in its entirety before impact is taken at about I-2 s and will have a scale of about
20 cm/pixel. Chances are larger than 50:50 that images taken after I-10 s could
be lost due to dust impacts causing pointing errors. All ITS data are returned
compressed from 14 to 8 bits using one of the selectable lookup table conversions.

4.4. PRE-IMPACT FLYBY DATA

Flyby science data acquired during the last day before impact comprises sets of color
images and IR spectrometer scans taken at regular intervals as spatial resolution
increases. The nucleus will be one pixel in diameter for the MRI and IR spectrometer
at about I-16 h. Data are returned immediately after acquisition up until I-5 h. VIS
data up to that time are all returned uncompressed. IR data prior to I-5 h are a mixture
of compressed and uncompressed data, some binned 2×2 and some unbinned.

At I-5 h, the science storage memory is erased completely in preparation for
storing the highest resolution, critical data bracketing the impact event. This erasure
is done under sequence control to ensure it happens on time; data played back in
the last couple of hours prior to this erasure will be at some risk since there will
be no time to request a replay if they are not successfully downlinked on the
first try. Almost all data between I-5 h and closest approach are stored for post-
flyby playback, some redundantly and some not. A few selected data sets taken
prior to I-1 h are designated for real-time playback with immediate erasure from
memory after they are played back. Hourly clear-filter long-exposure frames are
planned in an attempt to detect the dark limb of the nucleus for improved shape
modeling.

The frequency of data sampling accelerates as resolution improves and impact
approaches. Scans of the IR slit across the nucleus are performed to generate spectral maps. The ends of the IR slit typically extend well beyond the nucleus to give
spectral measurements of the coma (albeit likely to be at reduced signal-to-noise
ratio due to short exposures). Tables VI–VIII summarize the Flyby science data
taking scenarios for each instrument during the last day prior to impact. The column labeled "Prime" indicates data stored for post-flyby playback. The column
labeled "R/T" indicates the data that are scheduled for immediate playback. From
about I-13 min through closest approach, the queue of data for immediate playback contains a backlog. The table entries are color coded to indicate color sets,
uncompressed data, and data stored on only a single computer string. Just prior to
impact, the nucleus will fill about 1/3 of the HRI FOV and about 70 MRI and IR
spectrometer pixels. Except during IR scans, the instrument boresights are targeted
to the predicted impact point; no mosaicking is planned. Expected mean signal-to-
noise ratios (SNRs) and 3σ smear levels for the selected exposure times are given

TABLE VI

HRIVIS pre-impact data collection plan.

Time	Pixel scale (m)	Prime	R/T
I-25 hr	1830		3-clear + 7 color 512^2 set
I-23 hr	1697		3-clear + 7 color 512^2 set
I-21 hr	1540		3-clear + 7 color 512^2 set
I-19 hr	1400		3-clear + 7 color 512^2 set
I-17 hr	1250		3-clear + 7 color 512^2 set
I-15 hr	1100		3-clear + 7 color 512^2 set
I-13 hr	960		3-clear + 7 color 512^2 set
I-12 hr	890		3-clear 512^2
I-11 hr	815		3-clear + 7 color 512^2 set
I-10 hr			Calibration
I-10 hr	740		3-clear 512^2
I-9 hr	670		3-clear + 7 color 512^2 set
I-8 hr	600		3-clear 512^2
I-7.5 hr	560		3-clear 512^2
I-7 hr	525		3-clear + 7 color 512^2 set
I-6.5 hr	490		3-clear 512^2
I-6 hr	450		3-clear + 7 color 512^2 set
I-5 hr	380		3-clear + 7 color 512^2 set
I-4.5 hr	345	3-clear 512^2 + 256^2 long exp	3-clear 512^2 + 256^2 long exp
I-4 hr	310	3-clear + 7 color 512^2 set	3-clear + 7 color 512^2 set
I-3.5 hr	270	256^2 long exp	3-clear 512^2 + 256^2 long exp compr
I-3 hr	235	3-clear + 7 color 512^2 set	3-clear + 7 color 512^2 set
I-2.5 hr	200	256^2 long exp	3-clear + 7 color 512^2 set + 256^2 long exp compr
I-2 hr	163	3-clear + 7 color 512^2 set	3-clear + 7 color 512^2 set
I-1.5 hr	127	256^2 long exp	3-clear + 7 color 512^2 set + 256^2 long exp compr
I-1 hr	90	3-clear + 7 color 512^2 set	3-clear + 7 color 512^2 set
I-40m	66	128^2 8-color	128^2 8-color
I-27m	50	256^2 8-color +256^2 long exp compr	256^2 8-color + 256^2 long exp compr
I-18m	39	256^2 8-color	256^2 8-color
I-10m	29	8-color set, 7 @ 512^2, 1 @ 1024^2	
I-3.5m	21	8-color set, 7 @ 512^2, 1 @ 1024^2	
I-32s to I-7s	18	5-color set, 4 @ 512^2, 1 @ 1024^2	4-color set @ 512^2, 1 @ 1024^2
I-7s to I-3s	17	1024^2	

=>color set

=>one-string storage

=>uncompressed data

=>uncompressed and one-string storage

TABLE VII

MRIVIS pre-impact data collection plan (see Table VI for key to color coding).

Time	Pixel scale (m)	Prime	R/T
I-26 hr	9520		8 color 512^2 set
I-22 hr	8070		8 color 512^2 set
I-20 hr	7350		8 color 512^2 set
I-18 hr	6620		8 color 512^2 set
I-16 hr	5900		8 color 512^2 set
I-14 hr	5170		8 color 512^2 set
I-12 hr	4440		8 color 512^2 set
I-12 hr	4300		5 512^2 long-exposure gas filters
I-10 hr	3720		8 color 512^2 set
I-10 hr			Calibration
I-8 hr	2990		8 color 512^2 set
I-7 hr	2630		8 color 512^2 set
I-6 hr	2270		8 color 512^2 set
I-6 hr	2150		5 512^2 long-exposure gas filters
I-5 hr	1900	8 color 512^2 set	8 color 512^2 set
I-4 hr	1540		8 color 512^2 set
I-3 hr	1180	8 color 512^2 set	8 color 512^2 set
I-2.5 hr	1000		8 color 512^2 set
I-2 hr	815	8 color 512^2 set	8 color 512^2 set
I-1.5 hr	635		8 color 512^2 set
I-1 hr	453	8 color 512^2 set	8 color 512^2 set
I-40m	330	128^2 8-color	128^2 8-color
I-27m	250	256^2 8-color	256^2 8-color
I-18m	194	256^2 long exp	256^2 long exp
I-18m	194	256^2 8-color	256^2 8-color
I-10m	144	8-color set, 1024^2	3-color set, 1024^2
I-3.5m	105	8-color set, 7 @ 512^2, 1 @ 1024^3	
I-32s to I-7s	88	7-color set, 1024^2	4-color set @ 1024^2
I-7s to I-3s	87	1024^2	1024^2

in Table IX. For the IR spectrometer, the table lists several selected wavelengths and predictions at the longer wavelengths for scenes with surface temperatures of 200 K outside the anti-saturation filter that covers the central one-third of the slit and for surface temperatures of 300 K behind the anti-saturation filter (300 K surfaces outside the filter will saturate at these wavelengths). Note that when using the minimum exposure times for IR subframe modes, the smear and SNR will be somewhat lower (by roughly the square-root of the exposure time ratio).

4.5. FLYBY DATA AT IMPACT

The primary science objective during the impact event is to observe at high time resolution the physics of the event and the nature and geometry of the material

TABLE VIII

IR spectrometer pre-impact data collection plan.

Time	Binned pixel scale (m)	Prime	R/T
I-25 to I-11 hr	9160 - 4077		512-hi column x 50-wide scan every 2 hr; uncompressed
I-10 hr			Calibration
I-9 hr	3352		512-hi column x 50-wide scan
I-7 hr	2626		256-hi column x 50-wide scan; uncompressed
I-6 hr	2263		512-hi column x 50-wide scan
I-5 hr	1900	Clear NVM after last sample	256-hi column x 50-wide scan
I-4 hr	1440	128-hi column x 50-wide scan	256-hi column x 50-wide scan; uncompressed
I-3 hr	1080	128-hi column x 50-wide scan	512-hi column x 50-wide scan
I-2.5 hr	980		256-hi column x 50-wide scan; uncompressed
I-2 hr	810	128-hi column x 50-wide scan	256-hi column x 50-wide scan
I-1.5 hr	615		512-hi column x 50-wide scan
I-1 hr	440	128-hi column x 50-wide scan	256-hi column x 50-wide scan
I-36m	300	64-hi column x 40-wide scan over full nucleus	
I-19m	200	64-hi column x 64-wide scan over full nucleus; central 32 slits stored uncompressed	64-hi column x central 32-wide scan over full nucleus
I-14m	167	512-hi column unbinned on nucleus	
I-13m	166	512-hi column unbinned off edge of nucleus	
I-13m	160	256-hi column x 64-wide partial scan of coma	256-hi column x central 48-wide partial scan of coma
I-10m	146	512-hi column unbinned off edge of nucleus	
I-6m	125	64-hi column x 64-wide scan over full nucleus	
I-190s to I-66 s	95	128-hi column x 84-wide scan over full nucleus	One 128-hi column centered on impact site
I-35s to I-9 s	87	256-hi columns every 2.88 s	256-hi columns every 2.88 s

=>uncompressed, unbinned data
=>unbinned, compressed

initially ejected from the crater. Flyby instruments are trained on the predicted impact point and commanded to acquire data at high rate starting a few seconds before the expected time of impact and for some time thereafter. The predicted uncertainty in the actual time of impact calculated onboard by the autonav algorithms is about 3 s. MRIVIS images are taken at the maximum possible rate of 0.06 s by restricting its format to 64^2 in order to obtain the highest possible time resolution of the

TABLE IX

Predicted smear and nucleus signal-to-noise ratios for nominal exposures.

HRIVIS	Nominal exposure time (ms)	3σ smear (pixel)	Mean SNR
1-Clear	100	1.5	89
2-450 nm	500	2.75	69
3-550 nm	400	2.65	87
4-350 nm	700	3	48
5-950 nm	700	3	52
7-750 nm	400	2.65	74
8-850 nm	500	2.75	64
9-650 nm	400	2.65	85
MRIVIS			
1-Clear	100	0.4	174
2-514 nm	3000	0.8	144
3-526 nm	2500	0.75	89
4-750 nm	500	0.55	163
5-950 nm	2000	0.7	166
7-387 nm	3000	0.8	61
8-345 nm	3000	0.8	29
9-309 nm	7000	0.9	23
IR Spectrometer			
1.2 μm	2880	1.56	27
2 μm	2880	1.56	55
2.8 μm (300 K, filt)	2880	1.56	102
3.6 μm (300 K, filt)	2880	1.56	287
4.6 μm (300 K, filt)	2880	1.56	17
2.8 μm (200 K, no filt)	2880	1.56	42
3.6 μm (200 K, no filt)	2880	1.56	37
4.6 μm (200 K, no filt)	2880	1.56	55

impact event and its photometric light curve. The FOV should still be large enough to ensure capturing the impact site given the expected level of pointing errors and uncertainty in the impact site location (in fact, the selected FOV covers nearly the entire nucleus). This imaging rate is maintained up to 6 s after the nominal impact time. During this period, the HRI images are taken one-fourth as often but with a 256^2 image format. This approach yields comparable areal coverage as the MRI but with five times the spatial resolution. The IR spectrometer meanwhile is obtaining fixed-slit spectra centered on the predicted impact site every 0.72 s, its fastest frame rate, using a 64-pixel slit length.

The onboard data storage limitations prohibit continuing sampling at these high rates indefinitely, and we also wish to expand the instrument FOVs to capture more of the outflowing ejecta cone as time progresses. Therefore, we begin to step up the frame formats and decrease the sampling rate as time from impact increases. Tables X–XII list the data acquisition plan from impact until I+24 s for each instrument.

TABLE X

HRIVIS impact data (see Table VI for key to color coding).

Time	Pixel scale (m)	Prime	R/T
I-3s to I-0.5s	17	256^2 every 0.24 s	256^2 every 0.72 s
I-0.5s to I+1s	17	256^2 every 0.24 s	256^2 every 0.24 s
I+1s to I+7s	17	256^2 every 0.24 s	256^2 every 0.72 s
I+8s to I+9s	17	256^2 every 0.72 s	256^2 @ I+8s
I+9.5s	17	256^2	
I+10s	17	256^2	256^2
I+11s to I+15s	17	512^2 every 1.02s	512^2 @ I+13s
I+15s to I+20s	17	512^2 every 1.1s	512^2 @ I+16s
I+21s	17	512^2	
I+22s to I+24s	17	512^2 every 1.1s	512^2 @ I+22s

TABLE XI

MRIVIS impact data (see Table VI for key to color coding).

Time	Pixel scale (m)	Prime	R/T
I-3s to I-0.5s	86	64^2 every 0.06 s	64^2 every 0.24 s
I-0.5s to I+1s	86	64^2 every 0.06 s	64^2 every 0.06 s
I+1s to I+6s	85	64^2 every 0.06 s	64^2 every 0.24 s
I+7s to I+9s	85	128^2 every 0.12 s	128^2 every 1 s
I+10s	85	256^2	
I+11s to I+17s	84	256^2 every 0.72s	256^2 every 4 s
I+17s to I+20s	84	512^2 every 1.1s	
I+21s	84	512^2	
I+22s to I+24s	84	512^2 every 1.1s	512^2 @ I+22s

TABLE XII

IR spectrometer impact data (see Table VIII for key to color coding).

Time	Pixel scale (m)	Prime	R/T
I-6 s to I+4 s	86	64-hi column every 0.72 s	64-hi column every 0.72 s
I+4 s to I+15 s	85	64-hi column every 0.72 s	64-hi column every 1.44 s
I+15 s to I+24 s	83	128-hi column every 1.44 s	128-hi column every 2.88 s

A subset of the impact data is selected for real-time downlink by not selecting every time sample. All the data are compressed to 8 bits except for HRI and MRI images at I+10 s and I+21 s.

4.6. POST-IMPACT DATA

As the crater develops and ejecta continues to flow outward, we intend to monitor this continuing process (crater formation is nominally expected to be complete within about 4 min). Sampling with all the Flyby instruments continues; however, the sampling rate gradually decreases and the FOVs are progressively enlarged so that they extend off the edges of the nucleus to capture the near-nucleus coma. The first full-format images are taken at I+24 s. The instruments remain pointed at the nominal impact site until I+50 s when a small 11-slit-wide scan across the impact site is done to obtain two-dimensional IR spectrometer coverage. This scan is repeated again at I+87 s. Color imaging resumes with both the HRI and MRI at I+67 s, and periodic color sets are obtained thereafter interspersed with 1-color images all centered on the nominal impact site (no mosaicking). Most of the 1-color images are stored in only one computer string with alternate images going to string A and string B. A few selected images are saved uncompressed.

Starting at I+125 s until I+177 s, the IR slit is again nominally pointed and held at the predicted impact point. At the end of this period, a full nucleus scan of the IR slit is performed that lasts about 5 min. The first post-impact unbinned spectra are obtained at each end and once in the middle of this scan. HRI and MRI color sets centered on the impact site are obtained and returned in real time at the end of this scan. The nucleus fills the full HRI FOV at about I+550 s.

When these color sets are completed, the instrument boresights are pointed off the nucleus to obtain coma data. Three IR spectra with a 256-pixel slit are obtained with long exposures at different locations in the coma out to about 3 km from the center of the nucleus. Following this, the last highest-resolution HRI and MRI color sets are obtained pointed at the nominal impact site. Color imaging is avoided during the last 50 s to keep filter wheel vibration from degrading the highest resolution images. After completing the last color imaging, the last IR scan of the 300-m-wide region centered on the crater is performed. This scan is completed about 25 s before the spacecraft must go into its shield mode (SM) orientation to protect itself from dust impact during the closest approach period. During the course of this scan, one 512^2 image each from the HRI and MRI is selected for real-time return; these images are the last ones whose playback will be completed prior to closest approach. From SM-25 s until SM, the instruments are again pointed directly at the nominal impact site (no mosaicking). Regularly spaced images and IR spectra are taken until entering shield mode with the second-last images being stored uncompressed and the last, highest-resolution images being selected for real-time return as well as being stored for later playback. Their playback will not be completed until about

TABLE XIII

HRIVIS post-impact data (see Table VI for key to color coding).

Time	Pixel scale (m)	Prime	R/T
I+24s to I+36s	17	1024^2 @ I+24s, 512^2 every ~2.4s	512^2 @ I+33s
I+36s	16	1024^2	
I+39s to I+67s	16	512^2 every 10% of time since Impact; 1024^2 every 40% time since Impact	512^2 @ I+58s
I+67s to I+84s	16	4-color 512^2 + 1024^2	
I+84s to I+108s	15	512^2 every 10% of time since Impact	
I+108s to I+133s	15	4-color 512^2 + 1024^2	512^2 @ I+123 s
I+141s	14	1024^2	
I+155s to I+175s	14	4-color 1024^2 set	
I+185s to I+225s	13	1024^2 every 21s	
I+248s to I+273s	12	5-color 1024^2 set	
I+295s to I+329s	11	1024^2 every 29s	
I+356s to I+372s	10	4-color 512^2 set	512^2 @ I+364s
I+404s to I+449s	9 - 8	1024^2 every 40s	
I+465s to I+479s	7	4-color 1024^2 set	
I+508s	7	1024^2	
I+558s to I+578s	6	4-color 1024^2 set	
I+623s to I+637s	4.5	4-color 512^2 set	4-color 512^2 set
I+660s to SM-100s	4 - 3	1024^2 every 4s	
SM-100s to SM-68s	3	8-color 1024^2 set	
SM-68s to SM-40s	2.8	1024^2 every 2 s	
SM-40s	2	512^2	512^2
SM-40s to SM-28s	2	1024^2 every 2 s	
SM-28s to SM-4s	1.9 - 1.5	1024^2 every 2 s	
SM-2s	1.5	1024^2	
SM	1.4	1024^2	1024^2

158 s later (108 s after closest approach, about the time of crossing the orbit of Tempel 1). The nucleus will approximately equal the MRI FOV at the time of shield-mode entry.

Tables XIII–XV present the planned science data sets for each instrument during the post-impact period.

4.7. SHIELD-MODE AND LOOKBACK DATA

After entering shield mode, the spacecraft remains in this attitude until about 22 min after closest approach (CA) when the dust-impact hazard zone has been safely passed. While in this attitude, five long-slit IR spectra are obtained during the first

TABLE XIV

MRIVIS post-impact data (see Table VI for key to color coding).

Time	Pixel scale (m)	Prime	R/T
I+24s to I+36s	83	1024^2 @ I+24s, 512^2 every ~2.4s	512^2 @ I+33s
I+36s	82	1024^2	
I+39s to I+67s	81	512^2 every 10% of time since Impact; 1024^2 every 40% time since Impact	512^2 @ I+61s
I+67s to I+84s	79	4-color 512^2 + 1024^2	
I+84s to I+108s	76	512^2 every 10% of time since Impact	
I+108s to I+133s	74	4-color 512^2 + 1024^2	512^2 @ I+123 s
I+141s	70	1024^2	
I+155s to I+175s	68	4-color 1024^2 set	
I+185s to I+225s	63	1024^2 every 21s	
I+248s to I+273s	60	5-color 1024^2 set	
I+295s to I+329s	53	1024^2 every 29s	
I+356s to I+372s	49	4-color 512^2 set	512^2 @ I+364s
I+380s to I+449s	41	1024^2 every 40s	
I+465s to I+485s	38	4-color 1024^2 set	
I+508s	35	1024^2	
I+558s to I+578s	29	4-color 1024^2 set	
I+623s to I+637s	23	4-color 256^2 set	4-color 256^2 set
I+660s to SM-100s	18	1024^2 every 4s	
SM-100s to SM-52s	14	8-color 1024^2 set	
SM-52s to SM-40s	12	1024^2 every 2 s	
SM-40s	10	512^2	512^2
SM-40s to SM-28s	10	1024^2 every 2 s	
SM-28s to SM-4s	9 - 8	1024^2 every 2 s	
SM-2s	7	1024^2	
SM	7	1024^2	1024^2

minute after entering shield mode using long integration times to sample the near nucleus coma emissions. Since the spacecraft is traveling at 10 km/s relative to the comet and the pointing attitude is fixed, even with the minimum available integration time of 2.88 s, the IR spectra would be smeared over nearly 30 km of the coma in the along-track direction. Since good spatial resolution is not possible anyway, longer integration times of 7 and 14 s are planned to allow lower detection limits for any faint coma emissions that persist over many kilometers of the coma. The last such spectrum will be obtained when the spacecraft is near closest approach and the IR spectrometer FOV is looking at a point about 500 km behind the nucleus on a line parallel to the spacecraft flight direction and passing through the nucleus.

TABLE XV

IR spectrometer post-impact data (see Table VIII for key to color coding).

Time	Binned pixel scale (m)	Prime	R/T
I+24 s to I+32 s	83	128-hi column every 1.44 s	128-hi column every 2.88 s
I+32s to I+50s	81	256-hi column every 2.88 s	256-hi column at I+35s
I+50s to I+67s	80	128-hi column x 11-wide crater scan	One 128-hi column
I+70s	79	256-hi column	
I+77s	78	256-hi column	
I+84s	77	256-hi column	
I+87s to I+107s	76	128-hi column x 11-wide crater scan	One 128-hi column
I+110s to I+154s	71	256-hi column every 10% time from impact	
I+167s	69	256-hi column	
I+180s	68	512-hi column unbinned at edge of nucleus	
I+188s to I+246s	65	128-hi column x 40 wide scan of first part of nucleus	
I+250s	61	512-hi column unbinned at middle of nucleus	
I+255s to I+484s	60 - 38	256-hi column x 80 wide continuation scan of remainder of nucleus	
I+490s	37	512-hi column unbinned at edge of nucleus	
I+497s to SM-142s	37 - 21	256-hi column x 60 wide scan of central 1/4 nucleus	
SM-130s to SM-110s	18	Three 256-hi columns in coma; middle one uncompressed	One 256-hi column in coma
SM-98 s to SM-26 s	16 - 9	256-hi column every 2.88 s; 25-wide scan of 3x crater diameter	64-hi column every 2.88 s; 25-wide scan of 3x crater diameter
SM-23 s to SM-3 s	8	256-hi column every 2.88 s	
SM-3s to SM	7	256-hi column	256-hi column

At CA+22 min, the spacecraft begins an attitude maneuver to point the instruments back toward the nucleus. This maneuver takes about 9 min to complete. At that time, the first lookback science data of the nucleus and its surroundings are acquired. Eight-color sets are taken with both the HRI and MRI along with multi-frame sets (for co-adding) in the clear-filter using both normal and extended exposure times to image well both the nucleus and the coma as it evolves and to potentially locate and track any large debris from the impact that might be in orbit about the nucleus. The nucleus easily fits within the HRI FOV at this time and subtends only about 30 MRI pixels. These images should also map new areas of the nucleus surface at better resolution than on approach and

TABLE XVI

HRIVIS shield-mode and lookback data (see Table VI for key to color coding).

Time	Pixel scale (m)	Prime	R/T
SM+31m	36	11 1024^2 (3 long exp)	
SM+38m	44	6 1024^2 (3 long exp)	
SM+41m	48	6 1024^2 (3 long exp)	
SM+46m	56	6 1024^2 (3 long exp)	
SM+51m	60	6 1024^2 (3 long exp)	
SM+56m	66	6 1024^2 (3 long exp)	
SM+1 hr	73	6 1024^2 (3 long exp)	
SM+2 hr	145		18 1024^2 (5 long exp)
SM+3 hr	218		18 1024^2 (5 long exp)
SM+4 hr	290		18 1024^2 (5 long exp)
SM+5 hr	363		18 1024^2 (5 long exp)
SM+6 hr	435		18 1024^2 (5 long exp)
SM+12 hr	871		18 1024^2 (5 long exp)

help constrain the three-dimensional shape of the nucleus by mapping newly lit limbs as the nucleus rotates. IR long-slit spectra are taken at three coma positions, and a scan is done across the nucleus. The departure solar phase angle is about 116°.

HRI clear-filter sets of normal and extended exposures are acquired approximately every 5 min for the next 30 min (out to 1 h after CA). IR spectral sampling is repeated at about CA+40 min and CA+60 min. HRI color and multi-frame clear sets are repeated every hour out to CA+6 h and a final HRI sample is taken at CA+12 h. MRI color sets and multi-frame clear sets and IR coma samples at 3 positions (unbinned) are scheduled every 2 h out to CA+6 h and every 6 h out to CA+60 h to cover just over one full nucleus rotation.

Tables XVI–XVIII list the shield-mode and lookback data sets for each instrument. The data up to CA+1 h are stored as prime data for later playback. At this point the memory buffer is completely full of science data. Starting after the CA+1 h data collection, a spacecraft attitude maneuver is scheduled to allow playback and erasure of some of the initial lookback data from the memory buffer to free up space for the next lookback samples at CA+2 h. Attitude maneuvers are required for data return in this phase because the HGA is pointed away from the Earth in the lookback data-taking attitude. This pattern is repeated between each of the remaining lookback samples, which is why these data are listed only in the real-time column in Tables XVI–XVIII. After CA+6 h, the intervals between lookback samples increase to 6 h. The interleaved downlink sessions then become long enough to playback not only the previous lookback data set but also some of the prime data stored in memory. Playback of the prime data should be complete by CA+24 h.

TABLE XVII

MRIVIS shield-mode and lookback data (see Table VI for key to color coding).

Time	Pixel scale (m)	Prime	R/T
SM+30m	181	18 1024^2 (5 long exp)	
SM+2h	725		18 1024^2 (5 long exp)
SM+4h	1451		18 1024^2 (5 long exp)
SM+6h	2177		18 1024^2 (5 long exp)
SM+12h	4354		18 1024^2 (5 long exp)
SM+18h	6531		18 1024^2 (5 long exp)
SM+24h	8709		18 1024^2 (5 long exp)
SM+30h	10886		18 1024^2 (5 long exp)
SM+36h	13063		18 1024^2 (5 long exp)
SM+42h	15240		18 1024^2 (5 long exp)
SM+48h	17418		18 1024^2 (5 long exp)
SM+54h	19595		18 1024^2 (5 long exp)
SM+60h	21772		18 1024^2 (5 long exp)

5. Data Calibration

To be able to make full scientific use of the data gathered by the DI instruments during the mission, the raw data numbers (DNs) returned from each pixel in each image or spectrum must be converted to absolute scientific units of scene radiance or scene reflectance. The effective wavelength of the photons that generated the signal in each pixel must be known. In addition, it must be possible to determine from where in the scene or its surroundings the photons originated that produced the signal in each pixel. Accomplishing these functions is the goal of data calibration.

Figures 1 and 2 illustrate the planned DI radiometric calibration data processing pipeline. No correction for any instrument coherent noise is included in this plan since, although the final detailed analysis of the ground calibration data to search for evidence of coherent noise is not yet compete, no signs of such noise have yet been observed. While some evidence of electrical crosstalk between CCD quadrants has been observed for extremely high signal levels, this effect has not yet been fully characterized, and no noise removal algorithm has yet been derived. No correction for any spectral crosstalk in the IR spectrometer is currently included. While we found significant spectral crosstalk in the instrument during initial ground calibrations, steps were taken to eliminate its cause. Detailed analysis of any remaining crosstalk is not yet complete; however, no evidence of any obvious crosstalk was seen after the instrument corrections were implemented. The steps of the pipeline process are discussed in the following paragraphs.

To obtain the highest accuracy, calibration data are ordinarily returned uncompressed. Calibration correction files are then constructed using full 14-bit accuracy.

TABLE XVIII

IR spectrometer shield-mode and lookback data (see Table VIII for key to color coding).

Time	Binned pixel scale (m)	Prime	R/T
SM+3s to SM+21s	7	2 256-hi columns, 7-s exposure	
SM+24s to SM+1 min	5 - 30	256-hi columns every 14 sec, 14-s exposure	
SM+33m	190	256-hi columns at 3 coma positions plus 60-slit scan across nucleus	
SM+42m	250	256-hi columns at 3 coma positions plus 50-slit scan across nucleus	
SM+62m	375	256-hi columns at 3 coma positions plus 30-slit scan across nucleus	
SM+2 hr	744		512-hi columns at 3 coma positions
SM+4 hr	1470		512-hi columns at 3 coma positions
SM+6 hr	2190		512-hi columns at 3 coma positions
SM+12 hr	4370		512-hi columns at 3 coma positions
SM+18 hr	6550		512-hi columns at 3 coma positions
SM+24 hr	8725		512-hi columns at 3 coma positions
SM+30 hr	10900		512-hi columns at 3 coma positions
SM+36 hr	13080		512-hi columns at 3 coma positions
SM+42 hr	15260		512-hi columns at 3 coma positions
SM+48 hr	17435		512-hi columns at 3 coma positions
SM+54 hr	19615		512-hi columns at 3 coma positions
SM+60 hr	21790		512-hi columns at 3 coma positions

For flight data that are returned compressed to 8 bits/pixel, the first step in their calibration is to convert them back into their equivalent 14-bit values. A calibration step is also included here to correct for any possible non-uniformity in digital encoding DN intervals introduced during the initial instrument analog-to-digital data sampling (called uneven bit weighting). For the IR spectrometer, ground calibrations have shown that the output DNs do not increase linearly with integration time. Rather, the signal generation rate decreases as the total integrated signal increases.

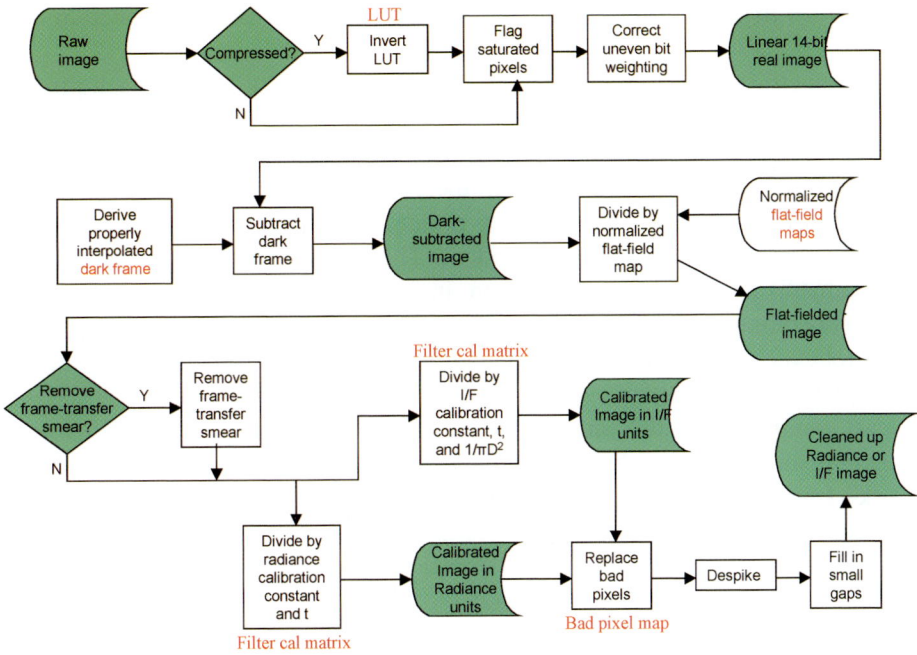

Figure 1. VIS radiometric calibration data processing pipeline.

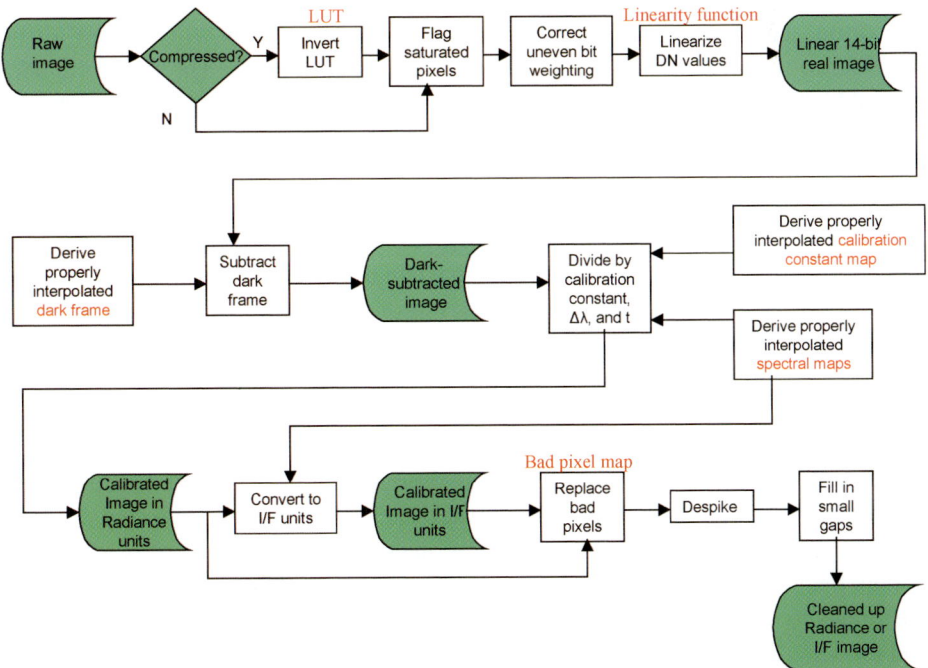

Figure 2. IR Spectrometer radiometric calibration data processing pipeline.

Therefore, a linearization step must be applied to IR data as the next step in calibration.

Radiometric calibration requires that the returned signal level in each pixel in the absence of any external scene input be determined. This offset level needs to be subtracted from the returned raw DN levels as a first step in isolating the scene-induced signal. For the IR spectrometer detector, the offset level varies with integration time, detector temperature, optical bench temperature, and to some extent the past history of detector readout. The offset level for each pixel in the CCD cameras can vary with integration time and detector temperature. For most camera operating modes, overscan pixels are returned that provide a measure of the offset for every line. However, ground calibration results analyzed to date suggest that the CCD bias level varies on time scales shorter than one line time by an amount at least comparable to the variation in the line-to-line average values. Therefore, CCD bias subtraction using averaged dark frames appears to be a better approach. "Dark" frames representing the instrument output in the absence of any scene input are acquired that span the range of independent variable values. A calibration frame properly interpolated to the parameter values that apply to a given image is subtracted as the next step in calibration.

For the VIS cameras, differences in the relative response rate of individual pixels within each detector are next removed by dividing by a normalized flat-field map. These maps will be initially derived from ground-based flat-field imagery. Updates may be performed based on the results of inflight images of the internal stimulus, the Moon, and/or the Earth; however, none of these sources will present a true flat-field to the detector, so the updates are likely to limited. The frame-transfer type of CCDs used in the VIS cameras allow some signal to be accumulated during the finite time required to shift the image from the active portion of the array to the masked storage region. This shift typically takes about 5 ms. For images that are acquired with short exposure times, this extra "smeared" signal could be detectable and a source of error in calculating scene radiance. Therefore, an optional step that removes the frame-transfer smear signal can be selected prior to converting the dark-subtracted, flat-field corrected data to radiometric units. This final radiometric conversion is done by dividing the image by a calibration constant unique for each filter position and by the commanded integration time.

For the IR Spectrometer, no single correction file can be derived to correct for pixel-to-pixel response variations because the mapping of wavelength to pixel location changes with optical bench temperature. Therefore, for the spectrometer, the flat-field and radiometric conversion steps are combined into one step. Files of wavelength, wavelength band, and calibration constant for each pixel as a function of bench temperature and instrument operating mode will be used to interpolate to the values to use for calibrating each flight image based on its bench temperature and mode.

After radiometric calibration is complete, other optional calibration steps can be applied. One possible additional correction is to clean up the image by interpolating over known bad pixel locations, removing cosmic-ray-induced noise spikes, and filling in any small gaps due to missing data. Another process that could be applied is a geometric correction for any optical distortions that might exist in the instrument. Finally, spatial resolution could potentially be improved by doing a correction for the instrument's point-spread function (or modulation transfer function) and scattered light response.

Besides the raw images, we intend to archive with NASA's Planetary Data System (PDS) the radiometrically calibrated versions of all images in both radiance and reflectance units. Any images that are selected for further calibration (cleanup, geometric correction, and/or resolution enhancement) will also be archived with PDS.

6. Data Processing Pipeline

6.1. INTRODUCTION

Each raw DI instrument frame is initially stored on the spacecraft as a single data "file." The data files are broken up into data packets, which are downlinked via the DSN to the Jet Propulsion Laboratory Advanced Multimission Operations System (JPL/AMMOS). AMMOS decommutates the packets to recreate the original raw observations. In data terms, the raw observations are simply a stream of data in a compact, mission-specific format unsuitable for scientific analysis.

6.2. SCIENCE DATA CENTER DESCRIPTION – OVERVIEW

The DI Science Data Center (SDC) is located at Cornell University, and its main tasks are to

1. Receive raw observations from JPL/AMMOS
2. Convert raw observations into products suitable for analysis and archiving
3. Make the products available to the Science Team and others.

To accomplish this, the SDC comprises hardware (computers and disk storage), software, and interfaces to the various suppliers and clients of the SDC. The SDC is designed for data handling and conversion only; science team members will perform scientific analysis at their home institutions using data obtained from the SDC. The Planetary Data System-Small Bodies Node (PDS-SBN) at the University of Maryland will archive SDC products as well as Science Team products (see The Data Archive section below).

6.3. SCIENCE DATA CENTER DESCRIPTION – HARDWARE

The hardware will be standard PC components for reasons of low cost, high reliability and simple sparing and maintenance. The SDC will be developed and built at Cornell University, but will be easily and inexpensively replicated at JPL for the encounter as well as at any other site desired.

6.4. SCIENCE DATA CENTER DESCRIPTION – SOFTWARE

The operating system software is Linux, chosen for its low cost, high reliability, and numerous, readily available, free, powerful development tools. The SDC-specific software will be a combination of standard Linux/Unix™ scripting tools and the Interactive Data Language (IDL) from Research Systems, Inc. Generally, the scripts will handle the file management tasks, and IDL software will handle the conversion of raw images into standard products. More broadly stated, the scripts will handle the data containers, and IDL will handle the data contents. The data will be organized in a database accessible to the Science Team using standard tools (PHP, a popular scripting language similar to Perl, and MySQL, a freely available database server and sequenced query language). As far as possible, the data acquisition, conversion, and delivery software will be automated.

6.5. SCIENCE DATA CENTER DESCRIPTION – INTERFACES

Each interface is designed around the supplier or client of the particular data or information passing through the interface.

All of the interfaces sit on top of the TCP/IP networking protocols and the Internet, which provides network access to the SDC at an arbitrary location to virtually any science team member. Interfaces transferring sensitive information will be password protected and/or encrypted via the Secure Sockets Layer (SSL) and Secure Shell (SSH) protocols.

The first interface transfers incoming data from the AMMOS to the SDC. Incoming data are pushed over the Internet by AMMOS scripts using SSH to encrypt the data. Data are run through relevant pipelines (see below), converted to standard products (see below), and made available to Science Team members and others via client interfaces in near-real time. Other data, such as commands and engineering telemetry, will be supplied by similar interfaces to this one.

The rest of the interfaces are client interfaces to post-pipeline standard products, and they all use the Hyper Text Transfer Protocol (HTTP), which makes them available to users on the Internet with a standard World Wide Web (WWW or "Web") browser.

The simplest client interface allows browsing of the directory structure of the SDC products. In this mode, the data are arranged hierarchically by mission phase (e.g. ground calibration, ground-based mission scenario tests, inflight calibrations, flyby), instrument, product type (e.g. raw, calibrated), date data were taken, and filename. The filename of each observation contains information specific to the observation in an encoded form. The main use of this interface will involve the

PDS downloading the entire dataset for archiving. It is not anticipated that this interface will get much other use unless one is trying to locate specific observations for which the filenames are already known.

The next client interface allows browsing of the data based on the synoptic environment unique to each observation. Specifically, this interface makes it possible for one to select observations using criteria from the FITS header (see Standard products below), and download one or more observations in a single step for scientific analysis. Given the large number of header keywords, this interface is very powerful but also very complex, requiring detailed knowledge of the keywords, which are restricted to sometimes cryptic mnemonics of eight characters. The interface will attempt to lessen the complexity by: (1) grouping the header keywords according to function, (2) sorting the keywords by priority assigned by the Science Team, and (3) making a description of each keyword available as part of the interface.

Another client interface is the inverse of the synoptic interface described above in that, where the synoptic interface looks backward in time at the observation environment, this interface looks at the data through the lens of the data set planning (see above). In this interface, the planned data set groupings (calibration, approach, pre-impact, impact, post-impact, etc.) will be part of the database, and the observations will be sorted as they arrive according to their original purpose or observational goal.

The details of the two previously described interfaces are described below in the section Data Library and Catalog.

Finally, Science Team products for Education and Public Outreach may be made available via a public interface to the SDC. This interface will be a typical website with the content comprising mostly images, figures and captions, generated by the Science Team and suitable for the audience.

6.6. PIPELINE PROCESSING FLOW

Figure 3 summarizes the flow of data to, through, and from the SDC. The figure shows the five basic functions the SDC performs on the data: convert, validate, calibrate, store, and distribute.

The main pipeline is as follows: decommutated spacecraft files and engineering telemetry from the AMMOS and commands from the Mission Operations Center (MOC) will be converted into standard raw products. The raw products will be validated and verified (see Data Validation below). Raw products that represent observations in instrument units (DN) will be calibrated, which converts them into physical units. The SDC will store all products on hard disk and distribute them to the Science Team and others via the WWW, or via alternate protocols where appropriate.

Additional inputs to the SDC include calibration files and procedures from the Spacecraft Instrument teams, ephemerides and other SPICE files from JPL/NAIF,

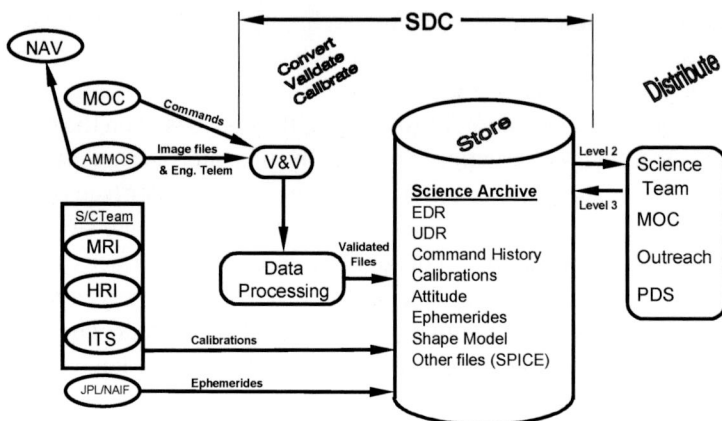

Figure 3. SDC Data Flow.

engineering telemetry from the spacecraft attitude control system (ACS), and products from the Science Team derived from the calibrated data and typically collapsible to a publication figure. Some of these additional inputs will become the basis for part of the pipeline, e.g., calibration files and procedures.

6.7. STANDARD PRODUCTS

SDC products come from several sources: Products from the instrument pipeline described above, ephemerides from JPL/NAIF, engineering telemetry from the ACS, and composite products from the Science Team (e.g., mosaics, shape models).

Instrument data, both images and spectra, will be available in both raw (DN) and calibrated (physical) units. Calibrated data may be available in either Intensity/Flux ratio or Radiance units, or both.

The images and spectra will be files in the Flexible Image Transport System (FITS) format, a standard data format used in astronomy (see http://fits. gsfc.nasa.gov for details). The FITS format combines each observation with header information in a single file. The header information is flexible and expandable; at a minimum, it contains a description of the observation's data. The SDC, at the direction of the Science Team, will also place mission- and observation-specific information in the header to aid in science analysis and archiving. The raw data headers will contain the information available in the raw spacecraft files such as epoch of the observation, instrument identification, state (e.g., temperatures in DN units, filters used), and parameter settings (e.g., exposure duration, gain states). The calibrated data headers will combine the header information from the raw data with external information such as ephemerides, attitude information, and calibrations to expand the headers to include temperatures in physical units, observational geometry, calibration files used, software versions used to generate the files, etc.

The product headers will also be used as the basis of the Data Library and Catalog described below.

Ephemerides in the form of SPICE kernel files from JPL/NAIF are not products of the SDC, but they will be stored on the SDC and distributed to clients of the SDC. Engineering telemetry from the ACS will be converted to SPICE C-kernel files and used in the calibration pipeline. They will also be distributed to clients of the SDC.

Members of the Science Team will provide composite products such as mosaics and shape models of Tempel 1 to the SDC in the form of data files. The format will typically be FITS format, but may be negotiated with the Science Team as necessary. The main function of the SDC for these files is as a central collection point for eventual archiving.

6.8. DATA VALIDATION

Most of the data validation and verification will take place when the AMMOS at JPL decommutates telemetry packets to re-create the spacecraft files. The result of that validation will be at most a single Data Quality Index (DQI) value in the FITS header of the data file. The purpose of the DQI is to alert science users to the possibility of corrupted data. As the data go through the SDC pipeline, there will be further verification to ascertain the original command sequence that created each observation. Once that is known, it will be placed in the FITS header and be available in the Data Library and Catalog (see below). The Data Library and Catalog may then be used for further verification, e.g., to look for missing data files. This will also allow SDC clients to locate data using the original purpose of each sequence as a search key.

6.9. DATA LIBRARY AND CATALOG – OVERVIEW

The workings behind two of the user interfaces, described functionally in the section SDC description – Interfaces above, will be described in more detail here.

The computer systems that make up the SDC keep track of observations by filename, but cryptic filenames such as dnivcaxf_20031013t22040001.fit are not much use to science team members and other clients of the SDC. As noted above, each observation, as a FITS file, has a header with mission- and observation-specific information. The catalog supports an interface for users of the SDC to search for, select, and retrieve observations based on the information in the FITS headers. To accomplish this, the SDC performs two tasks: convert the observational data into a database and provide a user-friendly interface to that database. Figure 4 summarizes these two tasks as connections between the data and the database and between the SDC user and the database. The next section details the pieces of the entire process.

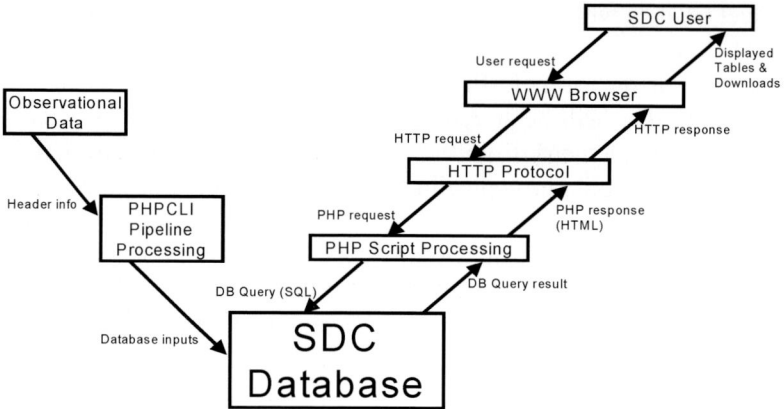

Figure 4. SDC User Interface Construction.

6.10. DATA LIBRARY AND CATALOG – DETAILS

The starting point for building the SDC database is the information stored in the FITS headers. That information is organized as keyword/value pairs as shown in Table XIX.

TABLE XIX

Subset of FITS header.

```
BITPIX  =                    16 / 16-bit signed integer per data pixel
NAXIS   =                     2 / Number of data axes
NAXIS1  =                  1024 /Number of positions along axis 1
NAXIS2  =                  1024 /Number of positions along axis 2
MISSION = 'DEEP IMPACT'         / Name of the spacecraft mission
MSNPHASE = 'Encounter'          / Phase of the spacecraft mission
EXPERMNT = 'LIGHT CURVE'        / Experiment type
ORIGIN  = 'CORNELL'             / Institution that originated this file
BUNIT   = 'DATA NUMBER'         / Physical units for data pixel
OBSERVAT = 'IMPACTOR'           / Observing platform (flyby or impactor)
INSTRUME = 'ITS'                / Instrument (HRI/MRI/ITS)
DETECTOR = 'VISUAL CCD'         / Detector Type
OBJECT  = '9P/TEMPEL 1'         / Target Name
EXPID   =                    40 / Exposure ID
INTTIME =                   100 / [msec] Total integration time
EXPTIME =                   100 / [msec] Exposure time
OBSDATE = '2003-163'            / Date at start of observation
OBSTIME = '12:34:56.678'        / Time at start of observation
```

TABLE XX

FITS headers' partial summary.

Observation	BITPIX	NAXIS1	NAXIS2	MSNPHASE	EXPERMNT
I030130_123117_827_A.FIT	16	1024	1024	Encounter	Lightcurve
I030130_132209_044_A.FIT	16	64	64	Encounter	Lightcurve
I030130_182853_344_A.FIT	16	1024	1024	Encounter	Lightcurve
I030130_205410_027_A.FIT	16	1024	1024	Encounter	Lightcurve

The leftmost one to eight characters on each record is the keyword. The set of keywords for each instrument is fixed for that instrument. To the right of each keyword is an "equals" sign that serves as the delimiter between the keyword and the value. The value follows the equals sign and may be numeric or text. Text values are delimited by single quotes. Everything to the right of the value is a comment, delimited by a "slash" character.

The first task of the catalog is to insert each instrument's keyword/value pairs into a database with the keywords as columns and each observation's values summarized as a single row in the table as shown in Table XX.

As each observation, bundled in a FITS file, enters the SDC, automated scripts extract the keyword/value pairs and insert them into the central database. Observations are afforded one line each, with one column per keyword/value pair. The scripts are written in PHP4. The database itself is composed of the MySQL package.

The second task of the catalog is to present the SDC user with one or more interfaces to search, review and download images that match user-specified criteria. Once the FITS headers have been inserted into the SDC database, the web-based interface can be dynamically generated through a straightforward application of standard tools and protocols (displayed on the right-hand side of Figure 4 above). Again, PHP scripts are the heart of the interface. The scripts relay user requests to the database, gather the requested information from the database response, and display the requested information as formatted web sites, viewable in any web browser. The interface is in a constant state of improvement through ongoing input from the Deep Impact Science Team.

7. The Data Archive

7.1. DATA PRODUCTS

Once science data files have been produced by the SDC, they will be transferred to the Principal Investigator team at the University of Maryland, where the content and format will be validated and the archive volumes will be prepared.

The DI archive will contain science data products from each of the instruments, instrument calibration data, command history data, navigation and ancillary data in the form of SPICE kernel files, software, and sufficient documentation of the data, software, and mission to enable scientists to understand and use the archive well into the future. To produce this archive, a number of steps need to be carried out, including design of the archive structure and contents, generation of the archive components and an archive interface control document (ICD), peer review with the PDS, and final packaging and delivery. The science data products form the core of the archive; a list of the expected data products from each of the instruments is given in Table XXI. The data set collections to be archived are expected to be several gigabytes (GB) in size. The archive will be on-line at the PDS Small Bodies Node (PDS-SBN), consistent with current PDS practice. Several copies of a hard

TABLE XXI

Data Products and Archive Components.

Archive component	Data type	Data volume (GB)
Imagers (HRIVIS, MRIVIS, ITS)	Raw images	66.7
	Calibrated images (pre- and post-flight)	
	Calibrations	
	Support data	
	Shape model	
Spectrometer (HRI-IR)	Raw spectra	20.13
	Calibrations	
	Calibrated spectra	
Radio Science	Trajectory estimates and supporting products, e.g., SPICE attitude files, used to determine the cumulative effect of comet dust particle impacts on the spacecraft	.05
Earth-based	Images (Lowell Obs, Mauna Kea Obs, etc.), Spectra	100
Supporting	Reprocessed data from IRAS, Images, photometry from 2000 apparition	5.02
MERGE	Shape models	.01
Ancillary	Mission history files, SPICE Kernels	.01
Software	Calibration Algorithms	.01
	Higher Level Software (as provided by Science team)	
Documentation		.05
Total archive		182 GB

media archive will also be produced for deep archive purposes, using compact disks (CD) or digital versatile disks (DVD).

7.2. ARCHIVE STRUCTURE

The DI archive will be broken down into data set collections – one for each instrument, one for data sets deriving from more than one instrument (the MERGE data set collection), and one for SPICE data. A typical volume will contain data from a specified time interval. The top-level directory of a volume will thus contain directories for each of the data set collections and directories for each of the additional components of the archive, as required by PDS. The MERGE data will be ordered by time first, then by instrument, and further divided by type of data, if relevant. Data types and volumes for each archive component and for the total archive are shown in Table XXI. PDS requires a number of documentation files for each archive volume. One is a readme.txt – a text file describing the contents of the volume. Also required is voldesc.cat – a catalog of all the files residing on the volume. Each of the sub-directories under the top-level directory also requires one or more files to document the contents of that directory. The details of these files are specified in the PDS Standards Reference (2002).

7.3. SAFED DATA

Packet data and some ancillary files will be assembled and placed on CD-R or DVD media for long-term safekeeping in the event problems are discovered with the formally archived higher-level data products. Copies of this safed data (and allied documentation) will be provided to the PI, the Cornell SDC, and the NAIF node of the PDS. These same data will be archived by the JPL InterPlanetary Network for a minimum of three years after end of mission.

7.4. ARCHIVE PREPARATION

Science data products will be generated in PDS–compatible formats. This approach requires that each data file (data table or image file) be in a format approved by PDS and be accompanied by a PDS "label", actually a detached descriptive header file describing formally the content and structure of the accompanying data file. Ancillary data describing the observing conditions and spacecraft state when science data were acquired will be extracted from the packet data and SPICE kernels and placed in these PDS labels.

Science data products also include an extensive Earth-based archive of Tempel 1 data that will be collected in both pre- and post-encounter phases.

Files documenting the archive components will be prepared by the parties generating the data. In general, all information necessary to interpret and use the data is to be included in the archive. Additionally, the source code of all software to be provided with the archive will be collected, documented, and included in the archive.

PDS standards call for the documentation of the mission, spacecraft, instruments, and data products with special files called "catalog objects." Since the catalog objects take the form of a template that must be filled out with prescribed information, they are often referred to as "templates" even when they are already filled out. The required templates are the "mission template" describing the Deep Impact mission as a whole, the "instrument host template" describing the spacecraft, one "instrument template" for each instrument, and one "data set template" for each data set. These templates are to contain much of the information necessary to document the archive, and should make it possible for scientists to make correct use of the data in the future when the mission personnel are not available to support them. The PDS-SBN will fill in portions of the catalog objects, requiring only text descriptions of the mission, spacecraft, instruments, and data sets from DI personnel.

Figure 5 illustrates the flow of data from the Distributed Object Manager (DOM) at JPL to the SDC at Cornell University and the building of the archive through the

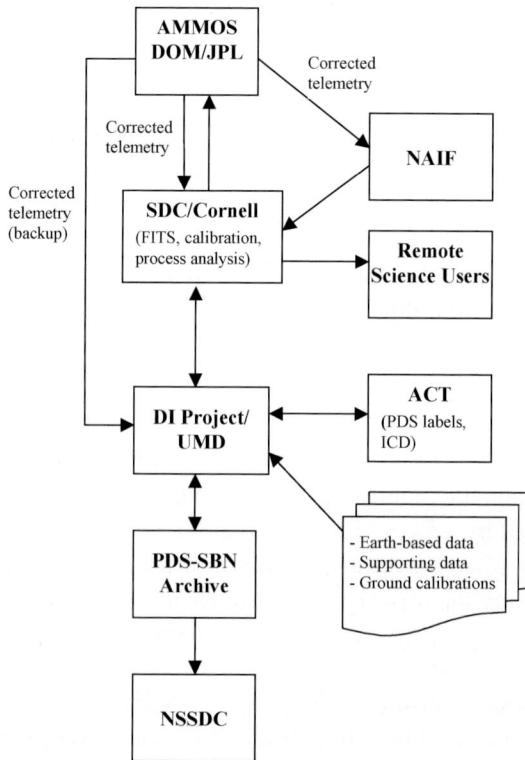

Figure 5. Building the Deep Impact archive.

TABLE XXII

Timeline for DI Project Archiving.

Delivery date	Archive Products
10/31/2003	IR model data (IRAS)
05/21/2004	Supporting data (Earth-based images, photometry from 2000 apparition)
07/2/2004	Ground calibration files
09/30/2004	Earth-based pre-encounter spectra and images
03/31/2005	Calibration files and payload tests during checkout
12/31/2005	Earth-based data leading up to impact. Spacecraft measurements through impact
03/31/2006	Earth-based post impact data

DI Project with its subcontractor, the Applied Coherent Technology (ACT), and the PDS Small Bodies Node at the University of Maryland.

7.5. ARCHIVE VALIDATION

Data validation falls into two types, validation of the data itself and validation of the compliance of the archive with PDS archiving requirements. The first type of validation will be carried out by the Science Team, and the second will be overseen by the PDS-SBN, in coordination with the Science Team. The delivery schedule, with separate delivery dates for different portions of the mission (Table XXII), will facilitate validation by ensuring that problems in the early deliveries are resolved by the time of the later deliveries.

The formal validation of data content, adequacy of documentation, and adherence to PDS archiving standards is finalized with an external peer review. The peer review will be coordinated by the PDS-SBN. The peer review process may result in "liens" – actions recommended by the reviewers or by PDS-SBN personnel to correct the archive. All liens must be resolved by the data set provider: the SDC and PI team personnel for raw and calibrated data, the science team for higher-level data products and calibration algorithms. PDS will do a final validation prior to packaging and delivery.

7.6. ARCHIVE PACKAGING AND DELIVERY

Data delivery will take place in stages, as specified in the timeline in Table XXII. Each delivery will be made to PDS via the appropriate medium. The final data delivery will incorporate the entire archive, including the earlier data deliveries.

7.7. SCHEDULE FOR ARCHIVE GENERATION, VALIDATION, AND DELIVERY

The principal archive elements, namely the science data products defined in Table XXI, will be generated during the course of the mission, as will many ancillary products such as SPICE files. The general guideline for Discovery missions is that they deliver archive quality volumes to PDS at intervals not exceeding six months after receipt of the data used to make the products contained on the volume.

The planned timeline for archive delivery to PDS is shown in Table XXII. The final archive delivery is to be made 90 days after the end of the nominal mission in order to allow time for PDS review and lien resolution before the end of operations.

Following the delivery of archive volumes to PDS, the data will be peer reviewed by PDS over a several-month-long period. Any liens that are identified by the peer review process will be rectified by the Project and the appropriate science team members before they cease operation (expected to be 90 days after the end of mission for the science team). The DI project is responsible for resolving all liens against the final archive delivery. Final acceptance of the data by PDS will occur only after all liens have been cleared. The delivery of post-launch checkout data to the PDS will help to identify early in the mission any potential problems that can be addressed before the final archive is generated, thus avoiding liens on the data that require significant resources to correct.

There are no proprietary data rights for the DI Mission. We anticipate raw and calibrated, single-instrument data to be available six months after impact through the PDS-SBN. These data will be in the process of being peer-reviewed. We expect higher-level products, such as shape models, to be available about nine months after impact.

8. Summary

The planned DI observing sequence and ground processing of the returned data will result in a unique and comprehensive archival science data set that will permit researchers to begin to unlock some of the mysteries of those fascinating and spectacular solar system wanderers – comets.

Appendix: Abbreviations

ACS   Attitude control system
ACT   Applied coherent corporation
AMMOS   Advanced multimission operations system
CA   Closest approach
CCD   Charge-coupled device
CD   Compact disk
Dec   Declination
DI   Deep Impact

DN	Data number, digital (raw) instrument output
DOM	Distributed object manager
DQI	Data quality index
DSN	Deep space network
DVD	Digital versatile disk
FITS	Flexible image transport system
FOV	Field of view
Gb	Gigabits
GB	GigaByte (10^9)
HGA	High-gain antenna
HRI	High-resolution instrument
HRI-IR	High resolution instrument – infrared spectrometer
HRIVIS	High resolution instrument – VISible imager
HTTP	Hyper text transfer protocol
I	Impact
ICD	Interface control document
IDL	Interactive data language
IPN	Interplanetary network
IR	Infrared
IRAS	Infrared astronomical satellite
ITS	Impactor targeting sensor
JPL	Jet Propulsion Laboratory
L	Launch
MB	MegaByte (10^6)
MOC	Mission Operations Center
MRI	Medium resolution instrument
MySQL	"My" Structured Query Language
NAIF	Navigation Ancillary Information Facility
NAV	Navigation team
NSSDC	National Space Science Data Center
NVM	Non-volatile memory
PC	Personal computer
PDS	Planetary data system
PDS-SBN	Planetary data system-small bodies node
PHP	PHP: Hypertext Preprocessor
PHPCLI	PHP Command Line Interface
RA	Right ascension

R/T	Real time data return
S/C	Spacecraft
SBN	Small bodies node
SDC	Science Data Center
SM	Shield mode
SNR	Signal-to-noise ratio
SPICE	NAIF toolkit
SSH	Secure shell protocol
SSL	Secure sockets layer protocol
UDR	User data record
UMD	University of Maryland
VIS	Visible
V_{mag}	Visual magnitude
V&V	Validation and verification
WWW	World Wide Web

References

A'Hearn, *et al.*: this volume.

Belton, *et al.*: this volume.

Blume, *et al.*: this volume.

Hampton, *et al.*: this volume.

Planetary Data System Standards Reference, October 15, 2002, Version 3.5, JPL D-7669, Part 2. (http://pds.jpl.nasa.gov/documents/sr/).

Richardson, *et al.*: this volume.

Schultz and Ernst: this volume.

Sunshine, *et al.*: this volume.

Thomas and Veverka: this volume.

EDUCATION AND PUBLIC OUTREACH FOR NASA'S DEEP IMPACT MISSION

L. A. MCFADDEN[1,*], M. K. ROUNTREE-BROWN[1], E. M. WARNER[1],
S. A. MCLAUGHLIN[1], J. M. BEHNE[2], J. D. RISTVEY[2], S. BAIRD-WILKERSON[3],
D. K. DUNCAN[4], S. D. GILLAM[5], G. H. WALKER[6] and K. J. MEECH[7]

[1]*Department of Astronomy, University of Maryland, College Park, Maryland 20742, U.S.A.*
[2]*McREL, Aurora, Colorado 80014, U.S.A.*
[3]*Magnolia Consulting, Louisa, Virginia 23093, U.S.A.*
[4]*Fiske Planetarium, Department of Astronomy, University of Colorado*
[5]*Jet Propulsion Laboratory, Caltech, Pasadena, California, U.S.A.*
[6]*American Museum of Natural History, New York, New York, U.S.A.*
[7]*Institute for Astronomy, University of Hawaii*
*(*Author for correspondence: E-mail: mcfadden@astro.umd.edu)*

(Received 20 August 2004; Accepted in final form 31 January 2005)

Abstract. The Deep Impact mission's Education and Public Outreach (E/PO) program brings the principles of physics relating to the properties of matter, motions and forces and transfer of energy to school-aged and public audiences. Materials and information on the project web site convey the excitement of the mission, the principles of the process of scientific inquiry and science in a personal and social perspective. Members of the E/PO team and project scientists and engineers, share their experiences in public presentations and via interviews on the web. Programs and opportunities to observe the comet before, during and after impact contribute scientific data to the mission and engage audiences in the mission, which is truly an experiment.

Keywords: solar system exploration, education public outreach

1. Introduction

The Deep Impact mission sends a pair of spacecraft to study the interior of comet 9P/Tempel 1. An impactor spacecraft separates from the flyby spacecraft 24 h before impact. Traveling at a relative velocity of 10.2 km/s the comet overtakes the impactor at 5:53 UT July 4, 2005. The impactor carries an imaging system that will study the surface of the comet upon approach, relaying data to the flyby spacecraft and then back to the Earth. The flyby carries two imagers and an infrared spectrometer that will record the impact and its aftermath (A'Hearn *et al.*, this volume). Images and spectra will allow the science team to address questions related to the nature and structure of comets:

1. What are the basic properties of a cometary nucleus?
2. How do comets evolve?
3. What are the primordial ices in comets?

Space Science Reviews (2005) 117: 373–396
DOI: 10.1007/s11214-005-6062-9

4. Do comets become extinct or only dormant?
5. Are comets made of discrete cometessimals?
6. Do cometary nuclei have a discernable impact history?

The science team anticipates that answers to the above questions will contribute to understanding the formation of the solar system in its early stages and be useful to determining how one would mitigate a cometary impact on Earth today.

1.1. DEEP IMPACT E/PO PROGRAM

NASA's Discovery Program is charged with sharing the excitement of space science with the public, contributing to the improvement of science, technology, engineering and math (STEM) education in the United States while inspiring the next generation of the 21st century scientific and technical workforce. To this end each mission plans and implements an Education and Public Outreach (E/PO) program. A series of three publications provide the guidelines and description of the infrastructure of E/PO programs for NASA missions (NASA, 1996, 2003, 2004). Rosendhal *et al.* (2004) provide a summary of the program, its history, philosophy and policy.

The E/PO team implemented a program built upon six principles comprising the Deep Impact's E/PO program philosophy:

1. Audiences will learn with opportunities to participate in many ways.
2. Partnerships with other education and outreach programs leverage the efforts of all programs.
3. Educators will respond to Deep Impact products and programs that meet the standards-based criteria to which they teach.
4. Learning is enhanced when inquiry and discovery are embedded in the process.
5. Educators look for real-world examples of teamwork and problem solving for their classrooms and mission work provides valuable lessons.
6. The internet enables delivery to targeted audiences.

The mission's E/PO program has four primary components: formal education, outreach programs including professional development, informal education and partnerships with other organizations, outreach products and the web. Two programs for undergraduate students were developed after the project was under way and were not in our original plan. They provide hands-on engineering and science opportunities for students in both high school and university.

1.2. ANTICIPATED OUTCOME

The anticipated outcome of the E/PO program includes: providing access to the excitement of a NASA mission and to the scientific, engineering and mathematical themes of the mission. The team strives to engage students, teachers and the interested public in the mission in order to learn and understand scientific investigations, their nature, process and results. The expected outcome, in the long term, is to inspire the next generation of scientists, engineers and educators, and to promote scientific literacy among the general public. To this end, a program evaluation is conducted to evaluate the extent to which target audiences access and use the E/PO materials and to examine the experiences of teachers and students using them. (See Section 5).

2. Formal Education Products and Programs

Educational units for school-aged children from elementary, middle and high school have been developed. Materials for grades K-4 are called activities. For middle school and high school, we developed educational modules that consist of guides for both students and teachers, and include alignment with science and/or math educational standards. All materials can be used in both formal classrooms and informal settings such as community or after school programs. The modules were reviewed by scientists for scientific accuracy before being tested with the intended audience. They were reviewed by educators, revised and reviewed again and posted on the project's web site. Some of the materials were reviewed by the Solar System Education Review process and recommended for distribution through NASA's CORE program. Major distribution is via the world wide web. All materials described here are available at http://deepimpact.jpl.nasa.gov/educ/index.html.

2.1. ELEMENTARY SCHOOL ACTIVITIES

Two principles drove development of materials for K-4:

1. Younger children begin by being curious about the world, efforts were made to build on that curiosity to prevent their losing it.
2. Children have different learning styles and will engage differently in educational activities.

Activities were built around things in which children naturally engage in during their early years – singing, listening to stories, using their imagination, and making crafts. A brief description is given below.

2.1.1. *C-o-m-e-t-s*

A simple song, written by E/PO team member, Maura Rountree-Brown, contains some basic facts about comets. The song spells out the word "comets" with an easy tune. Actions such as hand gestures and clapping are included so that the brain is recording the information through three different modes, with music, rhythm, and kinesthetic action. After singing the song, the students and presenter go over it line by line and ask questions such as, "If the song says a comet is very, very cold, what might a comet be made of?" The children easily come up with the concept of ice. By repeating the lines from the beginning every time a new line is added, the information is worked deeper and deeper into a student's memory. Teachers report that children are heard singing the song on the playground weeks after it was initially introduced to them. http://deepimpact.umd.edu/educ/storysong.html#cometsong.

2.1.2. *Comet on a Stick*

A hands-on activity on comet modeling called Comet on a Stick works well with young students following the song activity. This activity has been used with all age groups, however. Background information is provided to students about the reasons for developing a model of the comet for planning and design purposes. The students then form teams. Using an $8'' \times 11''$ piece of paper, the team shapes a comet nucleus and attaches it to a stick. They choose two or three facts about comets to model them using re-cycled materials: tin foil, black paper, beads, fiber fill for pillows and other items that a teacher can collect. Once their models are built, each team has a chance to show them, while other teams determine the subject being communicated. Misconceptions can be caught because they are physically showing what they learned. Modifications to this activity include selecting a candy bar that best represents what might be found beneath the surface of a comet and defending the reason for choosing it (layers, homogenous, rocky). http://deepimpact.umd.edu/educ/CometStick.html (Figure 1).

Figure 1. Team member Jochen Kissel shows a version of a modeled comet.

2.1.3. *Comet Sisters*

M. Rountree-Brown also wrote a mythical comet story after taking a class at the International Storytelling Center on incorporating story into education. "I'm going to tell you a story," she starts out. "Some of what I tell you is true about comets, some of it might be true and we will find out what happens when we execute the Deep Impact mission. Some of it – I just made up." Students who listen to the story can either draw their impressions on paper or discuss it afterward to find out what they know to be true or to perhaps be true about comets.

2.1.4. *Make a Comet and Eat It*

This activity starts with a basic ice cream recipe and suggests that students put in foods such as peppermint, ground cookies, peanuts, gummy bears, all of which represent components of a comet. The teams trade ice cream bags and then start using their individual senses to research what is in the ice cream: look at it, smell it, feel a separate cup of it, taste it. The process is compared to that of a spectrometer in using different wavelengths to collect different kinds of information about what is in the comet. http://deepimpact.jpl.nasa.gov/educ/IceCream.html.

2.2. MIDDLE AND HIGH SCHOOL MODULES

Materials for middle schools were built upon interdisciplinary themes combining social education (team work and communication) with science education. Workshops for teachers at the National Science Teachers' Association were held during which these materials were introduced and feedback was collected and incorporated into revision of the materials. High school material focuses on physics and includes inquiry-based processes.

2.2.1. *Collaborative Decision-Making*

Collaborative Decision Making is designed to engage students in grades 7–12 in activities that focus on collaboration and communication strategies. These activities strengthen students' understanding of and ability to use collaborative processes and communication practices to clarify, conceptualize, and make decisions. Students first consider cases in which they have to make decisions that are important in their life. They discuss how they arrive at a course of action or a decision. They are then presented with a problem that the Deep Impact team confronted, that of deciding the best time for the impactor to hit the comet. Students take on the roles of different project members, the principal investigator, the project manager, and engineer. After the risks are identified, they gather and convey evidence supporting and refuting the viability of these actions. The module's strategies rely primarily on student investigation into the background information that is necessary to support arguments; make quantitative risk analyses; engage in debate, role-playing, and practice persuasive writing and communication processes; and practice group decision-making procedures. The material is aligned with national science education standards (Appendix I). It was selected by the OSS, peer review process for wide distribution throughout NASA Resource Centers. http://deepimpact.umd.edu/collaborative_ed_module/index.html.

2.2.2. High Power Activity

An activity that takes less time to research and is not as technically involved as Collaborative Decision Making is called High Power Activity. It focuses on the same decision-making processes forming the basis of Collaborative Decision Making. Students play the role of different members of the project and are confronted with a problem that is similar to one that arose in the course of the Deep Impact mission. A company offered to pay to place extra cameras on the spacecraft to watch the impactor separate from the flyby spacecraft. After the students consider the pros and cons of this proposal, they decide what to do. They can compare their decision making with that of the Deep Impact team. http://deepimpact.jpl.nasa.gov/high_power/index.html.

2.2.3. Designing Craters

Designing Craters is a two-to-three week inquiry-based module addressing the question: "How do you make a 7–15 stories deep, football stadium-sized crater in a comet?" The lessons are designed for students in grades 9–12 and provide them with experience in conducting scientific inquiries, making measurements, displaying data and analyzing it to gain a greater understanding of scientific modeling while involving students in the excitement of a NASA mission in development. This unit was designed as part of a Masters degree in Science Education at University of Maryland. After studying the physics of crater formation based on the work of Melosh (1996), the graduate student then developed guidelines for student-designed experiments. The activities are designed to model one path that a scientific inquiry might take. The students begin by brainstorming what factors might influence crater size and doing some initial experimentation and exploration. They evaluate each other's suggestions and describe their initial ideas about cratering phenomena. Next, they design their own experiments to test one of the possible factors influencing crater formation. Emphasis is placed on experiment design, limiting the test to one variable, and quantifying the experiment. After analyzing the data for patterns that might be used to predict crater size from initial variables, the students test those predictions, use the results to refine their methods of prediction, and try again. The students discuss the advantages and limits of scientific modeling as they compare their own low-energy simulations, the work of Deep Impact Science Team cratering experts, and cratering on a Solar System scale. Finally, the students use current information about comets and the patterns they derived from their own investigations to give their best answer to the initial question. These answers can be submitted to the Deep Impact Education and Outreach Team.

Science team members reviewed the material while it was being tested in the classroom. When science team member Jay Melosh pointed out that laboratory experiments in high schools do not represent hypervelocity impacts in space, a section was added in which students discuss and compare their experimental conditions to those in space. They are led to an understanding of the limitations of laboratory experiments as well as knowledge that conditions in space are different. Science standards addressed in this unit are available in Appendix II. This unit is available at: http://deepimpact.umd.edu/designing_craters/index.html.

2.2.4. *Math Challenges*
A series of math problems was developed from computations that are necessary to carry out the Deep Impact mission. Algebra and geometry are required. Called Mission Challenges, they have been aligned to National Math standards and can be found at: http://deepimpact.jpl.nasa.gov/disczone/challenge.html.

2.3. UNIVERSITY PROGRAMS

Two higher education programs were developed after the initial planning of the Deep Impact E/PO program. Both programs offered undergraduate university students the opportunity to gain valuable research or engineering experience that will allow them to participate in the mission. The Deep Impact mission is a catalyst for long-term education and public outreach collaborations.

2.3.1. *Deep Impact Project Schools Technology Collaboration (DIPSTIC)*
DIPSTIC is a joint venture of the Jet Propulsion Laboratory (JPL), Los Angeles City College (LACC), University of Texas El Paso (UTEP), Digital Media Center (DMC) and the Deep Impact project at the University of Maryland. Participants in the program collect telescopic observations of 9P/Tempel 1 using a CCD camera built by the students and staff at LACC. This is a unique opportunity for undergraduates at a community college to participate in a NASA space mission. Data are gathered at Table Mountain Observatory (TMO) and analyzed by students. The results of their observations are distributed to the public through a web site built by students at the Digital Media Center (DMC) at UTEP. A cooled CCD detector system is built and operated by LACC students for comet spectroscopy.

The main goal of DIPSTIC is to provide a research experience for undergraduates. It is a focused effort reaching a small number of students and community college professors in an in-depth way. The team is on a journey to define the project's scientific goals, build the CCD camera, plan and execute novel astronomical observations, analyze data and finally, report results. NASA/JPL technology expertise has been transferred to LACC by developing the CCD.

Nine students between the ages of 19 and 25 have participated. They are physics, electrical engineering and computer science majors. None had previous experience in astronomy. Grades were not considered. Selection was based upon enthusiasm and interviews. In the CCD construction phase physics and engineering students were chosen. When the main task became software development, computer science students were selected. Physics and engineering students will observe the comet.

The educational benefits to students are several.

1. They learn that the real scientific world is about finding solutions to problems.
2. They learn to work independently and confidently.
3. They learn to set goals and meet them.
4. They learn to think "outside the box".
5. They also learn teamwork.

Their experiences encouraged the students to ask more questions in class, which probably carried over into their other classes and benefited their academic studies.

DIPSTIC provided one LACC instructor (Mike Prichard) with first-hand experience with unusual devices like thermoelectric coolers and thermistors. He promptly replaced experiments using more mundane electrical components with these devices in his electrical engineering classes. This required considerable modification of his curriculum. Experience in the DIPSTIC program gave him a new experimental perspective from which to teach his electronics classes.

The DIPSTIC laboratory and fabrication facility at LACC is essentially a college physics experiment. LACC science professors conducted tours of the lab for their students. DIPSTIC provides an example of the real world to students to motivate their academics.

A large draw for prospective DIPSTIC students is the opportunity to work with a NASA/JPL scientist who worked at LACC every Friday between May 2002 and June 2004. The program is now sustained by instructors; Mr. Dean Arvidson, Physics, Ren Colantoni and Mike Prichard, custom CCD electronics, Mike Slawinski, CCD cooling system.

Results are reported on the DIPSTIC web site at http://dml.nmsu.edu/dipstic/. The DMC built and maintains the web site. This aspect of the program brought together the expertise of the Digital Media Center (DMC) with the students in need of presenting their results to the public. In 2003 the DMC team filmed video interviews with DIPSTIC students. They gave knowledgeable presentations, without scripts, after participating for less than a year. The web site also presents the student's experimental logs and theoretical investigations on Microsoft™ Excel spreadsheets, PowerPoint presentations to the LACC physics club by DIPSTIC students and an HTML version of a poster-paper reporting the experimental work of the first year on the construction of the CCD cooling system. There are also photos and biographical sketches of all the participants. The DMC also produced the paper as a laminated 44 × 44 inch poster. It was presented at the 2003 Southern California Conference on Undergraduate Research.

2.3.2. Observing the Impact

High School and undergraduate students around the country are preparing to observe the comet both before and after impact. Many school educators who have been participating in training workshops over the duration of the mission will be working with students to observe the comet as part of the Small Telescope Science Program (Section 3.4) which requires a CCD camera and a telescope, or the Amateur Observers Program (Section 3.5) which requires just a dark sky and the ability to describe and possibly draw, what is seen. Professional scientists at colleges and universities around the country will be participating with observers at telescopes around the world in monitoring the behavior of the comet before, during and after impact. There are opportunities for students to observe using remote telescopes such as those connected with the TIE program, and the Faulkes Telescope, http://www.faulkes-telescope.com/.

3. Outreach Programs

Outreach extends the reach of materials and programs from the formal classroom setting to community and professional organizations with common interests and objectives to the E/PO program. It includes professional development for teachers, museum coordinators, and amateur astronomers. The programs engage the public and interested community organizations in the excitement of the mission. This is accomplished through lectures conducted via teleconference, telescope observation programs and by providing informative materials that explain and engage people in the mission.

3.1. PROFESSIONAL DEVELOPMENT

In 2000, Deep Impact, with other missions, partnered to provide professional development for educators through the Solar System Educator program (SSEP – see http://solarsystem.nasa.gov/ssep/) with as many as 75 teachers training 100 or more teachers each year. The program consists of training at JPL or another location, telecon updates providing the educators with information, written materials, videos and on-site training. In the final year of the Deep Impact mission, members of SSEP are forming their own programs and events for encounter (see Penny Kids http://deepimpact.jpl.nasa.gov/community/pennies.html). In addition, the Deep Impact outreach team has participated in professional development workshops and classes at NASA's Educational Resource Centers and other locations.

3.2. INFORMAL/PUBLIC PROGRAMS

Beginning in 2000, Deep Impact also combined with other missions to bring comet science and the technology of the Deep Impact mission to families, adults and communities through the Solar System Ambassadors (SSA – see http://www2.jpl.nasa. gov/ambassador/usstates.html). These 450 speakers participate in public events across the country reaching millions with information on NASA missions including

Deep Impact. The outreach team provides materials, training via teleconference, and mentoring as needed to meet their needs for information and expertise. Many of these ambassadors are also partnered or employed by informal institutions, school systems or are amateur astronomers further leveraging the reach of the Deep Impact mission into those outreach areas. As Deep Impact partners with museums and in-formal institutions across the country through the Museum Alliance at JPL, there is another leveraging effort to combine these different partners with special events hosted at informal institutions during encounter in July 2005. Many of their plans are self-sustaining bringing to fruition the goal of the Deep Impact outreach plan to provide opportunity for real participation by others. Deep Impact has also par-ticipated in several national and regional training for the leads of the Challenger Centers, many of whom would like to host encounter events in July. It is a goal within the Deep Impact outreach plan to leverage new opportunities off existing materials or programs, ViewSpace and Discovery Channel are producing videos and exhibits to reach an even larger section of the public through television and informal institutions.

3.3. UNDERSERVED AND UNDERREPRESENTED AUDIENCES

The Deep Impact outreach team focuses on the needs of several underrepresented and underserved audiences: young women, rural and inner city audiences, Native, Polynesian and Latin Americans, children with different learning styles. Efforts to reach these audiences are described below.

3.3.1. *Young Women*
The Deep Impact EPO team encourages young women to explore STEM careers by working with the Girl Scouts of the USA. Contributing comet materials to science kits, training Girl Scout master leaders who train other Girl Scout leaders in their council, participating in events for Girl Scouts in the geographical areas in which the outreach team is centered (California, Maryland and Colorado) the story of the mission and its scientific objectives are conveyed to girls and young women. In addition, the outreach team participates in conferences for young women interested in engineering and science careers. The Deep Impact mission also contributes to the Girl Scout/NASA Solar System Nights Kit bringing a traveling science discovery night to the families of an individual community.

3.3.2. *Rural and Inner City*
The outreach team has participated since 2000 in L.A.'s Best, an inner city after school program for middle school students. Mission team members visit schools teaching comet science, technology and using hands-on activities. Deep Impact team members have participated in the Challenger Center's Voyage through the Universe program, where scientists visit schools in the District of Columbia every year. Deep Impact supported a comet section and bulletin board exhibit distributed

to the 900 rural and inner city libraries of the Space Place Program bringing knowledge to young people in informal settings (http://spaceplace.jpl.nasa.gov/en/kids/deepimpact/deepimpact.shtml). In addition, the outreach team participates in both video and on-site national trainings for the Explorer school educators (many of whom are from underserved schools) and will be engaging some of those teachers in encounter events, particularly those in Hawaii.

3.3.3. *Native, Polynesian and Latin Americans*

The Deep Impact plan focused in particular on Native, Polynesian and Latin American cultures, especially in building early relationships and combining the science of Deep Impact with their cultural past. The project funded the Native American Initiative while it was active at JPL from 2000 to 2003 and participated in trainings for Native American educators and students. In 2004, the Deep Impact project participated in the first conference for Native American colleges held at JPL in California. The outreach team participates in conferences for the underserved such as Society for the Advancement of Chicano, Native American Scientists (SACNAS) giving workshops and distributing educational materials.

Successful collaborations are the key to effective education and public outreach. In Hawaii and the Polynesian islands, partners working to promote the NASA Deep Impact Mission include the University of Hawaii's Institute for Astronomy in Hilo, the Maui and Kauai Community Colleges, the Hawaii Public Libraries and Aloha Airlines. Nationally, the Carnegie Science Education Center and the NASA Astrobiology central EPO office supported programs. International partners include the Faulkes Telescope Foundation in England and an educational consortium in Iceland.

The Deep Impact mission serves as a perfect venue to promote public interest in astronomy and astrobiology in Hawaii with Deep Impact's Small Telescope Science Program (Section 3.4) since the impact will be visible in Hawaii. Information about the mission is incorporated into the Carnegie Institution of Washington, DC, traveling exhibit entitled: "Astrobiology: Discovering New Worlds of Life", on a 3-month tour to Oahu, Hawaii, Maui and Kauai, with support from Aloha Airlines. Lectures and teacher workshops were held in conjunction with this display. The exhibit introduced the local population to astrobiology and extremophiles living in hydrothermal vents. We transitioned to comets by using the University of Hawaii NASA Astrobiology Institute's (NAI) focus on water. The activities and materials on making observations of Comet Tempel 1 provide relevant applications at the workshops. We are recruiting teachers from the two NASA Explorer Schools on Kauai and Hawaii.

Since we have developed a cadre of master teachers through the NSF Toward Other Planetary System (TOPS) program, 1999–2003, these teachers served as instructors at our workshops. Deep Impact materials were presented and used at TOPS workshops. Having Karen J. Meech, a member of the Deep Impact science team, as well as the TOPS director, has given our Hawaii teachers extensive background for Deep Impact. Pacific Island teachers from Micronesia, Yap, Marshall Islands and Guam who were given 6″ Dobsonian telescopes through the TOPS program are encouraged to observe comet Tempel 1. Follow up with these teachers has been difficult, but will continue to be pursued for observing Tempel 1.

Our Hawaii teachers and students are also fortunate to have access to the Faulkes Telescope located on Haleakala, Maui. The Deep Impact mission is providing a perfect resource for students to conduct scientific investigations supporting the STSP. An educator from Iceland and another from England attend the Maui workshop to learn about Deep Impact and comets. Teachers from both the countries participate via a polycommunication video system. After the initial four workshops on each island, follow up sessions will be done via polycommunication system that the Institute for Astronomy has on three islands. The goal for formal education is having students enter comet projects in the 2006 science fairs.

3.3.4. *Convening Organizations for Outreach*

A year before launch, the E/PO team convened groups participating in the mission, including those from observatories, universities, outreach and institutional press officers. Personnel from science centers located close to mission activity, including Denver Museum of Science, Maryland Science Center, The Bishop Museum, Honolulu and Onizuka Science Center, Hilo, HI, are participants in developing programs for launch and encounter. The important role of Earth orbiting observatories, Hubble Space Telescope, Spitzer and Chandra and ground-based observatories provide a number of active and enthusiastic partners to engage in public outreach. Broker facilitators join in this planning for launch and encounter public outreach events.

3.4. THE SMALL TELESCOPE SCIENCE PROGRAM

The Small Telescope Science Program (STSP) was established in early 2000 to provide baseline information about comet Tempel 1 and to complement scientific data acquired at large, professional telescopes. The STSP is a network of advanced amateur, student, and professional astronomers who use telescopes, often <1.0 m in diameter, equipped with charged-coupled devices (CCDs) to make continuous, scientifically meaningful observations of comet 9P/Tempel 1. This effort will continue for several months after encounter in July 2005. See http://deepimapact.umd.edu/stsp.

Since 1999, the science team has conducted a vigorous program of ground-based observations of Tempel 1 (see Meech *et al.*, this volume) to characterize the nucleus and dust environment of the comet in terms of volatile outgassing, dust coma development and production rates, dust tail and trail development, and jet activity and outbursts. These characteristics vary as the comet moves along its orbit. However, observers at large telescopes have limited access to telescope time, and the baseline observations have gaps in temporal coverage. Advanced amateurs with access to small telescopes and private observatories can perform valuable, long-term and continuous monitoring of Tempel 1 to fill the gaps in the temporal coverage of the comet. Often the smaller telescopes used by these observers are fast, have large fields of view and are well suited for measuring the magnitude of the comet from which dust production can be measured. It is also important to look for evidence of jet activity as the comet approaches perihelion.

The STSP is the brainchild of Gary Emerson, an avid amateur astronomer and an engineer at Ball Aerospace and Technologies, the project's aerospace partner. After discussions with several team members and E/PO manager Lucy McFadden in 1999, the program was developed as a professional-amateur collaboration for the mission. The STSP was launched in early 2000, about 4 months after Tempel 1 passed perihelion. In 2000, the network of over 40 STSP observers delivered over 700 broadband VRI and over 300 unfiltered CCD images from observers in 12 countries, spanning 6 continents. The data were analyzed and a subset of the broadband, Red (R) filter images were used to calculate dust production (McLaughlin, 2001). Results supplemented data taken at large telescopes (McLaughlin *et al.*, 2003).

The program was in hibernation from 2001 through 2003, as the comet receded to aphelion. During the interim participants observed comet targets for NASA's Deep Space 1 and Stardust missions as well as several comets for dust tail and solar magnetic field interactions. The program re-launched in late 2004 when Tempel 1 returned to the inner solar system and became favorable for observing. The 2004–2005 observing campaign straddles Deep Impact's encounter with comet Tempel 1 on July 4, 2005. The program seeks the following types of observations:

- Broadband-RI CCD images and photometry for the duration of the campaign.
- Broadband-VRI CCD images from March 2005 through August 2005 to search for jet activity.

  During the 1983 apparition, Tempel 1 appeared to have at least two jets during the months before perihelion (see Lisse *et al.*, this volume). This needs verification in 2005. The impact is expected to make a new active area on the nucleus and may cause new jets or outbursts days or weeks later and evidence of any new activity needs to be documented.
- Broadband-VR and unfiltered images, in wide format, to look for interactions of the dust tail with the solar magnetic field.
- Any narrowband photometry and spectroscopy that STSP observers provide will be accepted; these types of data are within the capabilities of some advanced amateur astronomers.

Advanced amateur and student astronomers with access to small telescopes have the necessary equipment, experience, and skills to acquire scientifically meaningful data in support of the Deep Impact Mission.

3.5. AMATEUR OBSERVER'S PROGRAM

The Amateur Observer Program (AOP) is aimed at the more casual observer who may be hearing about the mission and the opportunities to observe comet 9P/Tempel

1 later rather than sooner. The material on the web site, http://deepimpact.umd.edu/ amateur is setup so that even the non-astronomer and new-comers to the field of astronomical observing can go out and observe. While all of the basics of observing cannot be taught via the world wide web alone, the goal is to convey enough so that a new comer can observe Tempel 1. Information that guides visitors through the observing process including some simple observing activities related to the project (tracking the comet on charts), and information "teasing" them into more advanced concepts are presented on the web site. Charts for finding the comet at different latitudes are also provided. AOP is designed to get people to observe and enjoy the event using the naked eye, binoculars or a small, portable telescope. The opportunity exists to present descriptions and images (sketches, film and digital pictures) on the web site. No specific equipment is needed since participants can attend events hosted by clubs and space places where telescopes are available.

While the web has become one of the easiest ways to disseminate materials, team members go to the amateurs giving presentations at astronomy club meetings and star parties (regional gatherings under dark skies). Early presentations concentrated on describing the mission and the role that amateurs could play. Later presentations focus on what the amateurs should look for just before, during and after the encounter.

4. Outreach Products

Products for outreach include a fact sheet describing the mission, a poster with educational activities on the back, paper models of the spacecraft, artwork depicting the comet and spacecraft (Figure 2), a digital animation, and a planetarium show. Their availability on the web and for distribution at launch and encounter in hard copy to organizations that serve the public contributes to successful outreach.

4.1. PLANETARIUM SHOW

The planetarium show was designed to provide the following information about the comets:

- They date to the beginning of the Solar System and contain pristine material.
- Parts of a comet.
- What we know about them.
- What we do not know about them.
- Where comets fit in the solar system.
- Some historical perspective on comets.

It touches on the following topics related to impacts, comets, and their effect on Earth.

Figure 2. Pat Rawling's rendition of comet Tempel 1 after impact.

- The K-T boundary impact.
- Impact effects elsewhere in the Solar System.
- The Solar System's dynamical nature; not a "calm" and unchanging place as a casual look might suggest.

The show describes the Deep Impact mission's scenario and shows the spacecraft being built at Ball Aerospace in Boulder, CO. Interviews with project scientists, engineers and managers are included. Emphasis was given to the process of science, including the fact that there is a wide range of predictions about what will happen on impact. The show emphasizes that the outcome of the experiment is not known, and that different scientists have different ideas about the outcome.

An effort is made to convey some awe with respect to comets, that they are beautiful and that they can display a range of presentations depending on their activity and geometry with respect to the observer. In providing an historical context, the point is made that in the distant past, with little knowledge about the nature of comets, they were thought to portend ill fates or victory in history. In sharing the passion that scientists have about comets, the show's objective is to inspire the audience to see a comet for themselves.

The Deep Impact E/PO team provided scientific oversight and visuals for the show. Ball Aerospace and Technologies Corp also provided visuals of the spacecraft under construction and interviews with project engineers. Fiske Planetarium staff wrote and produced the show that was distributed on CD-rom through the International Planetarium Society's monthly publication to approximately 800 member-planetaria world-wide in August 2004.

5. Developing the Web Site

The web site is the major communication channel to achieve the goals of the E/PO program. A planned effort was made to form a "community" of Deep Impact followers vested in the mission during our encounter in July 2005. The web site houses all Deep Impact products and activities as well as posts updates about the mission. Certain areas of a NASA web site are given: a place for educators and students to go to find formal education activities, *Education*. There is also a place where young people can go to have fun with the mission and where some informal activities are, *Discovery Zone*. There are places for learning the basic facts about the mission, *Mission, Science, Technology*, and there is a place to which the Press can go, *Press*. Images and animations of the spacecraft being built, experiments run in support of the mission, and of major mission milestones are posted in the *Gallery*. In developing the site, the team started with basic information and, over time, added new material to encourage people to return for updates. The writing style is intended to be friendly and casual and the team has been given compliments for its success in that regard. A section called "*Your Community*" was added a year before encounter. This section and additions to team biographies highlight the work of individuals who have become involved with the mission in some way. Examples are:

- Up Close and Personal – interviews with project members.
- Your Community – teachers, speakers, community organization leaders who have found a unique way to involve themselves with Deep Impact through research or special projects.
- Deep News newsletter – a monthly newsletter written by the outreach team with special features, games and articles.

In approaching launch and encounter, there is a building focus on extending the original mission to the many others who are making a "Deep Impact" with their participation and contributions to the project.

6. Evaluation and Review

Two types of evaluations were carried out, review and an evaluation study. A core planning team was convened early in the mission. Products were reviewed by science team members and by members of their intended audiences. Educational materials were reviewed by NASA's review panels. An independent evaluation was initiated in the final year of the project.

6.1. CORE PLANNING TEAM

The Deep Impact mission convened a 2-day Core Planning Team (CPT) retreat in the design phase (Phase A/B) to review and discuss the E/PO team's plans.

Representatives from proposed audiences including teachers, museums, planetaria and media organizations were invited. Further evaluation was carried out in classrooms and at teachers' workshops where the intent of the material and how it is communicated could be reviewed and revised.

6.2. EVALUATION STUDY

An evaluation study is carried out in the last year of the program with the purpose of evaluating the extent to which target audiences access and use the Deep Impact materials and examining the experiences of teachers and students using them. An independent sub-contract was issued to carry out this study. The evaluation design and its implementation span the course of 12 months during 2005. Two activities, Designing Craters and Modeling Comets for Mission Success will be evaluated in-depth during this project period, whereas the two other activities, Collaborative Decision Making and High Power Activity, will ensue with an exploratory evaluation of their visibility and reach to target audiences before pursuing a more in-depth inquiry as to their use in classrooms. Evaluations of Observing the Impact (Section 2.3.2) and the Amateur Observer Program (Section 3.5) as used in informal science settings will be conducted. The evaluation will be conducted in two phases. Phase One will focus on product dissemination and reach, and will include the development of appropriate methodologies for further examination of quality, utility, and impact in Phase Two.

The evaluation will address the following questions:

1. To what extent do target audiences access the Deep Impact education materials?
2. What are effective dissemination mechanisms for the Deep Impact materials?
3. How do teachers/facilitators use the materials in their education settings?
4. What are teacher/facilitator and student perceptions of and experiences with using the materials?

Study participants will include teachers who have previously used the Deep Impact materials or who will use them as part of the case study during the spring of 2005. The evaluation of the "Eyes on the Skies" and the Amateur Observer program will include educators and students in informal settings, such as astronomy clubs or summer camps. It is intended that participant recruitment efforts will identify three teachers/facilitators for each of the modules to be evaluated. As an incentive for participation, teachers/facilitators will receive the materials for free as well as a stipend of $50 for their time and contributions to the evaluation. Participant confidentiality and anonymity and will adhere to district processes and requirements for human subjects review.

A combination of quantitative and qualitative methods is included in the design to allow for a full understanding of the nature and extent of use of the Deep Impact materials. The evaluator will collect descriptive, implementation, and outcome data throughout the project period. Descriptive information characterizes the nature of the sample involved in the study and includes

- student and teacher demographic information,
- student attendance records, and
- document review.

Product implementation and monitoring measures the use of the Deep Impact materials in the classroom and documents any local history events that occur over the course of the study. Data collection methods to assess implementation include

- site visits for observations and interviews with teachers and students, and
- teacher implementation logs.

Evaluators will conduct site visits to a sample of sites. Evaluators will conduct interviews and observations with teachers/facilitators in order to identify any implementation issues and challenges and to gain in-depth information about implementation, instruction, teacher/facilitator perceptions of student engagement and learning, and student behaviors and perceptions of the materials. Teachers in formal education settings will complete an online implementation log that captures their personal, day-to-day experiences using the materials.

Outcome data reveals how teachers/facilitators and students have been impacted by education materials. Evaluators will inquire about the availability of existing school- or district-level student performance data in science, and will incorporate these data into the final analyses, if feasible. Teacher/facilitator and student outcome data will be based on self-report feedback through

- a teacher/facilitator survey, and
- a student interest survey.

The results of this evaluation will be published in a refereed journal upon its completion.

7. Conclusion

The Deep Impact E/PO program has grown out of a plan, a schedule and a budget that have been in place since the proposal-writing phase of the project. Materials specific to the Deep Impact project take the form of short informative material, audio-visual products, educational modules for the classroom and informal education venues. The breadth of the program is leveraged through partnerships with organizations that reach audiences including under represented and under served minority cultures, school children in urban and rural areas, and teachers and science center staff across the country. With enthusiasm generated by the project scientists, engineers and educators, the E/PO program and the project itself stands poised to meet its objectives of sharing the excitement of space science discoveries with the public, enhancing STEM education in the United States, and inspiring the next generation of scientists and engineers.

Appendix I. National Science Standards Alignment: Collaborative Decision Making

National Science Standards Addressed
    Grades 5–8
        Science as inquiry
            Abilities necessary to do scientific inquiry
                Identify questions that can be answered through scientific investigations
                Recognize and analyze alternative explanations and predictions
                Use appropriate tools and techniques to gather, analyze and interpret data
                Understandings about scientific inquiry
                Develop descriptions, explanations, predictions, and models using evidence
                Think critically and logically to make the relationships between evidence and explanations
                Communicate scientific procedures and explanations
                Use mathematics in all aspects of scientific inquiry

        Science and technology
            Understandings about science and technology

        Science in personal and social perspectives
            Science and technology in society
            Risks and benefits

        History and nature of science
            Science as a human endeavor
            Nature of science

        Earth and space science
            Earth in the solar system

    Grades 9–12

        Science as inquiry

Abilities necessary to do scientific inquiry

    Identify questions and concepts that guide scientific investigations

    Recognize and analyze alternative explanations and predictions

    Formulate and revise explanations and models using logic and evidence

    Communicate and defend a scientific argument

Science and technology

    Understandings about science and technology

    Use technology and mathematics to improve investigations and communications

Science in personal and social perspectives

    Environmental quality

    Science and technology in local, national, and global challenges

History and nature of science

    Science as a human endeavor

    Nature of scientific knowledge

Language Arts Standards

Standard: 8 Demonstrates competence in speaking and listening as tools for learning

Level III Grades 6–8

    Listening and speaking

        Conveys a clear main point when speaking to others and stays on the topic being discussed

        Presents simple prepared reports to the class

    Uses listening and speaking strategies for different purposes

        Uses strategies to enhance listening comprehension

        Listens in order to understand topic, purpose, and perspective in spoken texts

Level IV Grades 9–12

    Listening and speaking

        Adjusts message wording and delivery to particular audiences and for particular purposes (e.g., to defend a position, to entertain, to inform, to persuade)

        Makes formal presentations to the class (e.g., includes definitions for clarity: supports main ideas using anecdotes, examples, statistics, analogies, and other evidence; uses visual aids or technology)

Responds to questions and feedback about own presentations (e.g., defends ideas, expands on a topic, uses logical arguments)

Writing

Gathers and uses information for research purposes
Uses card catalogs and computer databases to locate sources for research topics
Uses a variety of resource materials to gather information for research topics
Determines the appropriateness of an information source for a research topic

Uses the general skills and strategies of the writing process
Drafting and revising: uses a variety of strategies to draft and revise written work

Life Skills Standards

Thinking and reasoning
Applies decision-making techniques
Secures factual information needed to evaluate alternatives
Predicts the consequences of selecting each alternative
Makes decisions based on the data obtained and the criteria identified
Identifies situations in the community and in one's personal life in which a decision is required

Effectively uses mental processes that are based on identifying similarities and differences
Compares different sources of information for the same topic in terms of basic similarities and differences

Principles and Standards for School Mathematics Grades 6–8, 9–12

Problem solving
Solve problems that arise in mathematics and in other contexts
Apply and adapt a variety of appropriate strategies to solve problems

Data analysis and probability
Formulate questions that can be addressed with data and collect, organize and display relevant data to answer them
Develop and evaluate inferences and predictions that are based on data

National Educational Technology Standards Grades K-12

Standard 5: Technology research tools

Students use technology to locate, evaluate, and collect information from a variety of sources

Standard 6: Technology problem-solving and decision-making tools

Students use technology resources for solving problems and making informed decisions

Students employ technology in the development of strategies for solving problems in thereal world

Grades 6–8

Standard 7

Collaborate with peers, experts, and others using telecommunications, and collaborative tools to investigate curriculum-related problems, issues, and information, and to develop solutions or products for audiences inside and outside the classroom

Standard 10

Research and evaluate the accuracy, relevance, appropriateness, comprehensiveness, and bias of electronic information sources concerning real-world problems

Appendix II. Designing Craters

National Science Education Standards Addressed
Grades 5–8

Science as inquiry

Abilities necessary to do scientific inquiry

Identify questions and concepts that guide scientific investigations

Design and conduct a scientific investigation

Communicating scientific procedures and explanations

Develop descriptions, explanations, predictions and models using evidence

Understandings about scientific inquiry

Technology used to gather data enhances accuracy and allows scientists to analyze and quantify results of investigations

Scientific investigations sometimes result in new ideas and phenomena for study

Earth and space science
Earth's history

Earth history is also influenced by occasional catastrophes, such as the impact of an asteroid or comet

Physical science

Properties and changes of properties in matter

A substance has characteristic properties, such as density, a boiling point, and solubility, all of which are independent of the amount of the sample

Motion and forces

The motion of an object can be described by its position, direction of motion, and speed. That motion can be measured and represented on a graph

Transfer of energy

Energy is a property of many substances and is associated with heat, light, electricity, mechanical motion, sound, etc.

Energy is transferred in many ways

Grades 9–12

Science as inquiry

Abilities necessary to do scientific inquiry

Identify questions that can be answered through scientific investigation

Design and conduct scientific investigations

Communicate and defend a scientific argument

Recognize and analyze alternative explanations and models

Formulate and revise scientific explanations and models using logic and evidence

Understandings about scientific inquiry

Scientists usually inquire about how physical, living, or designed systems function

Scientists rely on technology to enhance data

Results of scientific inquiry emerge from different types of investigations and communications between scientists

Principles and Standards for School Mathematics Addressed

Grades 6–8

Data analysis and probability

Develop and evaluate inferences and predictions that are based on data

Algebra

Understand patterns, relations, and functions

Analyze change in various contexts

Use mathematical models to represent and understand quantitative Algebrarelationships

References

Melosh, H. J.: 1996, *Impact Cratering: A Geologic Process, Oxford Monographs on Geology and Geophysics*, no. 11, Oxford University Press, Oxford.

McLaughlin, S. A.: 2001, *Deep Impact Mission's Small Telescope Science Program: CCD Broadband Photometry of the Target Comet 9P/Tempel 1*. Master's thesis, University of Maryland.

McLaughlin, S. A., McFadden, L. A., and Emerson, G., 2003, Science with Very Small Telescopes (<2.4-meters): Oswalt, T. D. (ed.), *The NASA Deep Impact Mission's Small Telescope Science Program in The New Millenium, Astronony and Space Science Library*, Vol. 289, Kluwer Academic Publishers, Dordrecht, p. 57.

NASA: October 15, 1996, *Implementing the Office of Space Science Education/Public Outreach Strategy*.

NASA: March 21, 2003, *Implementing the Office of Space Science Education/Public Outreach Strategy: A Critical Evaluation at the Six-Year Mark*. A Report by the Space Science Advisory Committee Education and Public Outreach Task Force.

NASA: March, 2004, *Explanatory Guide to the NASA Office of Space Science Education & Public Outreach Evaluation Criteria Version 3.0*.

Rosendhal, J., Sakimoto, P. L., Pertzborn, R., and Cooper, L.: 2004, *Adv. Space Res.* **34**(10), 2127.